Freeman Laboratory Separates in General Chemistry

Each exercise in this manual is available as a Freeman Laboratory Separate, numbered below in the order in which they appear in the manual.

The separates are self-bound, self-contained exercises. They are 8½ inches by 11 inches in size, and are punched for a three-ring notebook. They can be ordered in any assortment or quantity at 30¢ each. Order through your bookstore, specifying number and title. (For a complete listing of other Freeman Laboratory Separates in chemistry, see the last page and inside back cover of this manual.)

W. H. FREEMAN AND COMPANY
660 Market Street, San Francisco, California 94104
58 Kings Road, Reading, England RG1 3AA

THIRD EDITION

FRANTZ/MALM'S
Essentials of Chemistry in the Laboratory

James B. Ifft
Julian L. Roberts, Jr.

UNIVERSITY OF REDLANDS

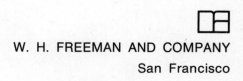

W. H. FREEMAN AND COMPANY
San Francisco

To
Fern, Frances, Evelyn, and Jane

CONTENTS

PREFACE

Today's society is confronting unprecedented challenges. Almost every day newspapers feature major items on the problems we face—energy shortages, overpopulation, pollution of our air and water, food shortages, and so on. To be sure, scientists themselves have contributed to some of these problems, but they—along with politicians, economists, and sociologists—will be called upon to help resolve all of them.

Chemistry is one of those natural sciences being applied to the quest for a better environment and better living standards for all people. Whether chemists are devising new and safer pesticides, more effective birth control measures, techniques to monitor and reduce air pollution, or a cure for cancer, their expertise is urgently needed by society.

We assume that the student is taking a laboratory course in chemistry, first because he thinks an acquaintance with this particular science can contribute to his understanding of the physical world and, second, because the skills he acquires can help him to earn a living. But we hope that he is taking it also because he realizes that a knowledge of this subject will enable him to make a positive contribution to the solution of some of the grave problems currently facing all of us.

Purpose. Some general chemistry instructors are dropping the laboratory portion of their courses. Most of these decisions are pragmatic, resulting from the very real difficulty of providing a laboratory experience for large numbers of students. It costs both money and time. We believe that it is worth every dollar and every hour.

In this new edition for the Frantz/Malm series of laboratory manuals, the basic premise continues to be that the laboratory part of the course must be a central aspect of the student's learning experience. The relationship between the lecture and laboratory portions of the course should therefore be closely coordinated and mutually supportive.

One reason is that chemistry *is* an experimental science. Although some theoretical calculations can be performed with complete rigor, vast areas of kinetics, quantum mechanics, and descriptive chemistry rest principally on the work of men and women conducting experiments at the benchtop.

Also, laboratory work is fun. The appearance of a precipitate at the right stage in a qualitative analysis experiment, the striking rainbow of colors from a series of indicators at a series of pH's, or the resolution of the light from a lamp into a series of lines with the use of a spectroscope one has made himself—all are stimulating and rewarding experiences in themselves.

Finally, we believe that conducting experiments in the chemistry laboratory is the best way to learn descriptive chemistry. Few students will forget the surprise with which they watch their results the first time they add base to a sodium dichromate solution, or add silver nitrate to a solution containing chloride ion, or reach the end point in the titration of an acid with a base using phenolphthalein as the indicator. One's own visual experiences, then, are definite aids to learning and retaining the inorganic and organic material described in the standard chemistry text.

Organization and Content. The organization of this manual has been revised somewhat in order to reflect changing course content and student needs. First, the arrangement of the experiments is different. For instance, all of the experiments on gases have been combined into one section, and all acid-base experiments are contained in a single section ("Chemical Equilibrium").

The order of the experiments has been altered also. The manual begins with a section on the structure of atoms and ends with a selection of experiments on some of the largest molecules known, the proteins. In the intermediate chapters we develop those topics that are treated in most general chemistry laboratory manuals—descriptive inorganic chemistry, the behavior of gases, thermodynamics, kinetics, chemical equilibrium and oxidation-reduction. The sequence in this manual conforms quite closely to that in a new general chemistry textbook, *Contemporary Chemistry* by John E. Hearst and James B. Ifft (W. H. Freeman and Company, in press), but the sequence of experiments can be easily altered to fit that of virtually any general chemistry text.

In addition to the reorganization, some important new material has been added. The section on atomic and molecular structure has been updated. It consists of the original experiment on crystal structure, from the previous editions, plus two new experiments that should complement today's

increased emphasis on attempts to provide the correct quantum-mechanical description of the atom. A new experiment on qualitative analysis with the use of paper chromatography has been added to demonstrate some of the newer methods available for detecting the presence of ions in solution. For the many students interested in the biological sciences, two experiments in the field of biochemistry have been added to the section on organic chemistry.

A modest treatment of classical qualitative analysis has been preserved. The methods employed in this area of experimental chemistry illustrate the principles of chemical equilibrium. The five original experiments plus the two new experiments provide a sound foundation for an understanding of qualitative analysis.

This manual contains 39 experiments—approximately twice the number performed in the standard general chemistry course. The instructor can thus choose from an ample selection those experiments that will best complement the lecture portion of his course.

A large number of problems are presented in the individual study assignments, and in the exercises at the ends of the experiments. These problems present a challenge to the student's understanding of the chemical principles illustrated in the laboratory experiments. They vary in complexity from the completely straightforward problems (the first few in every grouping) to some that are quite difficult and for which the students may need to seek the help of the laboratory instructor.

This manual contains laboratory report forms. The advantages of these forms are that they provide explicit guidance on what data to tabulate and how to arrange them so that they can be treated expeditiously. In addition, the report forms show the student how to perform some of the more difficult calculations. For a few of the experiments, the student is required to provide his own report form. This exercise should help him to prepare for subsequent independent laboratory research.

Level. This laboratory manual is designed for an intermediate level course in general chemistry. By "intermediate," we mean a course suitable for students in the life and earth sciences, for those chemistry and physics students who have not had high-school chemistry,[1] and for students in the

---

[1]For students who have had a sound preparation in chemistry in high school, the more advanced manual in this series, *Chemical Principles in the Laboratory,* is recommended.

social sciences and humanities. This manual does not require mathematics beyond the second-year course in high-school algebra. If a student has difficulty with some of the calculations or graphical techniques, he will find Appendixes A and B helpful.

Acknowledgments. We acknowledge our gratitude to Harper W. Frantz and W. H. Freeman and Company for inviting us to be the continuing authors of this prestigious and highly successful series of laboratory manuals. Indeed, one of us (J. B. I.) initially learned the essentials of laboratory chemistry from the first edition of one of the manuals in this series. We continue to believe that this manual

will provide a thorough foundation in the principles of the chemical laboratory. We trust that the current revision has retained the clarity and effectiveness of earlier editions, as well as introducing some recent experiments that will expose students to the important developments that are taking place in modern chemistry.

We are also grateful to our expert typist and friend, Eleanor Scott, who has so beautifully transformed our often confusing drafts into a typed manuscript ready for publication.

January, 1975   *James B. Ifft*
                                   Julian L. Roberts, Jr.

FRANTZ/MALM'S
Essentials of Chemistry
in the Laboratory

INTRODUCTION

The chemistry laboratory can be a place of joy, discovery, and learning. It can also be a place of danger and frustration.

The danger arises from the nature of the chemicals and apparatus that are used in these experiments, and their effects on protein. Skin consists of protein. We hope that you do not have to experience firsthand the fact that even 6 F solutions of acids and bases, not to mention concentrated solutions, have a most adverse effect on protein. The eyes are especially sensitive to acids, bases, and oxidizing agents. In addition, the open flame of a bunsen burner presents a continual hazard to clothing and hair.

The frustration is generated by experiments that don't seem to work, or by data on your report form that seem unintelligible when you are attempting to do the calculations the night before the report is due. You can minimize both types of problems by careful recording of data and thoughtful consideration of the data while collecting it.

We strongly advise you to learn and observe at all times the following laboratory rules and regulations in order that you will minimize the potential dangers and frustrations of laboratory work, and maximize the joy.

SAFETY RULES

These rules are designed to ensure that all work done in the laboratory will be safe for you and your fellow students.

1. The most important safety rule is that glasses must be worn at all times in the laboratory. Prescription glasses are adequate in almost all situations. If you do not wear glasses, obtain an inexpensive pair of safety glasses from your chemistry stockroom. In some procedures—such as heating a crucible to dryness or evaporating an acid solution—be sure to wear safety goggles or to carry out the experiment in a hood, which is provided in most laboratories for this purpose.

If any chemical comes in contact with the eye, the most effective first aid is the immediate flushing

ALWAYS smell a substance by wafting its vapor gently toward your face.

FIGURE i-1
The procedure for smelling a substance.

of the eye with copious amounts of tap water. You are seldom more than a few seconds from a faucet. Continue flushing for at least five minutes and then consult a physician at once. If your laboratory is equipped with eye fountains, familiarize yourself with their use and their location.

2. Fire is a constantly present danger. Learn where the nearest fire extinguisher is and how to use it. Your laboratory should also be equipped with a safety shower or fountain: if your hair or clothing should catch on fire, go to it at once and douse yourself.

3. Minor burns, cuts and scratches are fairly common injuries. However, you must report every such incident to your instructor, who will determine

what first aid is appropriate. If you or another student must report to the infirmary or hospital, be certain that someone else accompanies the injured person.

4. Bare feet are not allowed in a chemistry laboratory. Broken glass and spilled chemicals, such as concentrated acids, are all too common on the floors of chemistry labs. In addition, we recommend that bare legs, midriffs and arms be covered with old clothing or, preferably, with a laboratory apron or coat.

5. The vapors of a number of solutions are quite potent and can irritate or damage the mucous membranes of the nasal passages and the throat. Use the technique displayed in Figure i-1 when you need to sniff an odor.

6. In many experiments it is necessary to heat solutions in test tubes. Never apply heat to the *bottom* of the tube; always apply it to the point at which the solution is highest in the tube, working downward if necessary. Be extremely careful about the direction in which you point a tube; a suddenly formed bubble of vapor may suddenly eject the contents violently (an occurrence called "bumping"). Indeed, a test tube can become a miniature cannon.

7. Taste chemicals and solutions *only* when directed to do so. (Poisonous substances are not always so labeled in the laboratory.)

8. Beware of hot glass tubing—it *looks* cool long before it may be handled safely.

FIGURE i-2
Some important safety precautions.

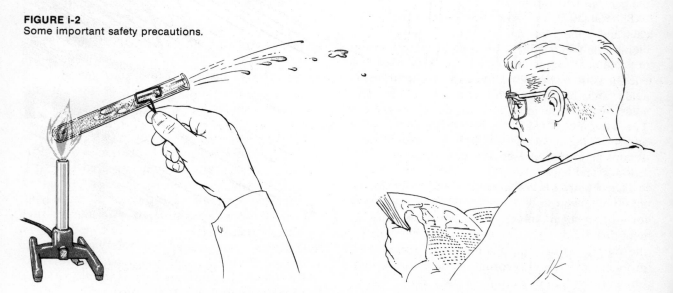

NEVER point a test tube of boiling liquid at your neighbor—it may bump.
SAFETY GOGGLES, worn regularly in the laboratory will protect your eyesight.

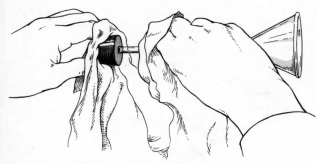

NEVER force a thistle tube or funnel into a stopper by grasping the large end. Use the stem and twist as you push. ALWAYS wrap your hands in a towel when putting a glass tube into a stopper.   Moisten with water and insert with a twisting motion.

FIGURE i-3
The procedure for inserting a glass tube into a stopper.

9. For reactions involving poisonous gases, *use the hood,* which provides suction to remove such gases or vapors.

10. Neutralize spilled acid or base as follows: (a) Acid on clothing; use dilute sodium bicarbonate solution. (b) Base on clothing; use dilute acetic acid. (c) Acid or base on the desk; use solid sodium bicarbonate for either, followed by water.

11. To insert glass tubing (including thermometers, thistle tubes, etc.) through a rubber stopper, first lubricate the tube and stopper with water or glycerine. Hold the tubing with a cloth *near the end to be inserted,* and insert with a twisting motion. (If you twist a thistle tube by the "thistle" end, it is easily broken.)

LABORATORY REGULATIONS

These regulations are designed to guide you in developing efficient laboratory techniques and in making your laboratory a pleasant place to work.

1. You must read each experiment thoroughly before entering the lab. If you do not, you will waste a great deal of time (both your own and your instructor's), you may expose yourself and others to unnecessary hazards, and you will probably not obtain reliable, useful data. (You will also routinely fail all pre-lab quizzes if your instructor chooses to use them.)

2. Discard solids into the waste crocks. *Never throw matches, litmus, or any insoluble solids* into the sink. Wash down liquids into the sink with much water; acids and salts of copper, silver, and mercury are corrosive to lead plumbing.

3. Leave reagent bottles at the side shelves. Bring test tubes or beakers *to the shelf* for transferring chemicals and carrying them to your desk.

4. Read the label *twice* before taking anything from a bottle.

5. Avoid using excessive amounts of reagent—1 to 3 ml is usually ample for test tube reactions.

6. *Never* return unused chemicals to the stock bottle. You may make a mistake from which other students' experiments will suffer.

7. Do not insert your own pipets or medicine droppers into the reagent bottles. Avoid contamination of the stock solution by pouring the solution from the bottle.

8. Do not lay the stopper of a bottle down. Impurities may be picked up and thus contaminate the solution when the stopper is returned. Hold the stopper as illustrated in Experiment 2, Figures 2-3 and 2-4.

9. Do not heat heavy glassware such as volumetric flasks, graduated cylinders, or bottles; they break easily and heating distorts the glass so that the calibrations are no longer valid. Test tubes may break if they are heated above the liquid level and liquid is then splashed over the hot glass. Evaporating dishes and crucibles may be heated red hot. Avoid heating any apparatus too suddenly; apply the flame intermittently at first.

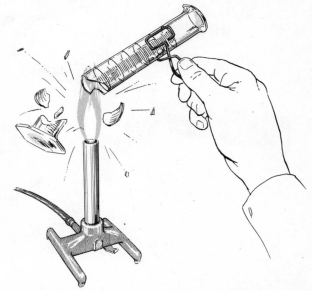

NEVER heat a graduated cylinder or bottle

FIGURE i-4
If heat is applied to the wrong type of laboratory apparatus, the outcome can be disastrous.

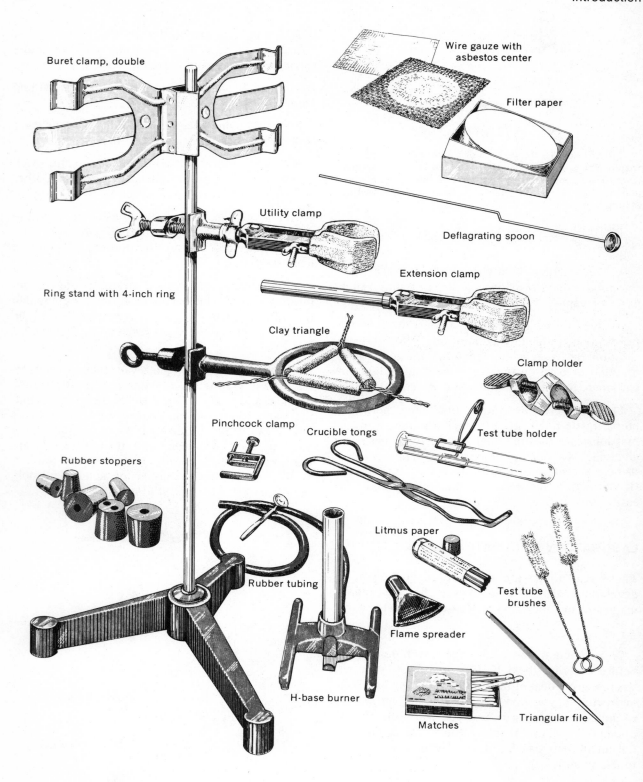

FIGURE i-5
Common Laboratory equipment. [From J. W. Hagen, *Empirical Chemistry*, W. H. Freeman and Company. Copyright © 1972.]

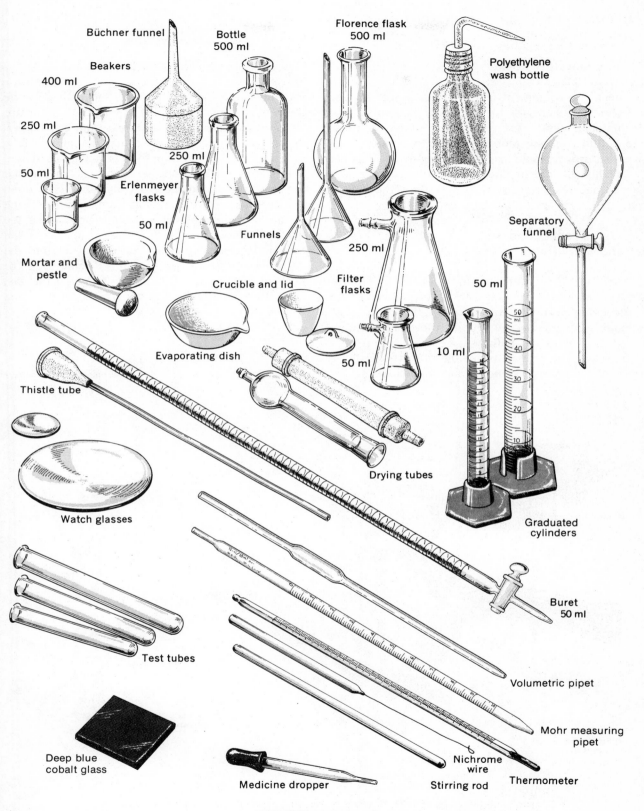

FIGURE i-5 (continued)

BASIC LABORATORY EQUIPMENT
AND PROCEDURES

The Laboratory Locker

Check the equipment in the locker assigned you. Refer to Figure i-5 for the identification of any unfamiliar items. Ascertain that all items are present, and examine them carefully to be sure they are in an acceptable condition. You are responsible for this equipment and will be charged for any breakage or shortage at the conclusion of the course.

Safety Precautions and
Laboratory Rules

At the outset of your work in the chemistry laboratory, familiarize yourself with the rules given in the preceding section. Obedience to these rules, as modified or added to by your instructor, is essential.

Your instructor will indicate the location and show you the proper use of the fire extinguishers, fire blanket, and first-aid cabinet and supplies.

Use of the Laboratory Burner

Examine your Bunsen burner[1] and note how the flow of air and the flow of gas to the burner are controlled. Connect the burner to the gas supply, and adjust the air control so that a minimum flow of air is supplied. Turn on the gas and hold a lighted match above and to one side of the gas flow, so that the first rush of air does not extinguish it. Now open the air intake slowly until a nonluminous blue flame results. Usually a blue inner cone is observed in the flame. The hottest part of the flame is just above the top of this inner cone. If the flow of air into the barrel is too great, the flame may "strike back" or burn *inside* the barrel. If it does, turn off the gas, close the air control, and relight the burner (see Figure i-6).

---

[1]In some laboratories, the Fisher burner may be used instead.

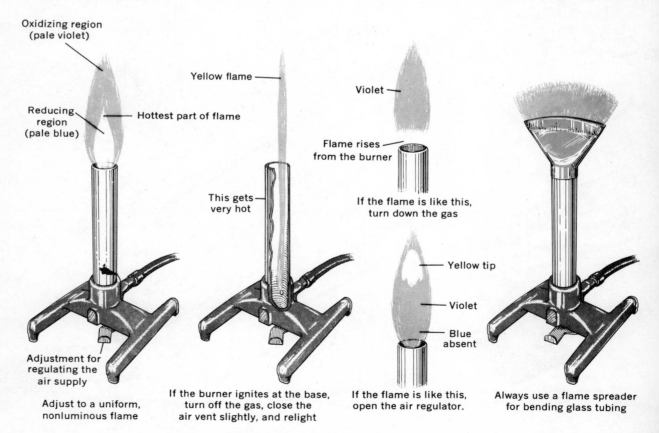

Oxidizing region (pale violet)

Reducing region (pale blue)

Hottest part of flame

Yellow flame

This gets very hot

Adjustment for regulating the air supply

Adjust to a uniform, nonluminous flame

If the burner ignites at the base, turn off the gas, close the air vent slightly, and relight

Violet

Flame rises from the burner

If the flame is like this, turn down the gas

Yellow tip

Violet

Blue absent

If the flame is like this, open the air regulator.

Always use a flame spreader for bending glass tubing

FIGURE i-6
Instructions for operating a Bunsen burner.

Operations with Glass Tubing

Glass is not a true crystalline solid and therefore does not have a sharp melting point. In this respect it more nearly resembles a solid solution or an extremely viscous liquid which gradually softens when heated. It is this property which makes glass working possible.

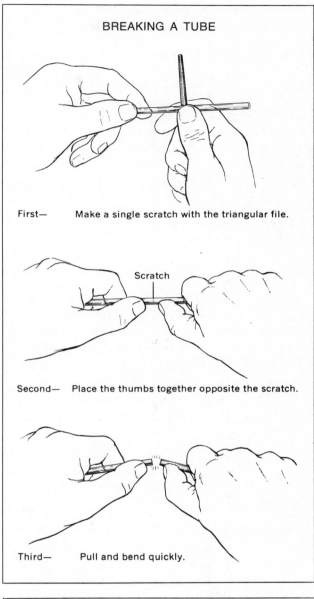

BREAKING A TUBE

First—   Make a single scratch with the triangular file.

Second—   Place the thumbs together opposite the scratch.

Third—   Pull and bend quickly.

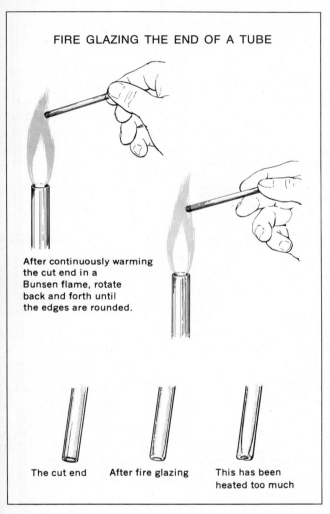

FIRE GLAZING THE END OF A TUBE

After continuously warming the cut end in a Bunsen flame, rotate back and forth until the edges are rounded.

The cut end   After fire glazing   This has been heated too much

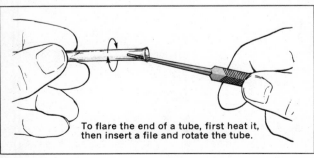

To flare the end of a tube, first heat it, then insert a file and rotate the tube.

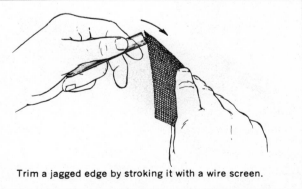

Trim a jagged edge by stroking it with a wire screen.

FIGURE i-7
Some elementary manipulations of glass tubing.

MAKING A BEND

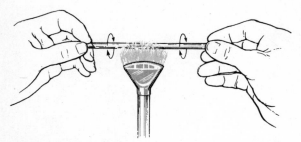

Roll the tube back and forth in the high part of a flat flame until it has become quite soft.

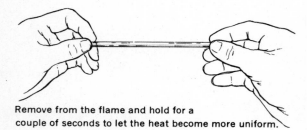

Remove from the flame and hold for a couple of seconds to let the heat become more uniform.

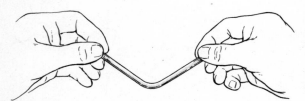

Bend quickly to the desired shape and hold until it hardens.

A good bend

Inadequate heating

Local overheating

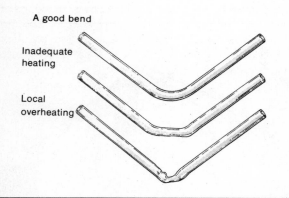

MAKING A CONSTRICTED TIP

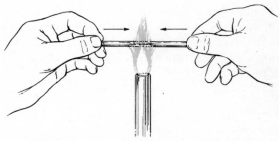

Allow the tube to become shorter as the walls thicken to about twice their original thickness.

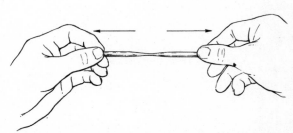

Remove from the flame and after a moment pull until the softened region is as small as desired.

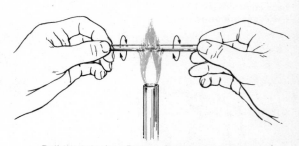

Roll the tube in a Bunsen flame until it softens. Don't use a flame spreader.

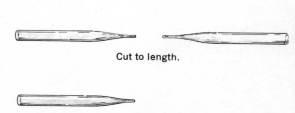

Cut to length.

Fire glaze, or file the tip.

FIGURE i-8
Additional manipulations of glass tubing.

Soda-lime glass, made by heating a mixture of soda (Na_2CO_3), limestone ($CaCO_3$), and silica (SiO_2), softens readily at about 300–400°C in the burner flame. Tubing of this glass is easily bent, but because of its high temperature coefficient of expansion it must be heated and cooled gradually to avoid undue strain or breakage. *Annealing* by a mild reheating and uniform cooling is often wise. Such glass must not be laid on a cold surface while it is hot, since this introduces strains and causes breakage.

Borosilicate glass (such as Pyrex or Kimax) does not soften much below 700–800°C and must be worked in an oxygen-natural gas flame or blow torch. Because it has a low temperature coefficient of expansion, objects made of it can withstand sudden temperature changes.

Figure i-7 shows the proper way to cut glass tubing and Figure i-8, the way to make a bend or constricted tip.

Care of Laboratory Glassware

Examine all glassware for cracks and chips. Flasks or beakers with cracks may break when heated and cause injury. Small chips in borosilicate glassware can sometimes be eliminated by fire polishing; otherwise chipped glassware should be discarded because it is easy to cut oneself.

The recommended procedure for cleaning glassware is to wash the object carefully with a brush in hot water and detergent, then rinse thoroughly with tap water, and finally rinse once again with a small quantity of distilled or deionized water. Then allow the glassware to drain dry overnight in your locker. If you must use a piece of glassware while it is still wet, rinse it with the solution to be used.

Cleaning solution (a solution of CrO_3 or $K_2Cr_2O_7$ in concentrated sulfuric acid) is sometimes used. Such solutions should be employed in the general chemistry laboratory only under the *direct* supervision of the instructor, because under certain conditions the use of concentrated acids can produce noxious gases, endangering you and others nearby.

Volumetric Measurement of Liquids

Volumetric measurements of liquids are made with graduated cylinders, burets, or transfer pipets. The graduated cylinder (Figure i-9) is usually used to

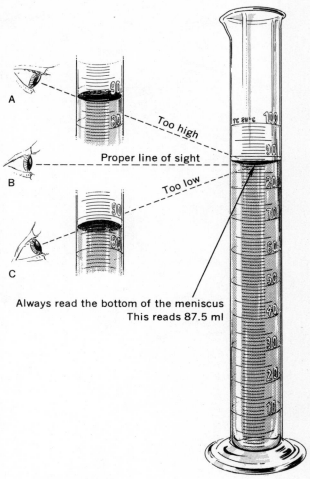

Always read the bottom of the meniscus
This reads 87.5 ml

FIGURE i-9
The proper method of reading a meniscus to avoid parallax error.

measure approximate volumes of liquids. Aqueous solutions wet the glass walls forming a concave meniscus; the bottom of the meniscus is used to indicate the volume of liquid. To avoid parallax error (caused by change of observational position), your eye should always be level with the meniscus when you are making a reading. The volume is estimated to one tenth of the smallest division. The graduated cylinder is calibrated to deliver (TD)[2] the volume that is read, and it actually contains slightly more than the volume read, so compensating for the thin film of liquid left on the walls when the contents are poured out.

-----

[2]The abbreviation TD designates glassware that is calibrated *to deliver* the volume specified: for example, a 50-ml TD pipet.

10

Using a small funnel, rinse a clean buret with
a few milliliters of the solution. Allow the buret
to drain.

Fill the buret to above the zero mark with the solution.

Open the stopcock wide for a few seconds to remove all air from the tip.

Refill to just *below* the 0.00 mark (somewhere between 0–1 ml). Take initial reading with eye level with meniscus. Do not attempt to set initial reading at 0.00 or 1.00 or any other specific reading.

FIGURE i-10
The use of a buret. [From J. W. Hagen, *Empirical Chemistry,* W. H. Freeman and Company. Copyright © 1972.]

Introduction

The buret (Figure i-10) is used for more precise volumetric work and for titrations. If it is clean, the solution will leave an unbroken film when it drains; if drops of solution adhere to the inside of the buret, it should be cleaned with a brush, hot water, and detergent until it drains properly (See Figure i-11). Absolute cleanliness is important because the volume of a 25- or 50-ml buret can ordinarily be estimated to the nearest 0.02 ml and the error caused by a single large drop adhering to the inside of a buret causes an error of about 0.05 ml. The presence of several drops would obviously result in poor measurement of the volume delivered.

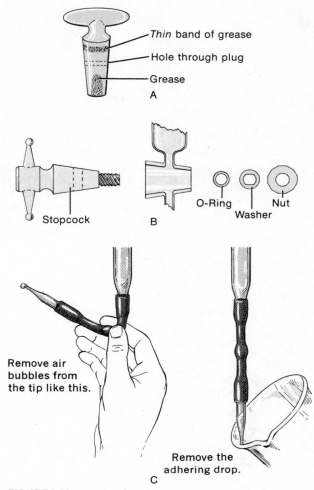

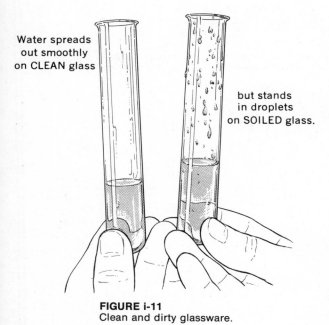

FIGURE i-11
Clean and dirty glassware.

Water spreads out smoothly on CLEAN glass

but stands in droplets on SOILED glass.

Remove air bubbles from the tip like this.

Remove the adhering drop.

C

Thin band of grease

Hole through plug

Grease

A

Stopcock

O-Ring

Washer

Nut

B

FIGURE i-12
Three varieties of stopcocks. (A) A glass stopcock: to grease a glass stopcock, remove old grease from both parts with organic solvent; wipe dry; apply a thin film of stopcock grease as shown. [From J. W. Hagen, *Empirical Chemistry*, W. H. Freeman and Company. Copyright © 1972.] (B) A teflon stopcock: no grease is used on a teflon stopcock. [From Hagen, 1972.] (C) A rubber-tubing, glass-bead stopcock: pinching the tube near the bead allows the solution to drain from the buret.

When filling a freshly cleaned buret with solution, add a 5- to 10-ml portion of the solution, being sure the stopcock is turned off, and tip and rotate the buret so that the solution rinses the walls of the buret completely. Repeat the procedure with at least two more fresh portions of solution; then fill the buret above the zero mark and clamp in a buret holder. Then open the stopcock wide to flush out any air bubbles between the stopcock and the tip of the buret. Next drain the buret below the zero mark, and take the initial reading, being careful to avoid parallax error. The smallest division on a 50-ml buret is ordinarily 0.1 ml; estimate the volume to the nearest fifth of the smallest division or the nearest 0.02 ml.

Burets may have stopcocks made of glass, which must be periodically cleaned and lubricated as shown in Figure i-12A. Teflon stopcocks (Figure i-12B) ordinarily require no lubrication, but they may have to be cleaned if they are plugged, or adjusted if the tension nut is too tight or too loose. A drawn-out glass tip—which is connected to the buret by a short length of rubber tubing containing a round glass bead—constitutes the simple yet effective stopcock shown in Figure i-12C.

The buret is calibrated to deliver (TD) and is capable of a precision of approximately ±0.02 ml when carefully used. The tip should be small enough that the delivery time is not less than 90 seconds for a 50-ml buret. This allows adequate time for drainage, in order that you can obtain a proper reading. If the delivery time of your buret is faster, wait a few seconds before taking a reading, so that you allow the buret to drain.

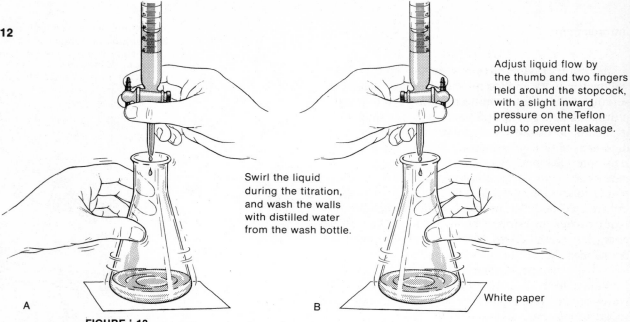

Adjust liquid flow by the thumb and two fingers held around the stopcock, with a slight inward pressure on the Teflon plug to prevent leakage.

Swirl the liquid during the titration, and wash the walls with distilled water from the wash bottle.

A

B

White paper

FIGURE i-13
Recommended technique for manipulation of a buret stopcock. Most left-handed students will manipulate the stopcock with the right hand (A), whereas most right-handed students will prefer to manipulate it with the left hand (B).

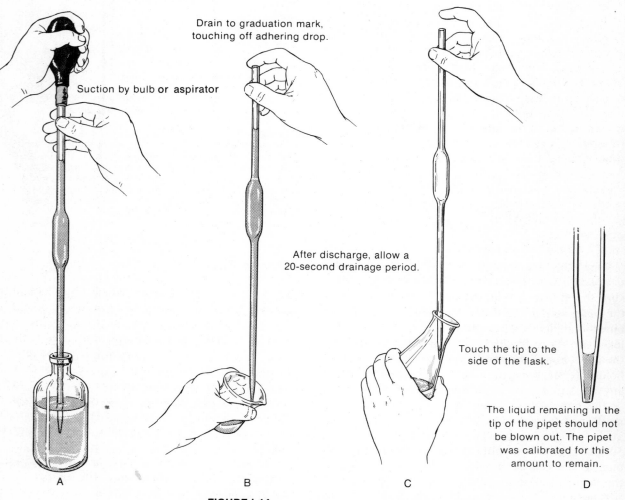

Drain to graduation mark, touching off adhering drop.

Suction by bulb or aspirator

After discharge, allow a 20-second drainage period.

Touch the tip to the side of the flask.

The liquid remaining in the tip of the pipet should not be blown out. The pipet was calibrated for this amount to remain.

A

B

C

D

FIGURE i-14
The procedure for using a transfer pipet.

Figure i-13 illustrates the recommended technique for manipulation of a buret stopcock. You may add the solution from the buret quite rapidly until it is close to the end point, but then reduce the flow until individual drops fall into the flask. As you add the last few drops slowly, swirl the flask to obtain thorough mixing.

Transfer pipets are designed to deliver a single fixed volume of liquid. The graduation mark is located on the narrow part of the pipet to assure good precision. They come in various sizes varying from less than 1 ml to 100 ml. They are calibrated to deliver (TD) the specified volume if they are handled in the prescribed manner.

Pipets are ordinarily calibrated at room temperature or close to it. In very careful work, temperature corrections are necessary if the solution temperature is markedly different from the calibration temperature of the pipet. Fill the pipet by placing the tip in a flask of the solution and using a suction bulb to draw the liquid up past the calibration mark (Figure i-14). Then slip off the bulb and, with the forefinger, quickly stop the flow before the solution drops below the calibration mark. Wipe the outside of the pipet with tissue or a clean towel, and then allow the liquid to flow out until the bottom of the meniscus is just at the calibration ring; to pick off the last drop adhering to the outside of the tip, touch the tip to the side of the flask. Then withdraw the pipet from the flask and hold it over the vessel into which the liquid is to be transferred. Allow the pipet to drain in a vertical position, with the tip against the side of the vessel. Allow 15 to 20 seconds for drainage after it appears that most of the liquid has drained out. The tip of the pipet will still contain some liquid. This has been accounted for in the calibration and should not be blown out. (Certain types of pipets are calibrated to be blown out—most of these will have a sandblasted ring at the top of the pipet). Like the buret, the pipet must be scrupulously clean if precise results are to be obtained. If the pipet is still wet from cleaning, rinse with several portions of the solution to be pipeted, using the same procedure as you would for rinsing a buret.

Filtration

A glass funnel and filter paper are most often used for filtering (Figure i-15). Fold a circle of filter paper in half and then fold again so that the edges do not meet but form an angle of about 10°. Tear off the top corner of the smaller half and open the

Fold and crease lightly.

Fold again.

About 10°

Tear off corner unequally.

Open out like this.

Fill with water and let it run until the air is washed out of the stem. When the water level drops to the top of the stem, add the mixture to be filtered.

Seal the moistened edge of the filter against the funnel.

The weight of this column of water hastens filtration.

The torn corner prevents air from leaking down the fold.

The filtrate should run down the walls of the beaker.

FIGURE i-15
The procedure for quantitative filtration.

larger half to form a cone, which you place in the funnel. The top of the paper cone should be at least 0.5 cm below the rim of the funnel. The paper should be firmly in place, and moistened with distilled water. With the fingers, press the upper edge of the cone against the funnel to form a seal. Then place it in a stand and add water until the stem fills with water. This column of water hastens the filtration process, but it will remain in the stem only if the paper is properly sealed at the top of the funnel.

When filtering a precipitate, first decant most of the supernatant solution into the funnel. Then, using a glass rod and wash bottle, wash the precipitate into the funnel as shown in Figure i-16. During this process the liquid level in the funnel should never be allowed to rise more than 0.5 cm below the top of the filter paper.

A more rapid filtration, particularly of fine precipitates, can be made using a Büchner funnel and a vacuum source as shown in Figure i-17. The Büchner funnel contains a perforated plate, which supports the filter paper. The filter paper should be slightly smaller than the plate. Place it in the funnel, drawing it evenly and smoothly over the plate, and moisten it with the solvent to be used. Now apply the vacuum. While the vacuum is on, pour the solution to be filtered into the funnel. Scrape solid material into the funnel with a spatula or glass rod. Some of the filtrate, or a fresh portion of solvent, can be used to flush the solid from the original container. You can wash the solid precipitate with a small amount of solvent, and then allow it to dry by drawing air through the funnel. You can remove the solid cake of precipitate from the filter by lifting one edge of the filter paper with a spatula or by inverting the funnel over a piece of clean glazed paper and rapping the funnel sharply against the paper. The trap is an important part of the apparatus: it keeps the filtrate from splashing over into the vacuum source, or, if an aspirator is used, it keeps any water that might back up in the aspirator out of the filter flask.

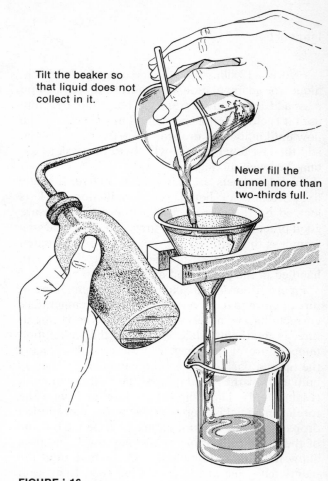

Tilt the beaker so that liquid does not collect in it.

Never fill the funnel more than two-thirds full.

FIGURE i-16
The quantitative transfer of a precipitate to a filter. [From J. W. Hagen, *Empirical Chemistry,* W. H. Freeman and Company, San Francisco. Copyright © 1972.]

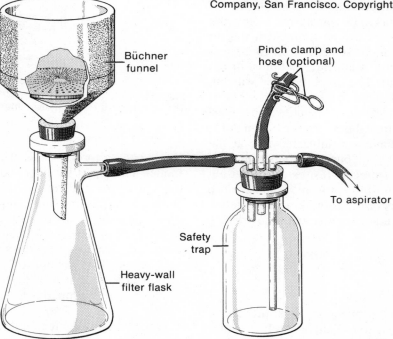

Büchner funnel

Pinch clamp and hose (optional)

To aspirator

Safety trap

Heavy-wall filter flask

FIGURE i-17
Filtration with the use of a Büchner funnel. [From J. W. Hagen, *Empirical Chemistry,* W. H. Freeman and Company. Copyright © 1972.]

INTRODUCTORY PRINCIPLES: PHYSICAL AND CHEMICAL PROPERTIES AND STOICHIOMETRY

MASS AND VOLUME RELATIONSHIPS

PRE-STUDY

Units of Measurement

Chemistry is distinctly an *experimental* science. The establishment of the truth of the fundamental laws and theories of the nature and behavior of matter depends on the careful measurement of various quantities—mass, volume, length, temperature, time, electrical magnitudes.

The metric system of units is especially convenient because it is a decimal system, like our system of numbers. The standard metric units were originally related to certain quantities in nature. For example, the meter was one ten millionth of the distance, on a meridian, from the earth's equator to the pole. Today the units are defined by the international prototype standards of the International Bureau of Weights and Measures, which preserves these standards in vaults near Paris, France. Each nation has its own metric standards, but all are based on and have been carefully compared with the prototypes. Practically all scientific research and development today uses metric units.

It is also true that most nations use metric units for everyday measurements. For example, in most cities of the world you would order half a kilogram (500 grams) of meat, if you wanted about a pound. If you wanted to know how far the next city was from the railroad station, you would be given the distance in kilometers (which you could then divide by 1.6 to obtain the number of miles). If the radio station reported the weather prediction as 30 degrees, you would head for the beach rather than the ski slopes because this temperature in degrees Centigrade is 86°F. You must buy milk (as well as wine) in liters, remembering that one liter is slightly larger than a quart.

Only the United States and a few small African nations remain on the foot–pound system rather than the metric system. However, a national commission has recommended that we convert to the metric system by 1980. This shift should be advantageous to almost everyone—from school children, to cooks, to sales persons. How long has it been since you encountered one of the following unpleasant problems?

If you add 13 gallons, 1 quart, 2 cups, and 13 liquid ounces *plus* 19 quarts and 19 liquid ounces, how many gallons of liquid in all will you have? The answer is 18.4 gallons.

$$13 + \frac{1}{4} + \frac{2}{16} + \frac{13}{128} + \frac{19}{4} + \frac{19}{128} = 18.4 \text{ gallons}$$

But if you needed to add quantities of four liquids that were expressed in metric units, the problem might be stated as follows: Add 51 liters, 40.0 milliliters, 190 deciliters, and 10 centiliters. The answer is 70.5 liters.

$$51 + 0.400 + 19.0 + 0.10 = 70.5 \text{ liters}$$

One has similar difficulties if he attempts to add 17 yards, 4 feet, and 10 inches; or 14 tons, 6872 pounds, and 4 ounces: these too would be eliminated by the metric system of weights and measures.

The metric unit of length is the *meter* (represented as m). Multiples and decimal fractions of this unit, as well as their symbols, are listed below.

1 km = 1 kilometer = 1000 meters

1 dm = 1 decimeter = 0.1 meter

1 cm = 1 centimeter = 0.01 meter

1 mm = 1 millimeter = 0.001 meter

1 μ = 1 micrometer = 10^{-6} meter

1 nm = 1 nanometer = 10^{-9} meter

Related units of length in common use in science are the *angstrom* (1 Å = 10^{-8} cm = 10^{-10} m) and the *millimicron* (1 mμ = 10^{-9} m = 1 nm).

The standard unit of mass is the *kilogram* (kg). The common submultiples of this mass are the *gram* (g) and the *milligram* (mg). The terms *mass* and *weight* are often used interchangeably by scientists although they represent different concepts. Mass is defined in terms of the Paris standard kilogram; weight is a *force*, actually a mass times the acceleration due to gravity. In chemistry one is concerned primarily with mass but, because virtually all scientific measurements are made on equal-arm balances, this dual terminology seldom causes problems.

The volume unit is the *liter*. This is a derived unit. It is the volume enclosed by a 1-decimeter cube. The *milliliter* (ml) is the most common volume unit in the chemical laboratory. Clearly, it is one-thousandth the volume of the liter. Recently, the milliliter was redefined slightly so that 1 ml $\equiv$ cm^3. (The triple equal sign means "is *exactly* equal to.")

A condensed table of the most commonly used metric units of length, mass, and volume, and their English equivalents, is included in Appendix C, Table 1.

Precision of Measurement

When measuring physical quantities it is important that the measuring devices be consistent with the precision desired. Thus, for a precision of 1% in the weighing of a 50-g sample, a balance which is accurate to only 0.5 g (1% $\times$ 50 g = 0.5 g) is required. For the same precision with a 1-g sample, a balance which is accurate to 0.01 g is necessary. In this experiment, platform balances which read to 0.1 g and graduated cylinders which can be read to about 0.2 ml should not cause uncertainty greater than about 1% in the calculated density. In a given determination, there is no point in using a greater precision for one measurement than for other measurements. Choose your measuring devices according to the precision desired.

The percentage error in an experiment is calculated by dividing the actual error by the accepted value, then multiplying the result by 100 to express it as percent, i.e., parts per hundred. See Appendix B for a more complete discussion of the treatment of experimental errors, the concept of precision, and related topics.

Density

The determination of this important physical property requires measurements of two quantities: the mass, M, and the volume, V, of a given amount of a substance. The ratio of these quantities, or the mass per unit volume, is the density, written $D = M/V$. In the metric system, this ratio is expressed as grams per cubic centimeter (g/cm^3) or grams per milliliter (g/ml). Study the relative densities of different substances as illustrated in Figure 1-1.

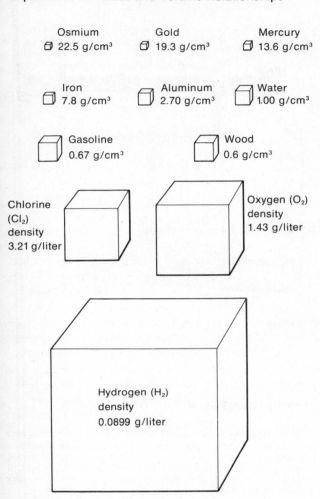

FIGURE 1-1
The relationship of density to volume. The cubes of
different sizes represent the relative volumes of equal
weights (about 0.2 g) of the various materials at 0°C.

The substances displayed in the figure visually
show the widely differing densities of a variety of
compounds and elements. It is apparent that liquids
and solids have much greater densities than gases.
Density is also influenced by temperature. The
densities of liquids and solids are affected only
slightly by changes in the temperature of the substance, but gas densities are quite sensitive to temperature changes.

The measurement of density is necessary for a
variety of important procedures in the science of
chemistry, such as the following: the calculation
of Avogadro's number from unit-cell dimensions
of crystals; the determination of the molecular

weight of a substance from its gas density (see
Experiment 12); the conversion of hydrostatic
pressure units (see Experiment 13); the conversion
from mass to volume (see Experiment 7); the measurement of densities of biopolymers in the ultra-
centrifuge; the determination of the concentration
of a solute from density measurements.

An understanding of density is also important
outside of the chemistry laboratory. For example,
the service station attendant determines the charge
of an automotive battery by measuring the density
(and hence the concentration) of the sulfuric acid
solution in the battery. Also, a winemaker (even
the amateur who makes his own wine at home)
measures the density of the grape juice to determine whether the sugar content is sufficient for
fermentation.

Measurement Techniques

Your instructor will demonstrate the correct techniques to use in reading the meniscus in a graduated
cylinder and in careful weighing with the balance.
See also Figures 1-2 and 1-3. For every weighing,

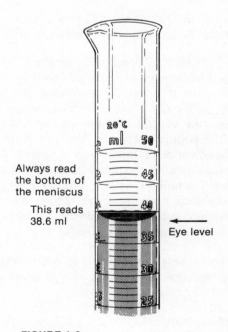

FIGURE 1-2
The proper method of reading a
meniscus (curved surface of a liquid)
in order to measure the volume of
the liquid.

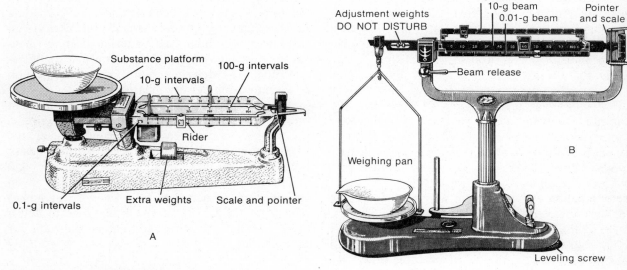

FIGURE 1-3
Laboratory balances. (A) A platform balance for crude weighing. (B) A triple-beam balance
for weighing quantities to ±0.01 g.

observe the following rules and precautions:

1. Keep the balance pans clean and dry. Clean up *immediately* any chemical which is spilled.

2. *Check the rest point of the empty balance.* To do this, first be sure all movable beam weights are at their zero position. Then release the fulcrum support, cause the balance beam to swing gently, and note the central position on the scale about which the pointer oscillates. Use this point as the reference-zero rest point in your weighings. (Never take readings with the beam and pointer at rest. Why?) If the pointer reading differs more than 2–3 scale divisions from the marked zero point, have your instructor adjust the balance. *Do not change the balance adjustments yourself.*

3. Never weigh an object while it is warm; the convection currents of warm air will affect the rest point.

4. If separate weights are used, always handle them with the weight forceps provided, never with the fingers. After weighing an object, return the weights to their proper place in the weight box, the beam weights to the zero position, and restore the fulcrum lift to its "rest" position.

EXPERIMENTAL PROCEDURE

Special Supplies: A metric rule. Liquid samples: various organic liquids, or solutions of unknown density prepared by dissolving inexpensive soluble salts, such as NaCl or $CaCl_2$, in water. Solid samples: coarse marble chips, a coarse silica sand, pieces of metal or metal shot, roll sulfur, or other solids (do not use any powdered material).

1. Density of a Liquid. Weigh a clean, dry, 150-ml beaker and watch-glass cover to the nearest 0.1 g, and record the weight in your experiment report. Place between 40 and 50 ml of the liquid into a graduated cylinder. Read the volume to a precision of 0.1 ml, then transfer as much of the liquid as possible to the beaker, and cover the beaker with the watch glass. Reweigh the beaker, its cover, and the contents. From these data, calculate the density of the liquid. Repeat all these measurements with a different volume of the same liquid, since the average of duplicate determinations will be more reliable, and will provide you with a check on gross errors in counting weights and reading volumes correctly.

2. The Density of a Solid. (a) Use the following technique to determine the density of an irregularly shaped solid. A sample will be designated by your instructor. Select 20 to 30 g of suitable-size pieces of the sample (avoiding fine powdered material), or such an amount as will give a volume increase in a graduated cylinder (see below) of a little less than 20 ml. Weigh a small beaker or evaporating dish to a precision of 0.1 g. Add the sample and weigh again. Place about 30 ml of water into the graduated cylinder and read the volume to 0.1 ml. Tilt the cylinder and slide the weighed sample pieces into it carefully, to avoid loss of water by splashing, then tap the sides to dislodge any adhering air bubbles. Again read the volume. The increase is the volume of the sample. (See Figure 1-4.) Calculate the density of the sample. If time permits, make a duplicate determination with a different weight of sample, to increase the accuracy of your determination.

(b) *Optional.* If you wish to determine the density of a regular-shaped solid, such as a rectangular block or a cylinder, the volume may be calculated from appropriate measurements of the dimensions. The weight may be obtained directly on the balance, and the density can then be calculated. In making these measurements, note how the quality and pre-

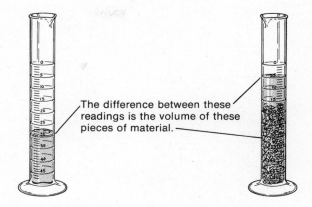

The difference between these readings is the volume of these pieces of material.

FIGURE 1-4
A method of measuring the volume of an irregularly shaped solid.

cision of the rulers and balances you use affect the number of significant figures you retain, and hence determine the degree of precision you can expect to obtain in your answer.

Mass and Volume Relationships

NAME	
SECTION	LOCKER
INSTRUCTOR	DATE

DATA AND CALCULATIONS

1. The Density of a Liquid

Identification of liquids used: 1 _____ 2 _____

Data	1	2
Weight of beaker, cover, and liquid	g	g
Weight of beaker and cover	g	g
Weight of liquid	g	g
Volume of liquid	ml	ml

Calculations[1]		1	2
Density from data		g/ml	g/ml
Density from literature		g/ml	g/ml
Percent error		%	%

[1]Always show the method of your calculations, including data with units, in the spaces provided, for all experiments throughout this manual. When two repetitive calculations are required (as in duplicate experiments), use this space to show only one of the calculations.

2. The Density of a Solid

Identification of irregular solids used: 1 _____ 2 _____

Data	1	2
Weight of beaker and solid	g	g
Weight of beaker	g	g
Weight of solid	g	g
Volume of liquid plus solid	ml	ml
Volume of liquid	ml	ml
Volume of solid	ml	ml

	Calculations	1	2
Density from data		g/ml	g/ml
Density from literature		g/ml	g/ml
Percent error		%	%

Optional. Enter below the measurements and weights of any regular-shaped solid that you measure, and calculate its density.

APPLICATION OF PRINCIPLES

NOTE: Study Appendix B on the use of dimensions and experimental errors in physical measurements. Always include the dimensional units and indicate all mathematical operations in the solution of problems, but omit actual arithmetic computation. Neatness is essential.

1. Exercises in the use of metric measurements.

   As an optional exercise, learn to approximate measurements in metric units at a glance.

   (a) Measure and record the length or diameter of laboratory utensils, such as test tubes (large and small), evaporating dishes, crucibles, etc.

   (b) Measure the precise volume in milliliters of test tubes, flasks, and beakers, and compare with the values stamped in the trademark.

   (c) Measure the length of this page in centimeters, and in inches, and calculate the conversion factor (ratio of centimeters to inches). Compare with the recorded value in Appendix C, Table 1.

   _____cm _____inch _____cm/inch

2. How would you modify the procedure of part 2 if you wished to determine the density of an irregular solid which is soluble in water, rock salt, for example, or sugar?

3. Would your density determinations be more precise if you weighed your samples to 0.01 g or 0.001 g, instead of only to 0.1 g, carrying out other measurements as before? Explain.

4. If 20 g of sulfur is used in the density determination, what is the limiting precision of measurement (%) if the weight is determined to a precision of 0.1 g?

_____%

5. What is the density of a brass sample if 50.0 g of coarse turnings, when placed in a graduated cylinder containing 10.3 ml of water, raises the level of the meniscus to a reading of 16.2 ml?

_____g/ml

6. What is the weight in pounds of a cubic foot of mercury, whose density is 13.6 g/cm³? (See Appendix C, Table 1, for any needed conversion factors.)

_____lb

THE LANGUAGE OF CHEMISTRY

INTRODUCTION

This study assignment provides you with a basic introduction to chemical nomenclature, and shows you how to write elementary chemical formulas, using the periodic table to correlate the ionic valences (or oxidation states) of the elements. It is necessary to establish rules for naming compounds in order to avoid the massive confusion that would exist if each person invented his own names and chemical symbols for the elements and their compounds. Consequently, chemists from all over the world meet periodically under the auspices of the International Union of Pure and Applied Chemistry (IUPAC) to agree on systematic rules for naming chemical compounds and writing formulas. We will present a few of these rules to help you begin to learn the language of chemistry.

As you study the behavior of the elements and their compounds you will begin to see how their properties are systematically correlated by the arrangement known as the *periodic table of the elements*. As you undertake the study of any given element, you should make frequent reference to the periodic table, printed on the inside front cover.

CHEMICAL SYMBOLS AND FORMULAS

We have already used chemical symbols and the formulas of substances in the first experiment, and we have written equations for a number of chemical reactions. Now we must pause to emphasize the exact meaning and correct usage of these and other terms that constitute the unique language of chemistry.

Each element is represented by a *chemical symbol*. The symbol consists of either one or two letters, such as C for carbon or Ba for barium. Several of the elements have symbols derived from their ancient Latin names: Cu for copper from *cuprum*; Fe for iron from *ferrum*; Au for gold from *aurum*; Pb for lead from *plumbum,* and Ag for silver from *argentum*. Several elements discovered after 1780 have names derived from Latin or Germanic stems: Na for Sodium from *natrium*; K for potassium from *kalium*; and W for tungsten from *wolfram*.

A *chemical formula* represents the composition of a given *substance,* which may be either an element or a compound. Thus H and O are the symbols for the elements hydrogen and oxygen, and they can also represent the atomic state of the elements, whereas H_2 (hydrogen gas) and O_2 (oxygen gas) represent the more stable molecular forms of the elements hydrogen and oxygen. When one speaks of the chemical properties of oxygen, it is usually the stable molecular form of oxygen, O_2, that is meant. In this manual we will always try to specify the chemical formula of a substance to avoid the possibility of any confusion or misinterpretation.

A *chemical compound* is a substance formed from two or more elements, such as H_2O (water), H_2O_2 (hydrogen peroxide), or NaCl (sodium chloride).

NOMENCLATURE, VALENCE, AND FORMULAS

In January, 1965, *Chemical Abstracts* (a periodical in which abstracts from all chemistry journals appear) started a compound registry index. This index contained more than 2 million different chemical substances by 1972 and was growing at the rate of 300,000 new substances each year. It is estimated that about 6 million chemical substances are known, the great majority of them being organic compounds (those containing carbon). Mastering the details necessary to name all of these substances would require a great deal of study. Fortunately, we work most of the time with a limited number of chemical substances, and therefore you will not need to learn more than a few hundred names. This task is made easier still because most of the names are established according to simple rules, and the elements are grouped in chemical families of the periodic table, those within each family bearing strong resemblances to one another. In addition, you do not have to learn all of the names at once. Just as you are able to learn the names of new friends, a few at a time, you will find it easy to learn the names of new chemical substances little by little, as you encounter them.

However, one problem cannot be avoided. The grand traditions of chemistry go back several centuries, and some compounds are commonly referred

to by names coined years ago, as well as by their more systematic names. Consequently, for some substances it is necessary to learn both the *common* or *trivial* name and the systematic name. Fortunately, the use of older common names is diminishing: for example, in industrial commerce it was once common to refer to sodium carbonate (Na_2CO_3) as "soda ash," but today you are not likely to find in a chemistry laboratory a bottle of sodium carbonate labeled as soda ash. Other substances, such as many of the organic compounds, are most often referred to by their common names. Thus the compound CH_3COOH is almost always called *acetic acid,* rather than by the more systematic name *ethanoic acid,* which reveals it to be a derivative of the parent hydrocarbon, ethane. For organic compounds of complex structure, common names are invented because the systematic names would be entirely too long and cumbersome for everyday use.

Binary Compounds and the Periodic Table

The periodic table can be roughly divided into two broad classes of elements, metals and nonmetals, by a zigzag line, which starts to the left of the element boron in Group III and runs down to the left of astatine in Group VII (refer to the periodic table on the inside front cover). The metallic elements lie to the left of the zigzag line and the nonmetallic elements to the right. When the metallic elements of Groups I and II react with the nonmetallic elements of Groups VI and VII, simple *binary compounds* are formed. Binary compounds contain only two elements. Their formulas are the simplest to build. In the naming of such compounds, the metallic element is given first; it is followed by the root of the name of the nonmetallic (or less metallic) element; the compound name is completed by the ending *-ide*. Some binary compounds are: NaCl, sodium chlor*ide*; CaO, calcium ox*ide*; $MgBr_2$, magnesium brom*ide*; H_2S, hydrogen sulf*ide*; AlI_3, aluminum iod*ide*; Mg_3N_2, magnesium nitr*ide*; K_2O, potassium ox*ide*.

Ionic Valence, Oxidation State, and the Periodic Table

Binary compounds may be approximately classified into two types: covalent compounds in which the chemical bonds are formed by sharing of the electrons and ionic compounds in which electrons are transferred from the metal to the nonmetal to form ions. The metallic elements form positive ions (thus having   positive electric charge) and the nonmetallic elements form negative ions (having a negative electric charge). In the formation of an ionic compound, each element in its ionic form achieves the electronic structure of an inert gas, an electron configuration that is notable for its stability. For example, when potassium reacts with oxygen, O_2, to form the compound potassium oxide (K_2O), each potassium atom loses an electron to become a potassium ion, K^+; in the process it achieves the same electron configuration as the inert gas argon. Each oxygen atom acquires two electrons to form an oxide ion, O^{2-}, which has an electron configuration like neon.

The charge possessed by an ion is sometimes called the ionic *valence*[1] because it determines the number of atoms of an element that will be contained in the smallest unit (or empirical formula) of a compound containing that element. For example, the potassium ion has an ionic valence of +1 and the oxide ion has an ionic valence of −2: therefore one unit of potassium oxide must contain 2 potassium atoms for each oxide ion, so that the number of negative and positive charges will balance exactly. Thus the formula for this oxide is K_2O.

The term *oxidation state* (or *oxidation number*) has come into widespread use and is often used instead of the term ionic valence. Neutral atoms of each element are by definition assigned oxidation number zero, represented by an Arabic 0. When one or more electrons are removed from a neutral atom that atom becomes a positively charged ion; when one or more electrons are added to the neutral atom, it becomes a negatively charged ion. For ions containing a single atom, the ionic charge is the same as the oxidation state. The metals of Group I give up one electron easily to form ions of charge +1. Thus we assign an oxidation number (or valence) of +1 to the alkali metals in all of their simple binary compounds: Li^+, Na^+, K^+, Rb^+, and Cs^+. The most common oxidation state of hydrogen is +1. The Group II metals easily give up two electrons to form ions of oxidation state +2 in their compounds: Be^{2+}, Mg^{2+}, Ca^{2+}, Sr^{2+}, Ba^{2+}, and Ra^{2+}; and the Group III metals give up three electrons easily to form ions of oxidation state +3 in their compounds: Al^{3+}, Sc^{3+}, and so on.[2]

All of the metal ions are given the same name as the element: Na^+ is called sodium ion; Mg^{2+} is called magnesium ion; Sc^{3+} is called scandium ion, and so on.

Like the metals, the nonmetals also show strong family resemblances in their oxidation states or ionic valences. The halogens, or Group VII elements, each tend to accept one electron to form an ion with oxidation number −1 in their simple compounds: F^-, Cl^-, Br^-, I^-. (Hydrogen atoms can also behave like the members of this group in reacting with the most electropositive metals to form ionic metal hydrides in which the hydrogen has an oxidation state of −1.) An oxidation number of −2 is characteristic of the Group VI elements in their binary compounds: O^{2-}, S^{2-}, Se^{2-}, Te^{2-}. The elements of Group V are assigned an oxidation number of −3 in many of their compounds: N^{3-}, P^{3-}, and so on.[3] A few Group IV ions are assigned oxidation number −4: C^{4-} and Si^{4-}. Each of these ions has an electron configuration, which is like that of the inert gas element that immediately follows the element in the same period; thus F^- has the electron configuration of neon, S^{2-} has the electron configuration of argon, and so on.

Thus you can see that certain patterns or regularities correlate with the periodic table: (1) all of the elements in a particular Group tend to have the same oxidation state; and (2) each element tends to achieve a stable electron configuration like that of an inert gas — the metallic elements tending to lose electrons to become positive ions and the nonmetallic elements tending to gain electrons to become negative ions.

---

[1]The word valence may be associated with "bond forming capacity" and is a term which appears to be rapidly falling into disuse. Other aspects of the topic of bonding, such as covalence and coordination number, will be considered more thoroughly in Experiment 21. A more complete discussion of nomenclature is given in Study Assignment D.

[2]Because of its high charge and small size, B^{3+} is not usually present as a simple ion, but tends to form complex ions with another element, particularly with oxygen, as in boric acid (H_3BO_3), which contains the borate ion, BO_3^{3-}. In this complex ion, the oxygen atoms are assigned oxidation state −2 so that boron is still assigned oxidation state +3. For the same reasons, H^+ (bare proton) does not exist as a discrete entity in solids and liquids. It is always attached to another species, for example, to the water molecule (H_2O) in the hydronium ion, H_3O^+.

[3]The heavier group V elements As and Sb are on the borderline between the metals and nonmetals and show many metal-like properties. For this reason they are sometimes called semimetals. They do not as readily form ions with oxidation state −3, as do N and P.

Binary Compounds Containing Two Nonmetallic Elements (Covalent Compounds)

When elements in Groups I and II react with elements in Groups VI and VII to form simple binary compounds, it is reasonable to assume that electrons are transferred from the metallic element to the nonmetallic element to form ions. It is on this basis that we assign the ionic valence or oxidation number for these elements. However, we must recognize that as we move inward to Groups III, IV, and V we may assign a formal oxidation number on the basis of the assumption that the element has achieved an inert gas configuration; but we also know from extensive studies that these elements form many compounds in which the electrons are shared to form a *covalent,* rather than ionic, chemical bond. The electrons are not transferred completely from the less electronegative to the more electronegative element. In these compounds the valence can be thought of as a measure of the number of bonds that can be formed by the element. Carbon in Group IV can share its four valence electrons to form four chemical bonds with other atoms, rather than forming an ionic carbide. In binary compounds of the nonmetallic elements the oxidation number is assigned *in a purely formal way; it does not represent the actual charge on the atoms.* In the naming of the binary compounds containing two nonmetallic elements, the less electronegative element is named first; the more electronegative element is named last; the prefixes *mono-, di-, tri-, tetra-, penta-, hexa-, hepta-, octa-,* and so on are attached, if necessary, to indicate the number of atoms of each element (the prefix *mono-* is usually omitted for the less electronegative element). The following are examples (with the oxidation state of each element indicated above by a small Arabic numeral):

$^{(+2)}CO^{(-2)}$ carbon monoxide

$^{(+4)}CO_2^{(-2)}$ carbon dioxide

$^{(+4)}SO_2^{(-2)}$ sulfur dioxide

$^{(+6)}SO_3^{(-2)}$ sulfur trioxide

$^{(+3)}N_2O_3^{(-2)}$ dinitrogen trioxide

$^{(+5)}N_2O_5^{(-2)}$ dinitrogen pentaoxide

Note that we have assigned oxygen an oxidation state of -2 in each of these binary oxides. The less electronegative element has been assigned the oxidation state that it must have in order for the net charge on the molecule to be zero, since these are all neutral covalently bonded molecules. Note also that in these compounds the *oxidation states can be variable*: for example, sulfur has oxidation state of $+4$ in sulfur dioxide and $+6$ in sulfur trioxide.

Some compounds, such as water (H_2O), ammonia (NH_3), and phosphine (PH_3), have common names that were given to them before the nomenclature of compounds was systematized, and these names are still retained. We will consider the nomenclature of inorganic compounds in greater detail in Study Assignment D.

Multiple Oxidation States of the Nonmetals

In the foregoing discussion we have stressed that the metals lose electrons and the nonmetals gain electrons to form monoatomic ions with an inert gas electron configuration. However, most nonmetals also form species in which the nonmetal atom is assigned an intermediate oxidation state. In these combinations the inert gas electron configuration is achieved by sharing of electrons with other atoms to form *polyatomic* species. For example, the oxidation states of carbon can range from -4 in the carbide ion, C^4, to $+4$ in the compound carbon dioxide, CO_2, in which oxygen is assigned its characteristic oxidation state of -2 and carbon is therefore required to have oxidation state $+4$. Similarly, nitrogen can have oxidation states that vary from -3 in the nitride ion, N^{3-}, to $+5$ in the nitrate ion, NO_3^-. Note that the difference between the most positive and most negative oxidation states is exactly eight, because the $2s$ and $2p$ electron shells have a total capacity of eight electrons. The most positive oxidation state is obtained when all of the electrons are formally removed; the most negative when the $2s$ and $2p$ orbitals are completely filled. The names of the negative ions of the nonmetals have a characteristic *-ide* ending and are shown in Table A-1.

TABLE A-1
Negative ions of the nonmetals

Group IV	Group V	Group VI	Group VII
			H^-, hydride
C^{4-}, carbide	N^{3-}, nitride	O^{2-}, oxide	F^-, fluoride
Si^{4-}, silicide	P^{3-}, phosphide	S^{2-}, sulfide	Cl^-, chloride
		Se^{2-}, selenide	Br^-, bromide
		Te^{2-}, telluride	I^-, iodide

Metals with Variable Valence

Some metals, notably the *transition* metals (atomic numbers 22 through 32 and those below them in the periodic table), show variable oxidation states. For elements having just two common oxidation states, the ending *-ous* is used to show the lower oxidation state, and the ending *-ic*, the higher.[4]

The more common elements with just two oxidation states are the following.

-ous	Fe^{2+}	Co^{2+}	Cu^+	Au^+	Hg_2^{2+}	Sn^{2+}
-ic	Fe^{3+}	Co^{3+}	Cu^{2+}	Au^{3+}	Hg^{2+}	Sn^{4+}

The following are names and formulas of a few corresponding compounds.

$FeBr_2$,	ferr*ous* bromide	Hg_2O,	mercur*ous* oxide
$FeBr_3$,	ferr*ic* bromide	HgO,	mercur*ic* oxide
$CoCl_2$,	cobalt*ous* chloride	$SnCl_2$,	stann*ous* chloride
$CoCl_3$,	cobalt*ic* chloride	$SnCl_4$,	stann*ic* chloride

Oxygen Acids and their Salts

It will be to your advantage to memorize the names and formulas for a very few common oxygen acids, because by doing so you will learn the ionic charges of the corresponding negative radicals.[5] Note that if the name of an oxygen acid ends in *-ic*, the name of the corresponding negative radical ends in *-ate*.

$HClO_4$,	perchlor*ic* acid	ClO_4^-,	perchlor*ate* ion
H_2SO_4,	sulfur*ic* acid	SO_4^{2-},	sulf*ate* ion
HNO_3,	nitr*ic* acid	NO_3^-,	nitr*ate* ion
H_3PO_4,	phosphor*ic* acid	PO_4^{3-},	phosph*ate* ion
H_2CO_3,	carbon*ic* acid	CO_3^{2-},	carbon*ate* ion

There are also a number of oxygen acids in which the central nonmetal atom has a lower oxidation state than those shown above. The names of these oxygen acids end in *-ous*. The names of the anions (negative ions)—produced by removing a proton (H^+) from the parent acid, end in *-ite*.

$HClO_2$,	chlor*ous* acid	ClO_2^-,	chlor*ite* ion
H_2SO_3,	sulfur*ous* acid	SO_3^{2-},	sulf*ite* ion
HNO_2,	nitr*ous* acid	NO_2^-,	nitr*ite* ion

In general, acids and salts containing three different elements (ternary compounds) are named according to a different system than that used in naming binary compounds. This system is considered in detail in Study Assignment D.

To complete the writing of a formula, simply insert such subscripts after each element or radical as will balance the total positive and total negative charges; for example, $(Na^+)(Cl^-)$ or $NaCl$, $(Ca^{2+})(Cl^-)_2$ or $CaCl_2$, $(Mg^{2+})(NO_3^-)_2$ or $Mg(NO_3)_2$, $(Ca^{2+})(O^{2-})$ or CaO, $(Ca^{2+})(SO_4^{2-})$ or $CaSO_4$, and $(Ba^{2+})_3(PO_4^{3-})_2$ or $Ba_3(PO_4)_2$.

Some Additional Rules of Nomenclature

Hydroxides are named according to the system used for binary compounds, even though there are three elements present. The hydroxide ion (OH^-) is a rather stable group of elements and is named as though it were a single nonmetal ion such as chloride (Cl^-). For example, $NaOH$ is sodium hydrox*ide*, and $Al(OH)_3$ is aluminum hydrox*ide*.

The ion NH_4^+ behaves very much like the alkali metal ions (Na^+, K^+, Rb^+) and is given the name "ammonium ion." It is named as though it were a simple metal ion. For example, NH_4Cl is ammonium chlor*ide*, and $(NH_4)_2S$ is ammonium sulf*ide*.

The hydrogen compounds of some of the nonmetals (e.g., HF, HCl, HBr, HI, H_2S, H_2Se) are named in the usual way, with the electropositive hydrogen specified first: HCl (hydrogen chloride), HBr (hydrogen bromide), H_2S (hydrogen sulfide), H_2Se (hydrogen selenide), and so on. The aqueous solutions of the hydrogen halides form strong acid solutions when dissolved in water. These *aqueous solutions* are often known by their common names:[6]

HF (hydrogen fluoride) becomes *hydro*fluor*ic acid*;

HCl (hydrogen chloride) becomes *hydro*chlor*ic acid*;

HBr (hydrogen bromide) becomes *hydro*brom*ic acid*;

HI (hydrogen iodide) becomes *hydro*iod*ic acid*.

---

[4]You will learn more systematic systems of nomenclature subsequently. The *Ewens-Bassett* system represents ions by writing the net charge on the ion as a superscript, e.g., Fe^{2+} or Fe^{3+}. The *Stock* system uses Roman numerals to represent different oxidation numbers of the elements; thus Fe^{2+} is called iron(II) and may be symbolized by Fe(II) or Fe[II]; Fe^{3+} is called iron(III) and may be symbolized by Fe(III) or Fe[III]. See Study Assignment D.

[5]The term "radical" is applied to a group of atoms, either neutral or ionic, which often act as a unit in a chemical reaction.

---

[6]Note that these names are not systematic names because the ending *-ic* is usually reserved for oxyacids. The use of names like hydrosulfuric acid for an aqueous solution of hydrogen sulfide is discouraged. See "Nomenclature of Inorganic Chemistry, 2nd edition," Definitive Rules 1970, *Pure and Applied Chemistry,* **28**, 1–106 (1971), page 22.

WRITING CHEMICAL EQUATIONS

A chemical equation is the chemist's shorthand expression of a chemical reaction. Fundamentally, it expresses the *identity* of each reactant and product and the relative quantities of each, termed the *stoichiometry* of the reaction. The word stoichiometry is derived from the Greek words *stoikheion* (element) and *metron* (measure). In a balanced equation, all of the atoms of the reactants must reappear in the products; that is, the equation is not balanced until there is the same number of each kind of atom on both sides of the equation. Thus, the reaction

$$Zn(s) + HCl(aq) \rightarrow ZnCl_2(aq) + H_2(g)$$

is not balanced. (The labels in parentheses specify the physical state of the reactants and products: that is, *g* refers to gas, *s* to solid, and *aq* to aqueous solution.) If the reaction is to be balanced, it is necessary to have 2HCl, since two atoms each of hydrogen and of chlorine are present, as shown on the right hand side of the equation. The balanced reaction is written as

$$Zn(s) + 2HCl(aq) \rightarrow ZnCl_2(aq) + H_2(g)$$

The following procedures are important for writing a correct, balanced equation:

1. Be sure that you know the correct formulas of all the reactants and products. You cannot proceed to write a proper chemical equation until you know the identity of each reactant and product. You must not change the subscripts of the formulas in order to balance the equation — that would be equivalent to changing the identity of the reactants or products.

2. Determine the coefficients that must appear in front of the formulas in order for the same number of atoms of each element to be present on each side of the equation. The coefficients in most simple reactions can be determined by inspection. A good policy is to start with the most complicated formula.

As an example, when steam is passed over red hot iron, analysis shows that the oxide formed is Fe_3O_4, magnetic iron oxide.[7] We first write the correct formulas of the reactants and the products (which have been determined by experiment):

$$Fe(s) + H_2O(g) \rightarrow$$
$$Fe_3O_4(s) + H_2(g) \qquad \text{(unbalanced)}$$

To balance the equation, we start with the Fe_3O_4. We note that $3\,Fe$ and $4H_2O$ are required on the left. This gives 8H on the left, so we must balance with $4H_2$ on the right:

$$3Fe(s) + 4H_2O(g) \rightarrow$$
$$Fe_3O_4(s) + 4H_2(g) \qquad \text{(balanced)}$$

---

[7]Note that if the conventional oxidation state of -2 is assigned to oxygen in Fe_3O_4, the oxidation state assigned to iron would have to be $+\frac{8}{3}$. It could also be regarded as a mixed valence compound $FeO \cdot Fe_2O_3$ with $\frac{1}{3}$ of the Fe atoms assigned oxidation state $+2$ and $\frac{2}{3}$ of the Fe atoms assigned oxidation state $+3$. This would give a weighted average oxidation state of $+\frac{8}{3}$ since $\frac{1}{3} \times 2 + \frac{2}{3} \times 3 = \frac{8}{3}$.

The Language of Chemistry

NAME

SECTION   LOCKER

INSTRUCTOR   DATE

EXERCISES ON FORMULAS AND NOMENCLATURE

NOTE: These exercises will help you learn how to write correct formulas and name compounds. Check your answers, if necessary, with your instructor.

1. Name the following.

FeI_2 _____ H_2CO_3 _____

I_2 _____ $CaCO_3$ _____

$FeCl_3$ _____ Be_2C _____

$Fe_2(SO_4)_3$ _____ $SnSO_4$ _____

FeS _____ $(NH_4)_2S$ _____

NCl_3 _____ N_2O_4 _____

PCl_5 _____ BaO _____

$La(NO_3)_3$ _____ $Sr(OH)_2$ _____

Mg_3N_2 _____ $Hg_2(NO_3)_2$ _____

K_2SO_4 _____ $HgCl_2$ _____

KNO_2 _____ $Ba(NO_3)_2$ _____

2. Write the correct chemical formulas.

Barium chloride _____   Ammonium sulfate _____

Stannous nitrate _____   Barium carbonate _____

Stannic nitrate _____   Sodium carbonate _____

Aluminum carbide _____   Sodium hydrogen carbonate _____

Magnesium phosphate _____   Calcium hydrogen carbonate _____

Nitrogen dioxide _____   Disulfur dichloride _____

Ferrous oxide _____   Cesium dihydrogen phosphate _____

Ferric sulfide _____   Cesium monohydrogen phosphate _____

Cobaltous chloride _____   Mercurous chloride _____

Cobaltic nitrate _____   Calcium nitride _____

3. The spaces below represent portions of some of the main groups and periods of the periodic table. In the proper squares, write the correct formulas for the chlorides, oxides, and sulfates of the elements of main groups I, II, and III, respectively. Likewise, write the formulas of the compounds of sodium, calcium, and aluminum with the elements of main groups VI and VII. Two of the squares have been completed as examples.

	Group I	Group II	Group III		Group VI	Group VII
Period 2	LiCl Li$_2$O Li$_2$SO$_4$		(omit sulfate)			
Period 3					Na$_2$S CaS Al$_2$S$_3$	
Period 4						
Period 5						

4. Give the ionic charge (including + or −) for the italicized element or radical in each of the following.

Cu_2O _____ $La_2(SO_4)_3$ _____ NH_4OH _____ H_3PO_4 _____ SnS_2 _____

$CuSO_4$ _____ Ca_3N_2 _____ SnO _____ $MgNH_4PO_4$ _____ $TiCl_4$ _____

Cr_2O_3 _____ H_3AsO_4 _____ $Mg(NO_3)_2$ _____ XO_2 _____ CaZ_2 _____

5. From the valences that you found for certain elements, including the hypothetical elements X and Z above, determine the formulas of the following by filling in the proper subscripts.

$H_( )Z_( )$ $X_( )(OH)_( )$ $Cr_( )(SO_4)_( )$ $Cr_( )Z_( )$ $X_( )S_( )$

$(NH_4)_( )Z_( )$ $Mg_( )(AsO_4)_( )$ $As_( )Z_( )$ $X_( )(PO_4)_( )$ $X_( )Z_( )$

6. Balance the following chemical reactions by writing the proper stoichiometric coefficient in front of each reactant and product.

_____ Ba(s) + _____ HNO$_3(aq)$ → _____ H$_2(g)$ + _____ Ba(NO$_3$)$_2(aq)$

_____ Fe(s) + _____ O$_2(g)$ → _____ Fe$_2$O$_3(s)$

_____ S$_8(s)$ + _____ O$_2(g)$ → _____ SO$_2(g)$

_____ Na$_3$PO$_4(aq)$ + _____ CaCl$_2(aq)$ → _____ Ca$_3$(PO$_4$)$_2(s)$ + _____ NaCl(aq)

SOME COMMON
PHYSICAL PROPERTIES OF
SUBSTANCES

PRE-STUDY

In this experiment the emphasis is on observation. Our responses to a substance we observe can be primarily sensory—so that our observation is based on the usual spontaneous perceptions of smells, colors, shapes, weight and textures. Who can forget seeing mercury for the first time, its mobile shiny surface perfectly described as "quicksilver," or its surprisingly heavy weight? But much more remains to be discovered about mercury—the variety of its chemical reactions, as well as its potent toxicity.

The trained scientific observer will perceive much more, noticing subtleties that escape the untrained senses. In addition the trained observer may structure his own experience by making observations in a controlled way—this procedure being the essence of the experimental method. Each experiment may produce new results that must be incorporated into some framework of interpretation. The step of interpretation is very im-

portant. Without it, our observations would amount only to a jumble of facts.

Among your goals in a first course in chemistry are to train your senses to observe more than what is immediately perceived, to train your mind to draw inferences from the observations, and to learn to construct a framework for interpretation of these experiences. To provide this framework in the broadest sense, you will correlate physical and chemical properties with the chemical structure of atoms and molecules.

The Kinds of Matter

A chemist deals in substances, and in the changes which these undergo, both in nature and in the laboratory. Carefully study Figure 2-1, and also review in your text the definitions of the following terms, which classify and describe the kinds of matter.

By a *substance* the chemist means a material, all samples of which are identical in composition,

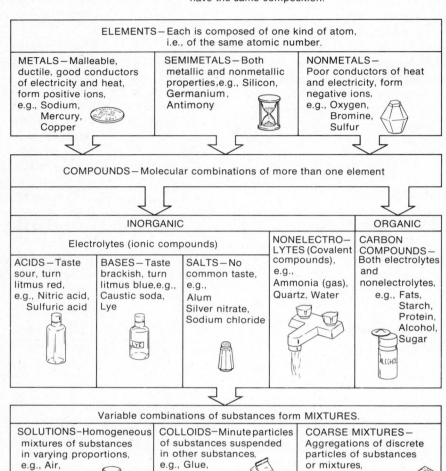

SUBSTANCES—All samples of a given substance have the same composition.

ELEMENTS—Each is composed of one kind of atom, i.e., of the same atomic number.

METALS—Malleable, ductile, good conductors of electricity and heat, form positive ions, e.g., Sodium, Mercury, Copper

SEMIMETALS—Both metallic and nonmetallic properties,e.g., Silicon, Germanium, Antimony

NONMETALS— Poor conductors of heat and electricity, form negative ions, e.g., Oxygen, Bromine, Sulfur

COMPOUNDS—Molecular combinations of more than one element

INORGANIC

ORGANIC

Electrolytes (ionic compounds)

ACIDS—Taste sour, turn litmus red, e.g., Nitric acid, Sulfuric acid

BASES—Taste brackish, turn litmus blue,e.g., Caustic soda, Lye

SALTS—No common taste, e.g., Alum Silver nitrate, Sodium chloride

NONELECTRO-LYTES (Covalent compounds), e.g., Ammonia (gas), Quartz, Water

CARBON COMPOUNDS— Both electrolytes and nonelectrolytes, e.g., Fats, Starch, Protein, Alcohol, Sugar

Variable combinations of substances form MIXTURES.

SOLUTIONS—Homogeneous mixtures of substances in varying proportions, e.g., Air, Glass, Brass, Syrup

COLLOIDS—Minute particles of substances suspended in other substances, e.g., Glue, India ink, Milk, Egg white, Smoke

COARSE MIXTURES— Aggregations of discrete particles of substances or mixtures, e.g.,Wood, Concrete, Granite

FIGURE 2-1
The kinds of matter.

such as water, baking soda, common salt, and copper. A *mixture* is made up of more than one substance. A mixture so intimate and homogeneous that its several components are present as one phase (either solid, liquid, or gaseous), without separate boundaries, is called a *solution*. A mixture containing two phases—in which one phase is a continuous medium (either a gaseous, liquid, or solid substance), and the other phase consists of particles (too small to be recognized by the naked eye) that remain suspended and do not settle out of the mixture—is called a *colloid*.

The first problem of the chemist is to recognize the various substances with which he deals. The qualities or characteristics that distinguish one substance from another are called **properties**. *The chemist identifies substances by their properties.* Some of these properties, such as color, odor, density, melting and boiling points, crystalline form, hardness, malleability, ductility, and thermal and electrical conductivity, are called *physical properties*. These can all be observed without changing the identity of the substance. In studying the *chemical properties* of a substance, however, the chemist notes the changes it undergoes in being transformed into one or more different substances. A property that is specific enough to identify a certain substance, particularly if interfering substances have been removed, is the evidence one looks for as a *test for a substance.*

In this experiment, we are concerned with some of the physical properties of certain substances. Experiment 3 will introduce some of the chemical properties of many of these same substances. Table 2-1 lists these substances, and includes some of their more important physical properties.

EXPERIMENTAL PROCEDURE

Special Supplies: Low-power magnifier for part 3.

Chemicals: Cu (turnings), Zn (mossy), S (roll, crushed), $I_2(s)$ HCl (dilute), $Ca(OH)_2(s)$, $CuCO_3(s)$, $Cu(NO_3)_2 \cdot 3H_2O(s)$, $KClO_3(s)$, and other salts for crystallization, as $NaCl(s)$, $NaNO_3(s)$, $KNO_3(s)$; $CS_2(l)$, sand.

NOTE: Reread the safety rules and laboratory regulations, given at the beginning of the Introduction. Study the suggestions in Figures 2-3 and 2-4 for handling and removing chemicals from reagent bottles without spilling them. Do not waste chemicals by removing from the bottle more than you need.

1. Preliminary Observations. (a) *General Characteristics, Density, Melting Point, and Solubility.* Study Table 2-1 in the pre-study section, and compare the listed properties with the actual appearance and general characteristics of the several chemicals *as you observe them.* Note the range of densities, melting points, and solubilities, and the variation in these depending on the type of substance. Note that certain substances decompose chemically when heated rather than melting as the same substance.

(b) *Odor and Taste.* Smell any substances cautiously by fanning the vapors toward the nose (see

TABLE 2-1
Physical properties of some substances

Substance	Some general characteristics	Density (g/cm³)	Melting Point (°C)	Solubility (g/100g H₂O, 20°C)
Copper, Cu (a metal)	Yellow-red, cubic crystals, good electrical conductor	8.93	1083	Insoluble
Zinc, Zn (a metal)	Bluish-white, hexagonal crystals, good electrical conductor	7.14	419	Insoluble
Sulfur, S (a nonmetal)	Yellow, rhombic crystals, non-conductor of electricity	2.07	113	Insoluble
Iodine, I_2 (a nonmetal)	Blue-black, rhombic, flat crystals, metallic luster	4.93	114.2[1]	0.029
Hydrogen chloride, HCl (acid in water solution)[2]	Colorless gas, very sharp odor	0.00163	−112	72 (475 liters/liter H₂O)
Calcium hydroxide, $Ca(OH)_2$ (a base)	White powder	2.34	dec. 580 −H₂O	0.165
Cupric carbonate, $CuCO_3$[3] (a salt)	Green powder	3.9	dec. 200	Quite insoluble
Cupric nitrate ($Cu(NO_3)_2 \cdot 3H_2O$)[3] (a salt)	Blue crystals, deliquescent	2.05	dec. 170 −HNO₃	55.6[4]
Potassium chlorate, $KClO_3$ (a salt)	Colorless, monoclinic crystals	2.32	368	7.4
Sodium chloride, NaCl (a salt)	Colorless, cubic crystals	2.16	801	35.8

[1]Iodine has a vapor pressure of 90.1 mm Hg at its melting point, hence it sublimes readily even below this temperature.
[2]Hydrochloric acid is a solution of the gas in water. The "dilute" laboratory reagent contains 21.9 g HCl per 100 ml solution, with a density of 1.10 g/cm³.
[3]Cupric carbonate is usually the basic salt, $CuCO_3 \cdot Cu(OH)_2$.
[4]Cupric nitrate crystallizes from solution as $Cu(NO_3)_2 \cdot 6H_2O$ below 26.4°C, and as $Cu(NO_3)_2 \cdot 3H_2O$ above that temperature. The solubility is given in grams of anhydrous salt per 100 g H₂O.

the Introduction, Safety Rule 5). Never taste chemicals except as directed, and then only by touching a particle or drop to the tongue. Do not taste copper salts, since they are somewhat poisonous. Other substances in this experiment may be tasted, but first dilute the hydrochloric acid—a few drops in 2 ml of water, then add more if you cannot taste the acid.

NOTE: Perform both or either parts 2 and 3 according to the time available, as directed by your instructor.

---

2. The Quantitative Analysis of an Unknown Mixture Based on Solubility.   Obtain from your instructor a 3- to 5-g sample of a mixture of sand and common salt, NaCl, of unknown composition. (Exchange one of your small test tubes for the unknown contained in a similar test tube.) Weigh a clean, dry 150-ml beaker and a clean, dry evaporating dish, each to ± 0.01 g on the semianalytical triple-beam balance. Also weigh your test tube containing the unknown sample. Without loss, transfer all of the unknown to the weighed beaker. Again weigh the empty test tube, and the beaker plus the unknown. (The differences in weight permit you to check on the weight of your unknown sample.)

Add about 10 ml of distilled water to the unknown in the beaker, gently warm this, with stirring, to an estimated 60–80°C, then let the mixture settle, and carefully decant the clear solution down the stirring rod into the weighed evaporating dish (see Figure 2-2). Add a second 10-ml portion of distilled water to the sand mixture, and repeat the extraction and decantation. Place the evaporating dish and solution over a wire gauze on a ring stand, and begin careful evaporation, using a small flame (see Figure 2-3). The complete extraction of any remaining salt from the sand will probably require that a third and possibly a fourth 10-ml portion of distilled water be added to the evaporating solution. Finally, decant as much water as possible from the sand (without any loss of sand), and gently (at first) heat the beaker of moist sand, until it is thoroughly dry. Let this cool, and weigh it.

As the solution in the evaporating dish becomes concentrated and crystals form, be very careful to avoid spattering. Pass the flame around the edges of the dish, but not directly under it. When the salt is completely dry, let the dish cool completely, and weigh it. (Your weights now give you a duplicate check on the weight of salt in the unknown.) From these data, calculate the percent of NaCl in your unknown mixture.

FIGURE 2-2
The separation of a heavy solid by decantation of the supernatant liquid.

Heat the mixture very gently.

FIGURE 2-3
The procedure for evaporating a liquid.

FIRST METHOD

First—   Roll and tilt the bottle until some of the contents
         enters the inside the plastic cap.

Second— Carefully remove the cap so that some
         of the contents remains in it.

Third—   Tap the cap with a pencil until the
         desired amount falls out.

SECOND METHOD

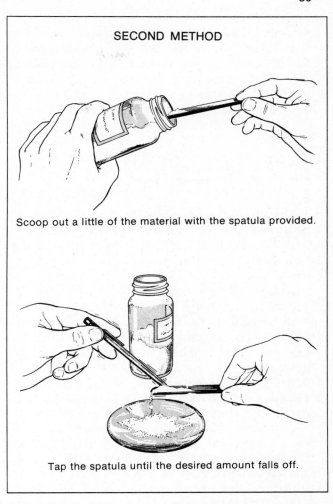

Scoop out a little of the material with the spatula provided.

Tap the spatula until the desired amount falls off.

THIRD METHOD

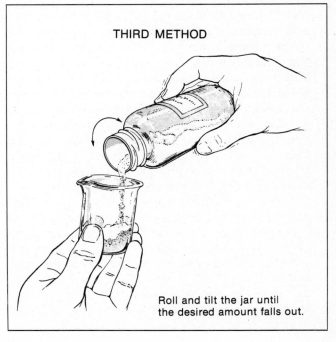

Roll and tilt the jar until
the desired amount falls out.

FIGURE 2-4
Methods for transferring powders and crystals.

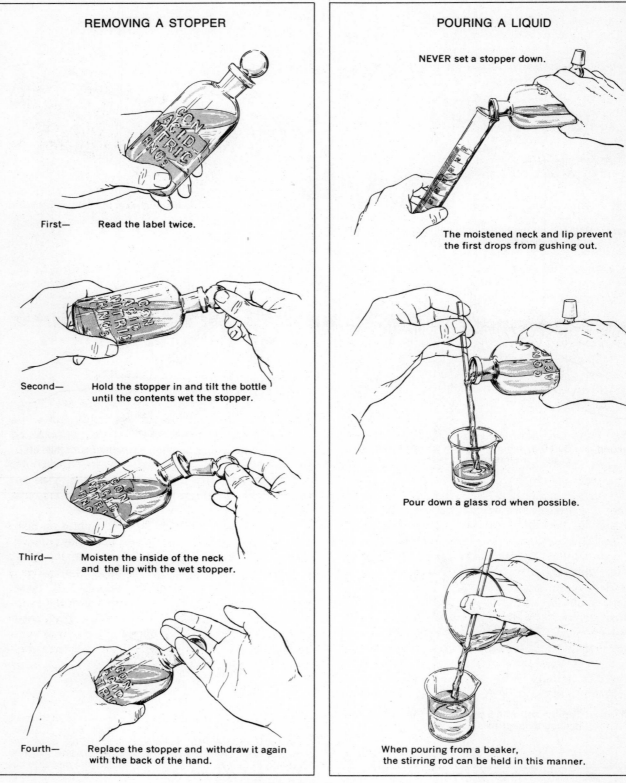

REMOVING A STOPPER

First—   Read the label twice.

Second—   Hold the stopper in and tilt the bottle
             until the contents wet the stopper.

Third—   Moisten the inside of the neck
             and the lip with the wet stopper.

Fourth—   Replace the stopper and withdraw it again
             with the back of the hand.

POURING A LIQUID

NEVER set a stopper down.

The moistened neck and lip prevent
the first drops from gushing out.

Pour down a glass rod when possible.

When pouring from a beaker,
the stirring rod can be held in this manner.

FIGURE 2-5
The method for transferring liquids.

3. Crystalline Form. Prepare crystals of as many of the following substances as your instructor directs. With a low-power magnifying glass, observe the characteristic shape of each type of crystal (you may also observe crystals of different substances prepared by your classmates), and describe each in your report sheet. If practical, include a sketch of the crystals. While observing the various crystals, be thinking of any evidence supporting the atomic theory which the formation and growth of crystals suggests.

(a) *Rhombic Sulfur.* Place about 2 g of crushed, roll sulfur[1] and about 5 ml of carbon disulfide in a 15-ml test tube. (**Caution:** *Carry out this operation in a hood. Do not handle carbon disulfide within five feet of any flame; its vapor, mixed with air, is very explosive.*) Mix gently for about 5 minutes to promote solution. Fold a filter paper in the usual manner, hold it (without a funnel) directly over a watch glass, and filter the mixture. Set the watch glass aside in a safe, quiet place until the liquid has evaporated. Observe the crystals.

(b) *Monoclinic Sulfur.* At an elevated temperature sulfur crystallizes from the molten material as monoclinic crystals. The equilibrium temperature of the two *allotropic[2] forms* is 95.5°C. Such a *transition temperature* is a definite physical property of a substance, analogous to the melting point or the boiling point.

Prepare a cone of filter paper and support it either in a funnel or a small beaker. Fill a 15-cm test tube about two-thirds full with sulfur. Holding it with a test tube clamp, heat it slowly and uniformly *in order not to superheat any portion of it.* The sulfur will darken if it is superheated; this can be avoided by moving the test tube in and out of the flame. When the sulfur is just melted, it should be a light yellow, straw-colored liquid. Pour it into the filter cone previously prepared, and with a match stick in hand watch for the formation of long needle-shaped crystals. Just as the surface of the liquid begins to solidify, break it open with the match stick, and quickly pour the remaining molten sulfur into a beaker of water. Let the filter cone cool, then break it open and observe the crystals.

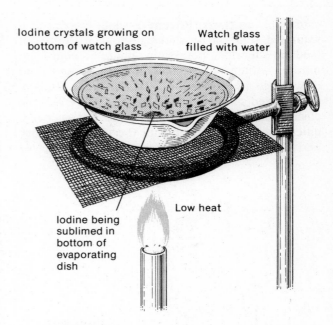

FIGURE 2-6
Apparatus for the recrystallization of iodine by sublimation.

(c) *Iodine.* Iodine has a very appreciable vapor pressure at temperatures considerably below its melting point; hence it may easily be *sublimed* (caused to crystallize directly from the vapor without first liquefying) from a warm surface to a cooler one. (Owing to the high cost of iodine, your instructor may designate a limited number of students to demonstrate this experiment.)[3]

Put about 1 g of iodine in an evaporating dish, and place this on a wire gauze and ring support. Over the evaporating dish, place a watch glass which is partially filled with water (this provides a cool, condensing surface; see Figure 2-6). Now, very gently and with a small flame, warm the evaporating dish. Do not rush the process, as the growth of larger crystals is favored by slow growth with no disturbance of the vessel. Finally, let the watch glass cool, and observe the crystals.

---

[1]Flowers of sulfur is not satisfactory for this purpose.

[2]Allotropes are different molecular forms of the same element such as oxygen (O_2) and ozone (O_3).

[3]As an alternative or supplement to this experiment we have found the observation of the sublimation of iodine in sealed tubes very effective. When sealed in a vacuum (*in vacuo*), the iodine travels long distances in the tube; when sealed under an inert gas, the subliming iodine molecules travel a much shorter distance. Such sealed tubes can be made in the laboratory, or they are available commercially (Sargent-Welch Scientific Co., Sci-Ed Catalog No. 4425).

(d) *Various Salts.* Prepare a small amount of a saturated solution by shaking about 3 g of a salt (sodium chloride, potassium chlorate, potassium nitrate, sodium nitrate, or other) with about 5 ml of water in a test tube for about 5 minutes. Let any undissolved salt settle, and decant (Figure 2-2) the clean liquid onto a clean watch glass. Set aside in a quiet place overnight or until the next laboratory period, to permit the water to evaporate. Observe the crystals. (Ordinary table salt frequently contains well-developed crystals, which may be observed under low magnification. Rock salt, also, may contain excellent crystals. *Optional*: Try crystallizing sodium chloride from a small amount of a 30% urea solution, instead of from water, to attempt to obtain the octahedral facial development of salt crystals.)

OBSERVATIONS AND DATA

1. Preliminary Observations

(a) Note your observations on general characteristics, density, melting point, and solubility. Comment on any specific features of special interest *as you observed them*.

(b) Tabulate and describe the odor and taste of any substances that have these characteristics.

2. The Quantitative Analysis of an Unknown Mixture Based on Solubility

Unknown sample No. _____   Weight of beaker and sample _____ g

Weight of test tube and sample _____ g   Weight of beaker and sand (extracted, dry) _____ g

Weight of test tube _____ g   Weight of beaker _____ g

Weight of sample (from test tube weighings) _____ g   Weight of sample (from beaker weighings) _____ g

Weight of evaporating dish and salt _____ g   Weight of sand _____ g

Weight of evaporating dish _____ g   Weight of salt (by subtracting the beaker weighings) _____ g

Weight of salt extracted _____ g

Calculations: Compare and average the weights of the salt, and also of the sample, if there are differences between the results of the two methods, and calculate the percent of salt in the sample.

_____ %

3. Crystalline Form

Characterize and sketch crystals of the various salts you have prepared. (Report the actual samples to your instructor as directed.)

EXERCISES AND PROBLEMS

NOTE: See Appendix A on the use of dimensional units in the solution of problems. Set up each problem solution *neatly*, using *good form*. Label all quantities with the appropriate units, and indicate all mathematical operations, inserting the actual values in the problem, but omit detailed multiplication and division. Record the answers in the space at the right.

1. What supporting evidence of the atomic theory does the formation and growth of crystals suggest? (Base your answer on the observations you noted in part 3 of the report).

---

2. An experiment calls for 50.0 g of concentrated nitric acid, density 1.42 g/ml. Suppose no balance is available, and you decide to use a graduated cylinder. What volume should you use?

_____ml

---

3. The dilute sulfuric acid on the laboratory desk has a density of 1.18 g/ml, and is 25.0% sulfuric acid, the remainder being water.[1]
   (a) How many grams does 15.0 ml of this acid weigh?

_____g

   (b) What is the weight of pure sulfuric acid in this 15.0 ml of solution?

_____g

---

4. Calculate the volume of magnesium, density 1.74 g/cm³, that would be equal in weight to 500 cm³ of lead, density 11.4 g/cm³.

_____cm³

---

[1]Percent composition means percent by weight unless percent by volume is specifically stated. The composition of gases is usually expressed as percent by volume.

SOME ELEMENTARY CHEMICAL PROPERTIES OF SUBSTANCES

PRE-STUDY

In this experiment we shall continue the study of properties of the same substances used in Experiment 2. Here we are interested in typical modes of *chemical* behavior, resulting in the formation of new substances.

Physical and Chemical Changes Taking Place on Heating

When a substance is heated in the air it may:

1. change its physical state from solid to liquid, from liquid to gas, or directly from solid to gas;

2. react chemically with a constituent of the air (usually oxygen) to produce one or more new substances;

3. decompose to produce simpler substances.

Elements, of course, ordinarily cannot decompose.[1] The products of a decomposition must contain all the elements originally present. None can be destroyed, and no additional elements can be present in the products, except possibly some constituent of the air with which there has been a reaction. These phenomena can be explained by the Law of Conservation of Mass, which states that mass can neither be created nor destroyed during a chemical reaction.

The physical changes that accompany a given chemical change often help us to identify it. Thus, mercuric oxide (a red powder) decomposes to form mercury (a silvery metallic liquid) and oxygen (a

---

[1] We are not considering here the unusual forces brought into play in a nuclear reactor, in which different elements, with different properties, are formed. Such changes are not "chemical changes" in the usual sense. They are called nuclear transformations. Radioactive elements also spontaneously undergo nuclear transformations.

colorless gas). If we heat calcium carbonate (limestone), two simpler compounds result—calcium oxide (quicklime), and carbon dioxide (a colorless gas). If carbon (e.g., coke) is heated in excess air, it forms carbon dioxide, the oxygen coming from the air.

Properties of and Tests for Some Common Gases

Use the descriptions below in identifying any gaseous products formed during the experiments that follow these descriptions.

Oxygen, $O_2(g)$: Colorless and odorless. Causes the glowing end of a wood splint to brighten greatly or even burst into flame. This test is made by thrusting the glowing end of a splint into the gas being tested. Only one other common gas, nitrous oxide, $N_2O(g)$, has similar properties.

Hydrogen, $H_2(g)$: Colorless and odorless. One of a number of gases that burn when ignited in the presence of air or explode when a mixture of the gas and air is ignited. You can detect the presence of hydrogen by holding a lighted match near the mouth of an inverted test tube containing a gas. If the tube contains a mixture of air and hydrogen, a slight "pop" will result.

Carbon dioxide, $CO_2(g)$: Colorless, odorless, a slight acid or sour taste if in solution. Recognized by the milky precipitate of calcium carbonate it forms when exposed to a calcium hydroxide solution (limewater). You may observe this reaction by holding a glass tube or rod, from which a drop of limewater is suspended, in the mouth of the test tube containing the substance being tested, and noting whether the drop turns milky. (Touch the glass rod to some limewater in a test tube. Never put a rod or medicine dropper into a reagent bottle.) A better test (see Figure 3-1) is to insert into the test tube containing the substance being tested a rubber stopper and bent delivery tube (see directions on glass bending, Figure 8 in the Introduction). The other end of the delivery tube is submerged in limewater in a 10-cm test tube. When the substance is heated, any $CO_2(g)$ expelled into the limewater will turn it milky. Or, in testing for $CO_2(g)$ when an acid is added to a substance, disconnect the rubber stopper just long enough to permit the addition of a little HCl; it is reconnected *at once,* and, if necessary, the test tube is warmed slightly to assist in expelling the evolved gas into the limewater.

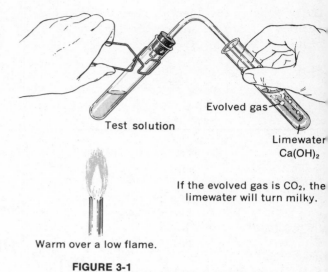

Test solution

Evolved gas

Limewater
Ca(OH)₂

If the evolved gas is CO_2, the limewater will turn milky.

Warm over a low flame.

FIGURE 3-1
The limewater test for carbon dioxide.

Sulfur dioxide, $SO_2(g)$: Colorless, with a sharp choking odor.

Hydrogen sulfide, $H_2S(g)$: Colorless, poisonous, with a very characteristic foul odor.

One of the tests you will use in this and a number of subsequent experiments to examine the chemical properties of a substance is to dissolve it in water and determine whether the properties of the solution are basic or acidic. An acid solution contains an excess of hydrogen ions, H^+, and is characterized by a sour taste. The taste of vinegar is a familiar example of a dilute, acid solution. Basic solutions contain an excess of hydroxide ions, OH^-, and feel slippery. A soap solution has this property and is an example of a basic solution. The acid-base nature of a solution is easily determined by means of an indicator that exhibits different colors in the two kinds of solutions. (The pre-study for Experiment 20 provides a more complete discussion of the way in which indicators function.) Litmus paper is a paper impregnated with the indicator litmus, which is red in an acidic solution and blue in a basic solution.

EXPERIMENTAL PROCEDURE

Chemicals: Cu (turnings), Zn (mossy), Zn (dust), S(s), I₂(s), dil. HCl, Ca(OH)₂ (sat. sol.), CuCO₃(s), Cu(NO₃)₂·3H₂O(s), KClO₃(s), Mg (ribbon).

1. Reaction to Litmus. Test solutions of each of the *soluble* substances in the above chemicals with red and blue litmus paper. (See also Table 2-1 in the preceding experiment. A blue color with iodine solution is due to starch in the paper. It is not a basic indication.)

2. The Effect of Heat on Certain Elements. Heat each of the following elements, as directed. For each, note any physical changes, and also note any evidence that a new substance may have been produced.

Copper. With the forceps, hold a small piece of copper sheet, or turnings, at the top of the Bunsen flame for a short time.

Sulfur. Heat a bit of sulfur the size of half a pea in a porcelain crucible *in a hood* until it melts and then catches fire as it becomes hotter (Figure 3-2). *Cautiously* note the odor. (Follow the procedure described in the beginning of the Introduction— Safety Rule 5—for sniffing gases.) Continue heating until all the sulfur is burned out of the crucible. *Keep the crucible under a hood.*

Iodine. Heat one or two small crystals of iodine in a 15-cm test tube, keeping the upper part of the tube cool.

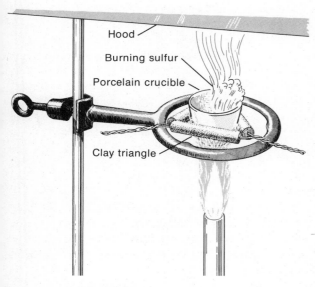

FIGURE 3-2
The ignition of sulfur in a crucible.

3. The Decomposition of Certain Compounds by Heat. Place small samples each of cupric carbonate and potassium chlorate in separate test tubes. Before you heat them, be prepared to test for any gases which you think may be formed. (See the pre-study for this experiment on properties of and tests for common gases.) Heat each substance, and note any changes. Identify the gaseous products by the tests or observations described in the preceding discussion. The white residue left on heating potassium chlorate contains no oxygen, and is potassium chloride.

4. The Effect of an Acid on Various Substances. Place small samples each of copper, zinc (use mossy zinc), sulfur, calcium hydroxide (solid), and cupric carbonate in separate test tubes. Be prepared to test for any gases you might anticipate. Add 3–5 ml of dilute hydrochloric acid to each test tube, and note changes. Identify any gases that form.

5. The Reaction of Zinc and Sulfur.[2] Mix thoroughly an estimated 2 g of zinc dust (about 1 cm deep in a dry 10-cm test tube) with about twice its volume (1 g) of powdered sulfur. Place half of the mixture in a 15-cm test tube, and add dilute hydrochloric acid. Make suitable physical and chemical observations to identify the principal gas formed. A small amount of another gas is present. What is it? Place the remainder of the mixture on a 7-cm square of asbestos paper, or in your evaporating dish, and ignite it as follows. Holding a 4-cm length of magnesium ribbon with your forceps, light it by heating the tip in the Bunsen burner, and at once touch the burning magnesium to the zinc-sulfur mixture. (*Exercise reasonable care, and do not get too close, because the mixture burns rapidly.*) For additional evidence of any chemical change that has taken place between the zinc and sulfur, put the residue into a test tube, and add a little dilute hydrochloric acid. Carefully note the odor. Identify the principal gas formed.

---

[2]To avoid fumes in the laboratory and to minimize potential hazards, it is recommended that the instructor perform this experiment as a demonstration.

**Some Elementary
Chemical Properties
of Substances**

NAME

SECTION   LOCKER

INSTRUCTOR   DATE

OBSERVATIONS AND DATA

1. Reaction to Litmus

List the substances whose solutions are:

Acidic_____

Basic_____

Neutral_____

2. The Effect of Heat on Certain Elements

List any changes in physical state, color, odor, and so forth, and name any new substances produced on heating.

Element	Observations	New substances
Copper		
Sulfur		
Iodine		

3. The Decomposition of Certain Compounds by Heat

List any changes in physical state, color, odor, or other properties that occur as the substance is heated. List any tests performed on gaseous products in order to identify them.

Compound	Observations, including tests for any gaseous products evolved	New substances
Cupric carbonate		
Potassium chlorate		

4. The Effect of an Acid on Various Substances

List any observable effects, such as solution of the solid and color of the solution, when hydrochloric acid is added. List any tests performed (and the results) on any gases evolved.

Substance	Observations, including tests performed	Identity of gas, if any
Copper		
Zinc		
Sulfur		
Calcium hydroxide		
Cupric carbonate		

5. The Reaction of Zinc and Sulfur

State clearly *all* the evidence in the experiment that leads you to believe that a chemical change did, or did not, take place when:

(a) the zinc and sulfur were mixed, before igniting;

(b) the zinc and sulfur mixture was ignited. (Caution: follow directions carefully.)

INTERPRETATION OF THE DATA

In drawing conclusions from experimental data, it is important that you consider all the observations made, but also *that you do not generalize beyond the data.* Consider a hypothetical case:

Example. Six metals are treated with dilute hydrochloric acid. In every reaction, a gas is evolved which, on mixing with air, explodes with a sharp report when it is ignited.

Possible generalizations and their validity:

1. Hydrogen can be formed by the action of hydrochloric acid on certain metals. *True.* (The behavior noted is characteristic of hydrogen. The number of occurrences justifies the conclusion.)

2. Every metal will displace hydrogen from hydrochloric acid. *Insufficient evidence.* (Not all metals were tried, nor each metal under all possible conditions.)

3. When a metal is treated with any acid, hydrogen is liberated. *Insufficient evidence.* (Generalizations cannot be made about acids that were not tried.)

4. The gas liberated is carbon dioxide. *False.* (Contrary to data. Though no test for carbon dioxide was made, compounds of carbon were not among the reactants.)

Problem. For each of the following statements, circle the T if sufficient evidence is presented in *Experiments 2* or *3* to justify the statement; circle the F if the statement is contrary to the data; circle the I if insufficient evidence is given.

1. All metals, when treated with an acid, liberate hydrogen.	1.	T	I	F
2. Most metals unite with oxygen when heated in air.	2.	T	I	F
3. All compounds may be decomposed by heating.	3.	T	I	F
4. All compounds containing oxygen decompose to yield oxygen gas when heated.[1]	4.	T	I	F
5. The determination of any one specific property is always sufficient evidence to identify a substance.	5.	T	I	F
6. Calcium hydroxide is more soluble by reaction with hydrochloric acid solution than by solution in water.	6.	T	I	F
7. The aqueous solution of every substance not an acid or base is neutral to litmus.	7.	T	I	F
8. All salts are soluble.	8.	T	I	F
9. The disappearance of sulfur on heating always indicates its complete change into different substances.	9.	T	I	F
10. The decomposition of potassium chlorate by heat is a good example of one of its chemical properties.	10.	T	I	F
11. Copper oxide will dissolve in hydrochloric acid.	11.	T	I	F
12. Crystals of a substance are shaken with water, and some crystals remain after shaking. The substance is insoluble.	12.	T	I	F

---

[1]The symbol for oxygen, O, with or without a subscript, appears in the formula of any compound containing oxygen.

CHEMICAL STOICHIOMETRY: UNITS OF QUANTITY AND CONCENTRATION

THE "PARTICLE SCALE" OF QUANTITY IN CHEMICAL MEASUREMENTS—THE MOLE

You are accustomed to measuring "amounts" of substances in terms of weight—grams in the metric system. Since the unit in a chemical reaction is always a particle (an atom, molecule, ion, or electron), weight is not as convenient for chemical purposes as a system of units in which we can compare directly the relative number of such reacting particles.

Because it is not practical to work in the laboratory with single atoms or molecules on account of their small size, a larger unit that contains a definite number of atoms or molecules is used. This unit is called the *mole* (sometimes abbreviated *mol*). We will define the mole and show how this unit is used in calculations involving chemical reactions.[1]

As you will discover in Experiment 4, the weights of the elements that react with one another are proportional to the atomic weights of the elements.[2] The reason is that each atomic weight unit (or mole) of an element (for example, 12.0 g of carbon, 16.0 g of oxygen, 63.5 g of copper, 107.9 g of silver) contains the same number of atoms as one mole of any other element. That number of atoms is thus a constant. This constant is called Avogadro's number, and it has a value of 6.02×10^{23}. In other words, every mole contains 6.02×10^{23} particles.[3] We therefore adopt the mole as our fundamental measure of quantity in chemistry: *One mole of a chemi-*

---

[1]Two highly recommended books dealing with the solution of many types of problems in general chemistry are A. B. Loebel, *Chemical Problem-Solving by Dimensional Analysis,* Houghton Mifflin, Boston, 1974, and C. Pierce and R. N. Smith, *General Chemistry Workbook,* 4th ed., W. H. Freeman and Company, San Francisco, 1971.

[2]The *atomic weight* of an element is the average weight of the natural isotopic mixture of atoms of that element relative to the mass of the stable, most abundant isotope of carbon, which is arbitrarily assigned a value of exactly 12. Before 1961 the standard was the natural isotopic mixture of oxygen atoms, with an assigned value of exactly 16. The present standard reduces the older atomic weight values by a factor of only 1.000043—a difference which is negligible except for precision greater than about 4 significant figures, and for elements whose atomic weight values are known to that precision.

[3]Avogadro's number is experimentally determined in Experiment 7.

*cal substance has a mass in grams numerically
equal to the formula weight of the substance and
contains an Avogadro's number of particles, defined
by the chemical formula of the substance.*

Thus, a mole of carbon contains 12.0 g of carbon
and 6.02×10^{23} particles (atoms) of carbon. A mole

of CO_2 contains 44.0 g of carbon dioxide and $6.02
\times 10^{23}$ particles (molecules) of carbon dioxide. A
mole of NaCl contains 58.5 g of sodium chloride
and 6.02×10^{23} particles of NaCl (where a particle
of NaCl consists of 1 sodium ion and 1 chloride
ion).

Helium (He)
Atomic weight = 4.003
Volume = 22.4 liters
1 mole = 4.003 g

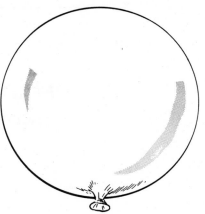

Oxygen (O_2)
Mol. wt. = 32.00
Volume = 22.4 liters
1 mole = 32.00 g

Carbon dioxide (CO_2)
Molecular weight = 44.01
Volume = 22.4 liters
1 mole = 44.01 g

Mercury (Hg)
Atomic weight = 200.6
Density = 13.546 g/ml
1 mole = 200.6 g

Water (H_2O)
Molecular weight = 18.016
Density = 1.00 g/ml
1 mole = 18.016 g

Carbon disulfide (CS_2)
Molecular weight = 76.13
Density = 1.261 g/ml
1 mole = 76.13 g

Carbon (diamond)
Atomic weight = 12.01
Density = 3.51 g/cm³
1 mole = 12.01 g

Salt (NaCl)
Formula weight = 58.44
Density = 2.165 g/cm³
1 gfw = 58.44 g

Sulfur (S_8)
Molecular weight = 256.5
Density = 2.07 g/cm³
1 mole = 256.5 g

FIGURE B-1
The mass and volume of 1 mole of some chemical substances. Although the weight in grams
is very different for different substances, the number of atoms or molecules per mole—namely
Avogadro's number—is the same for all. (Values given are for gases at 0°C, and for liquids
and solids at 20°C.)

In this laboratory manual we will apply the term *mole* to pure elements, molecular substances, ionic substances, or to any pure substance whose composition is expressed by a chemical formula, such as $Na_2SO_4 \cdot 10H_2O$. Some authors prefer to use the term mole only for molecular substances, and use the term *gram-atom* (abbreviated *g-at*) for a mole of an element, and the term *gram-formula weight* (abbreviated *gfw*) for ionic substances that do not exist as discrete molecules. This distinction does not seem necessary to us if care is taken to specify the chemical formula of the substance. Thus to speak of a mole of O or a mole of O_2 or a mole of $CuSO_4 \cdot 5H_2O$ produces no ambiguity: for each, we mean a mass (or weight) in grams numerically equal to the formula weight of the substance. Thus a mole of O would mean 16.0 g of oxygen, a mole of O_2 would mean 32.0 g of oxygen, and a mole of $CuSO_4 \cdot 5H_2O$ would mean 249.7 g of copper(II) sulfate pentahydrate. The mole of each substance contains an Avogadro's number of particles, the particle being defined by the chemical formula of the substance.

A chemical symbol or formula is often used by chemists to designate, not a single atom or molecule, but 6.02×10^{23} atoms or molecules (Avogadro's number), that is, a mole of a substance.[4] These units are large enough to be practical weighable amounts, and their use makes calculations of relative chemical quantities simple, logical, and direct. Thus, the chemical reaction

$$C + 2S \rightarrow CS_2$$

may be interpreted in two ways: (1) 1 atom of carbon reacts with 2 atoms of sulfur to form 1 molecule of carbon disulfide; or (2) 1 mole of carbon reacts with 2 moles of sulfur to form 1 mole of carbon disulfide. Thus, the symbol C may sometimes be used to represent 12.0 g (1 mole) of carbon, the symbol S used to represent 32.0 g (1 mole) of S, and the formula CS_2 used to represent 76.0 g (1 mole) of CS_2. The weight fraction of sulfur in carbon disulfide is

$$\frac{2S}{1CS_2} = \frac{64.0 \text{ g}}{76.0 \text{ g}} = 0.842 \text{ or } 84.2\%$$

Students sometimes confuse the number of

[4] Engineers often use the English system of weights, in which case the corresponding units are pound-atom, pound-mole, and pound-formula weight. These units would also contain equal numbers of particles, but not the above Avogadro's number, which is based on the metric system.

moles with the weight in grams it represents. The defining equation is

$$\text{moles} = \frac{\text{grams}}{\text{g/mole}} \qquad (1)$$

To calculate the weight of 10.0 moles of S, we transpose equation (1) and write

$$10.0 \text{ moles S} \times 32.0 \frac{\text{g}}{\text{mole}} = 320 \text{ g S}$$

Conversely, to calculate the number of moles in 38.0 g of carbon disulfide, from equation (1) above, we write

$$38.0 \text{ g } CS_2 \times \frac{1 \text{ mole}}{76.0 \text{ g}} = 0.500 \text{ mole } CS_2$$

For an example of this type of calculation that involves a chemical reaction, let us consider the typical data that might be obtained by a student from Experiment 4, which follows this study assignment. The loss in weight of the copper wire was 0.250 g, and the corresponding weight of silver metal precipitated by replacement from the silver sulfate solution was 0.860 g. From these data we may calculate the relative number of moles (and of individual atoms) that reacted:

$$\frac{0.250 \text{ g Cu}}{63.54 \text{ g/mole}} = 0.00393 \text{ mole Cu}$$

$$\frac{0.860 \text{ g Ag}}{107.9 \text{ g/mole}} = 0.00797 \text{ mole Ag}$$

This is a ratio of 1.00 mole of copper to 2.03 mole of silver, or, within experimental error, 1 "Cu" to every 2 "Ag." This corresponds to and justifies the equation for this reaction as described by the third chemical change occurring in Experiment 4, namely,

$$Cu(s) + Ag_2SO_4(aq) \rightarrow CuSO_4(aq) + 2Ag(s)$$

Observe that *the reacting proportions as expressed by the coefficients in an equation are the result of experimental observation*—not magical numbers picked out of the air in some mysterious manner. *Question:* Suppose that an aluminum wire is suspended in this copper sulfate solution. Assuming that excess aluminum is present, what will be the loss in weight of the wire due to the replacement reaction by this more active metal? The equation is

$$2Al(s) + 3CuSO_4(aq) \rightarrow Al_2(SO_4)_3(aq) + 3Cu(s)$$

The original 0.250 g of copper is all present in the solution as copper sulfate.

First, express this quantity of copper as gram-atoms, as before:

$$\frac{0.250 \text{ g Cu}}{63.54 \text{ g/mole}} = 0.00393 \text{ mole Cu}$$

Second, compare the reacting quantities from the equation for the reaction:

$$0.00393 \text{ mole Cu} \times \frac{2 \text{ mole Al}}{3 \text{ mole Cu}} = 0.00262 \text{ mole Al}$$

Third, convert the quantity of aluminum to grams:

$$0.00262 \text{ mole Al} \times \frac{27.0 \text{ g Al}}{3 \text{ mole Al}} = 0.0707 \text{ g Al}$$

As you gain experience in thinking in terms of "particles," you will generally eliminate the intermediate calculations and summarize these operations in one series of steps:

$$\frac{0.250 \text{ g Cu}}{63.54 \text{ g/mole Cu}} \times \frac{2 \text{ mole Al}}{3 \text{ mole Cu}} \times \frac{27.0 \text{ g Al}}{1 \text{ mole Al}} =$$
$$0.0707 \text{ g Al}$$

Always include the dimensional units in your work, because the fact that they cancel out properly is a check on the logic of your thinking.

THE MOLAR VOLUME OF A GASEOUS SUBSTANCE

Avogadro made the brilliant hypothesis, which is confirmed by a number of experiments, that equal volumes of gases at the same temperature and pressure contain equal numbers of molecules. This means that one mole of any pure gas, containing an Avogadro's number of molecules, will occupy a definite volume, independent of the molecular weight of the gas. The *molar volume* measured under conditions of standard temperature and pressure is 22.4 liters STP[5] per mole, and is the same for all *ideal*[6] bases. This relationship allows one to

---

[5]The standard temperature and pressure (STP) conditions are 0°C and 1 atmosphere (760 Torr).

[6]An ideal, or perfect, gas is one in which the molecules occupy no volume and there are no attractive forces between molecules. In real or actual gases, neither of those conditions is exactly correct. Therefore, real gases show a deviation from the molar volume of 22.4 liters, which must be taken into account in careful work. Such gases are called *nonideal*.

measure the number of moles of a gas by volume rather than by weighing the gas.

CONCENTRATION UNITS IN SOLUTION CHEMISTRY

It is convenient to study many chemical reactions in solution by dissolving the reactants in a solvent such as water and then measuring the quantity of solution rather than weighing out the reactants directly. The quantity of solution may be based on either a volume or a weight measurement.

Volume Units of Concentration

Concentration and Amount. Many student difficulties arise because of failure to differentiate these two terms. For example, if we dissolve 34.2 g of sugar ($C_{12}H_{22}O_{11}$) in enough water to make 100 ml of solution, we have a quite concentrated, sweet-tasting solution. If we dissolve 34.2 g of sugar in enough water to make 10 liters of solution, we have used the same *amount* of sugar, but the *concentration* is much less, and the solution hardly tastes sweet. Concentration means the amount per unit volume; that is,

$$\text{Concentration} = \frac{\text{amount of solute}}{\text{volume}}$$

Formality and Molarity. If one mole (or gram-formula weight) of a substance is weighed out, dissolved in water, and diluted to a total volume of 1 liter (not added to 1 liter of water), the resulting solution is called a one formal (1 F) solution. It may also be called a one molar (1 M) solution. Likewise, we may dissolve two formula weights in 2 liters of solution and still have a 1 F solution (see Figure B-2). The defining equation is

$$\text{Formality} = \frac{\text{moles (or gfw) of solute}}{\text{liters of solution}} \quad (2)$$

$$F = \frac{\text{no. moles (or gfw)}}{V \text{ (liters)}}$$

There is no universally accepted convention about whether equation (2) defines only formality or both formality and molarity. Some chemists use

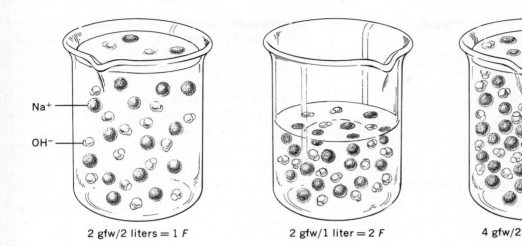

2 gfw/2 liters = 1 F   2 gfw/1 liter = 2 F   4 gfw/2 liters = 2 F

FIGURE B-2
"Amount" versus "concentration" of solutions. The first two beakers each contain the same
amount of sodium hydroxide but in different volumes and therefore at different concentrations—the first at 1 F and the second at 2 F. The second and third beakers contain different
amounts, but at the same concentration, each 2 F.

formality, F, and molarity, M, interchangeably. We prefer to make the distinction that formality is *prescriptive* and molarity is *descriptive*. For example, a 1 F solution of NaCl is made up by weighing out 1 mole of NaCl, dissolving it in water and diluting the solution to a total volume of 1 liter. We prefer not to call the resulting solution a 1 M solution of NaCl because the formula NaCl does not represent a definite molecular species that is present as such in the solid crystal or in solution. Because the solute NaCl is regarded as being completely dissociated in solution, we may speak of a 1 F NaCl solution as containing 1 M Na$^+$ and 1 M Cl$^-$.

Other situations in which the distinction between formality and molarity is useful may arise. A 0.100 F solution of acetic acid is made up by dissolving 0.100 mole of acetic acid in water and diluting to a total volume of exactly 1 liter. The formal concentration of acetic acid is therefore 0.100 F. However acetic acid dissociates slightly to give some acetate ion and hydrogen ion, so that the actual equilibrium

$$CH_3COOH \rightarrow CH_3COO^- + H^+$$

acetic acid   acetate ion   hydrogen
                                    ion

concentration of undissociated acetic acid in this solution is about 0.098 moles/liter. In this situation we can make a distinction between the formal acetic acid concentration, 0.100 F, and the equilibrium concentration of undissociated acetic acid which is 0.098 M.

Example 1. What are the concentrations of the sugar solutions mentioned above, which contain 34.2 g $C_{12}H_{22}O_{11}$ in 100 ml and in 10.0 liters of solution, respectively?

Since 34.2 g (0.100 mole) of sugar is weighed out, we can designate this as a "formal" concentration, but it would be equally correct to call it a "molar" concentration because the solution contains molecules of the composition $C_{12}H_{22}O_{11}$, which do not dissociate in solution. First we express 34.2 g as moles:

$$\frac{34.2 \text{ g}}{342 \text{ g/mole}} = 0.100 \text{ mole}$$

From the defining equation, we write

$$F = \frac{\text{moles}}{V} = \frac{0.100 \text{ mole}}{0.100 \text{ liter}}$$
$$= 1.00 \ F \text{ (or } M)$$

for the first solution, and

$$F = \frac{0.100 \text{ mole}}{10.0 \text{ liters}} = 0.0100 \ F \text{ (or } M)$$

for the second solution.

Example 2. How many moles are there, and what is the weight of H_2SO_4 in grams, in 250 ml of 0.300 F H_2SO_4?

Transposing the defining equation $F = \text{moles}/V$, we have

$$\text{moles} = F \times V = 0.300 \frac{\text{mole}}{\text{liter}} \times 0.250 \text{ liter}$$

$$= 0.0750 \text{ mole } H_2SO_4$$

To express this quantity in grams, we write

$$0.0750 \text{ mole} \times 98.1 \frac{g}{\text{mole}} = 7.36 \text{ g } H_2SO_4$$

Example 3. What volume of 0.300 F H_2SO_4 is required to react with 4.00 g NaOH? We have

$$\frac{4.00 \text{ g NaOH}}{40.0 \text{ g/mole}} = 0.100 \text{ mole NaOH}$$

From the equation for the reaction

$$2NaOH + H_2SO_4 \rightarrow Na_2SO_4 + 2H_2O$$

it is evident that 2 moles of NaOH will require 1 mole of H_2SO_4; therefore 0.100 mole of NaOH will require 0.050 mole of H_2SO_4. Transposing the defining equation $F = \text{mole}/V$, we have

$$V = \frac{\text{mole}}{F} = \frac{0.0500 \text{ mole}}{0.300 \text{ mole/liter}}$$

$$= 0.167 \text{ liter}$$

$$= 167 \text{ ml of } 0.300 \ F \ H_2SO_4$$

Normality. This unit of concentration is related to molarity and formality in the same way that the equivalent weight (see Experiment 13) is related to the mole. The normality of a solution expresses the number of equivalents of solute per liter of solution. The defining equation is

$$\text{Normality} = \frac{\text{equivalents of solute}}{\text{liters of solution}}$$

or

$$N = \frac{\text{no. equiv}}{V \text{ (liters)}}$$

The use of normality has the advantage that equal volumes of solutions of the same normality are equivalent chemically and just react with one another. This is not always true when concentrations are expressed in molarity or formality, since reactions do not always take place one formula weight with one formula weight. Normality has the disadvantage that it may be ambiguous unless the defining reaction is specified.

Equivalent Weights of Acids and Bases. In neutralization, one hydrogen ion reacts with one hydroxide ion to form one molecule of water. These amounts are therefore equivalent. We may calculate the equivalent weights of acids and of bases by dividing the respective formula weights by the number of potential hydrogen ions or hydroxide ions in the formula, *provided that all these take part in the reaction under consideration.* Consider the reactions

$$H_2SO_4 + Ca(OH)_2 \rightarrow CaSO_4 + 2H_2O \qquad (3)$$

$$H_2SO_4 + 2NaOH \rightarrow Na_2SO_4 + 2H_2O \qquad (4)$$

$$H_2SO_4 + NaOH \rightarrow NaHSO_4 + H_2O \qquad (5)$$

In equations (3) and (4), the equivalent weight of H_2SO_4 is one-half its formula weight ($\frac{1}{2} \times 98$ g or 49 g), since both hydrogen atoms are neutralized. The equivalent weight of $Ca(OH)_2$ is one-half its formula weight ($\frac{1}{2} \times 74$ g, or 37 g), and the equivalent weight of NaOH is its formula weight (40 g). In equation (5), only one of the two hydrogen atoms in H_2SO_4 has been neutralized by the one formula weight of NaOH, which is obviously one equivalent. The equivalent weight of H_2SO_4 in this reaction is therefore the same as its formula weight, 98 g.

Sometimes chemists refer to the equivalent weight of an acid or base without reference to a particular reaction, generally inferring the maximum number of equivalents per formula weight.

Example 1. How many equivalents, and how many grams, of NaOH are there in 200 ml of 0.300 N NaOH? Transposing the defining equation

$$N = \text{equiv}/V$$

we have

$$\text{equiv} = N \times V = 0.300 \frac{\text{equiv}}{\text{liter}} \times 0.200 \text{ liter}$$

$$= 0.0600 \text{ equiv}$$

To express this in grams, we write

$$0.0600 \text{ equiv} \times \frac{40.0 \text{ g NaOH}}{\text{equiv}} = 2.40 \text{ g NaOH}$$

Example 2. What is the normality of a solution that contains 1.11 g of $Ca(OH)_2$ dissolved in 2.00 liters of solution?

First we express 1.11 g $Ca(OH)_2$ as equivalents. The formula weight is 74.0 g. Since there are two replaceable OH groups, the equivalent weight is 74.0/2 or 37.0 g, and the number of equivalents is

$$\frac{1.11 \text{ g}}{37.0 \text{ g/equiv}} = 0.0300 \text{ equiv}$$

From the defining equation, we may then calculate the normality:

$$\frac{0.0300 \text{ equiv}}{2.00 \text{ liters}} = 0.0150 \ N \ Ca(OH)_2$$

Example 3. What is the normality of 3 F H_2SO_4, when used in the usual manner, where both hydrogen atoms in the formula react? We get

$$\frac{3 \text{ mole } H_2SO_4}{\text{liter}} \times \frac{2 \text{ equiv}}{\text{mole}} = \frac{6 \text{ equiv}}{\text{liter}}$$

$$= 6 \ N \ H_2SO_4$$

Weight Units of Concentration

Percent by Weight. This refers to parts of solute per one hundred parts of solution. Thus, a 5% solution contains 5 g of solute to every 95 g of solvent. (This terminology is frequently used in the medical and dental professions.)

Grams of Solute per 100 g of Solvent. Data are usually expressed in this manner in the reference literature, so that they will be independent of atomic weight values assigned at the time.

Molality. This term refers to the number of moles (or gram-formula weights) per 1000 g of solvent:

$$m = \frac{\text{no. moles (or gfw)}}{1000 \text{ g solvent}}$$

Mole Fraction. For a two-component system (solute + solvent), the mole fraction is given by dividing the moles of a component by the total number of moles in the system:

$$X_{\text{solute}} = \frac{\text{moles of solute}}{\text{moles solute} + \text{moles solvent}}$$

$$X_{\text{solvent}} = \frac{\text{moles solvent}}{\text{moles solute} + \text{moles solvent}}$$

Concentration units, such as molality and mole fraction, express the relative numbers of molecules in the system. Molality expresses the relative number of molecules of solute and solvent. For example, in aqueous solutions, 1 kg solvent contains 55.5 moles H_2O [1000 g H_2O/(18 g/mole)]. Molality is therefore the ratio of moles of solute to moles of solvent. The mole fraction expresses the ratio of the moles of one component to the total number of moles in the solution. Concentration units based on weight are independent of temperature, whereas the formal or molar concentrations change with temperature because the volume of a solution changes with temperature. The weight units are much used in problems based on the colligative properties: molecular weights in solution, vapor pressure relationships, and osmotic pressure problems.

Example 1. Express the composition of 3.00 F H_2SO_4 (density 1.180 g/ml) as (a) percent composition, (b) grams per 100 g solvent, and (c) molality

(a) *Percent Composition:* 1 liter of 3.00 F H_2SO_4 contains 3 moles/liter $\times$ 98.1 g/mole or 294 g of H_2SO_4, and weighs 1.180 g/ml $\times$ 1000 ml/liter or 1180 g. The percent composition is

$$\frac{294 \text{ g}}{1180 \text{ g}} \times 100 = 24.9 \text{ g/100 g solution}$$

$$= 24.9\% \ H_2SO_4$$

(b) *Grams per 100 g of Solvent:* From (a) we have 294 g of H_2SO_4 in 1180 g − 294 g, or 886 g of H_2O. Therefore we have

$$\frac{294 \text{ g } H_2SO_4}{886 \text{ g } H_2O} \times 100 = 33.2 \text{ g/100 g } H_2O$$

(c) *Molality:* From (b), multiplying by 10, we have 332 g of H_2SO_4 per 1000 g of solvent, or

$$\frac{332 \text{ g}}{98.1 \text{ g/mole}} = 3.38 \text{ moles/1000 g solvent}$$

$$= 3.38 \ m$$

CHEMICAL EQUATIONS AND STOICHIOMETRY

All stoichiometry problems involve essentially the same principles. For example, if we have a chemical reaction which we may represent in the following general way

$$aA + bB \rightarrow cC + dD$$

we may make the following statement: the coefficients (a, b, c, d) of the *balanced* reaction tell us that a moles of A and b moles of B react to give c moles of C and d moles of D. In order to solve a problem correctly you must first be sure that the equation for the reaction gives the proper reactants and products and is balanced to conserve atoms.

The amounts of reactants and products are often not given in moles. If they are not expressed in moles, most commonly they will be given in one of three ways:

1. In grams:

$$\text{grams } A \times \frac{\text{mole } A}{\text{grams } A} = \text{moles } A$$

2. In liters (if A is a gas):

$$\text{liters (STP)} \times \frac{1 \text{ mole } A}{22.4 \text{ liters (STP)}} = \text{moles } A$$

or the number of moles of a gaseous substance may be obtained from the volume measured at any temperature and pressure by application of the ideal gas law,

$$n = \frac{PV}{RT}$$

where n is the number of moles and the gas constant R has the value 0.0821 (liter-atm)/(mole-K).

3. In liters (if A is in solution):

$$\text{liters solution} = \frac{\text{moles } A}{\text{liters solution}} = \text{moles } A$$

These simple conversions should be learned so well that you can do them automatically. A con-

version of this type will be necessary in almost all problems of stoichiometry. We may represent this relationship in the following schematic way:

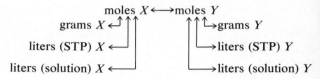

Aside from these general principles, we will not give more explicit directions for solving problems in stoichiometry, since the problems will all differ slightly in details. In general, follow these steps:

(a) Read the problem carefully to determine what is being asked for.

(b) Formulate in your mind a plan by which you can obtain the desired result from the given data. Do not be misled by extraneous or irrelevant data.

(c) Proceed to use the plan in a logical and orderly fashion, paying special attention to proper units or dimensions. If the units do not cancel to give the desired dimensions, the solution to the problem cannot be correct.

(d) After writing the unit equation, insert the numerical data and use a slide rule or calculator to obtain the exact numerical result. When using a slide rule, determine the decimal point or power of ten either in your head or on a sheet of scratch paper.

Examples. Consider the following reaction in which the hydrocarbon gas ethane, $C_2H_6(g)$, is burned in oxygen, $O_2(g)$, to form carbon dioxide, $CO_2(g)$, and water vapor, $H_2O(g)$.

The quantity relationships (weights and volumes) as indicated below the following balanced equation for this reaction are based on 1.0 mole of ethane.

(1) Equation:	$C_2H_6(g)$	+	$3\frac{1}{2}O_2(g)$	$\rightarrow$	$2CO_2(g)$	+	$3H_2O(g)$
(2) Moles:	1.0		3.5		2.0		3.0
(3) Weight:	$1 \text{ mole} \times \dfrac{30 \text{ g}}{\text{mole}}$		$3.5 \text{ moles} \times \dfrac{32 \text{ g}}{\text{mole}}$		$2 \text{ moles} \times \dfrac{44 \text{ g}}{\text{mole}}$		$3 \text{ moles} \times \dfrac{18 \text{ g}}{\text{mole}}$
(4) Volume (at STP):	$1 \text{ mole} \times \dfrac{22.4 \text{ liters}}{\text{mole}}$		$3.5 \text{ moles} \times \dfrac{22.4 \text{ liters}}{\text{mole}}$		$2 \text{ moles} \times \dfrac{22.4 \text{ liters}}{\text{mole}}$		$3 \text{ moles} \times \dfrac{22.4 \text{ liters}}{\text{mole}}$

Suppose the problem is to find all of these quantities when 75.0 g of ethane is burned. Since (75.0 g of ethane)/(30 g/mole) = 2.5 moles of ethane, the problem can be solved quickly by multiplying each of the given quantities by 2.5.

In most problems, all of the quantities (moles, weight, volume) are not needed in the solution, and so the calculation is further simplified. But the solution still requires the fundamental steps— conversions from weight to moles to weight (or volume) for the substances involved.

Example 1. Weight-moles-weight problem. (a) What weight of concentrated sulfuric acid solution, 96% pure H_2SO_4, will be required to dissolve 50.0 g of aluminum in a mixture with water? (b) How many grams of anhydrous aluminum sulfate will be produced, by evaporation of the resulting solution and drying of the salt?

(1) $2Al(s) + 3H_2SO_4(aq) \rightarrow Al_2(SO_4)_3 + 3H_2(g)$

(2) Stoichiometry
   2 moles   3 moles   1 mole   3 moles

(3) $\dfrac{50 \text{ g Al}}{27 \text{ g/mole}} = 1.85$ moles Al

(a) $1.85 \text{ moles Al} \times \dfrac{3 \text{ moles } H_2SO_4}{2 \text{ moles Al}}$

$\qquad = 2.78$ moles H_2SO_4

$2.78 \text{ moles } H_2SO_4 \times \dfrac{98 \text{ g}}{\text{mole}}$

$\qquad = 272$ g pure H_2SO_4

$\qquad = 272 \text{ g } H_2SO_4$

$\qquad \times \dfrac{100 \text{ g "concentrated" } H_2SO_4}{96 \text{ g } H_2SO_4}$

$\qquad = 284$ g "concentrated" H_2SO_4

(b) $1.85 \text{ moles Al} \times \dfrac{1 \text{ mole } Al_2(SO_4)_3}{2 \text{ moles Al}}$

$\qquad = 0.93$ mole $Al_2(SO_4)_3$

$0.93 \text{ mole } Al_2(SO_4)_3 \times \dfrac{342 \text{ g}}{\text{mole}}$

$\qquad = 318$ g anhydrous $Al_2(SO_4)_3$

Example 2. Volume-moles-weight problem. What weight of zinc would be required to produce 12.0

liters of hydrogen gas, measured at STP, by reaction with dilute hydrochloric acid?

(1) $Zn(s) + 2HCl(aq) \rightarrow ZnCl_2(aq) + H_2(g)$

(2) Stoichiometry:
   1 mole $Zn(s)$ produces 1 mole $H_2(g)$

(3) Moles:

$\dfrac{12.0 \text{ liters } H_2(g)}{22.4 \text{ liters/mole}} = 0.536$ mole $H_2(g)$,

$\qquad\qquad = 0.536$ mole $Zn(s)$ required

(4) Weight:

$0.536 \text{ mole } Zn(s) \times \dfrac{65.37 \text{ g}}{\text{mole}} = 35.0$ g $Zn(s)$

Example 3. Volume-volume problem. What volume of oxygen will be required to react with 3.50 liters of nitric oxide, $NO(g)$, and how many liters of nitrogen dioxide, $NO_2(g)$, will be formed? (All gases measured at the same conditions of temperature and pressure, but not necessarily STP.)

(1) $\qquad\qquad NO(g) + \tfrac{1}{2}O_2(g) \rightarrow NO_2(g)$

(2) Stoichiometry: 1 mole $\tfrac{1}{2}$ mole   1 mole

(3) From Avogadro's principle, recognize that the stoichiometry by moles, 1 to $\tfrac{1}{2}$ to 1, is also the relation between relative volumes of the gases. Therefore, 3.50 liters of $NO(g)$ will react with 1.75 liters of $O_2(g)$ to produce 3.50 liters of $NO_2(g)$.

Example 4. Weight-moles-volume at any T or P. What volume of oxygen will be produced at 27°C and 700 mm Hg pressure by the complete decomposition of 73.56 g of potassium chlorate, $KClO_3(s)$?

(1) $\qquad\qquad 2KClO_2(s) \rightarrow 2KCl(s) + 3O_2(g)$

(2) Stoichiometry: 1 mole   1 mole   1.5 mole

(3) moles: $\dfrac{73.56 \text{ g } KClO_3}{122.6 \text{ g/mole}} = 0.600$ mole $KClO_3$

$0.600 \text{ mole } KClO_3 \times \dfrac{1.5 \text{ moles } O_2}{1 \text{ mole } KClO_3}$

$\qquad\qquad = 0.900$ mole O_2

(4) $V = \dfrac{nRT}{P}$: $V = \dfrac{(0.900)(0.0821)(300)}{700/760}$

$\qquad\qquad = 24.07$ liters

Example 5. Volume (solution)-moles-weight problem. What volume of 3.00 F H_2SO_4 would be required to precipitate all of the Ba^{2+} ion as $BaSO_4$ from a solution containing 3.00 g of $BaCl_2$?

(1) $$H_2SO_4(aq) + BaCl_2(aq) \rightarrow BaSO_4(s) + 2HCl(aq)$$

(2) Stoichiometry: 1 mole   1 mole   1 mole   2 moles

(3) $\dfrac{3.00 \text{ g } BaCl_2}{208 \text{ g/mole}} = 0.0144$ mole $BaCl_2$

0.0144 mole $BaCl_2$ $\times \dfrac{1 \text{ mole } H_2SO_4}{1 \text{ mole } BaCl_2} = 0.0144$ mole H_2SO_4

(4) 0.0144 mole H_2SO_4 $\times \dfrac{1 \text{ liter}}{3.00 \text{ moles } H_2SO_4} = 0.00480$ liter 3 F H_2SO_4

$$= 4.80 \text{ ml } 3 \text{ } F \text{ } H_2SO_4$$

REPORT

B

**Chemical
Stoichiometry**

NAME		
SECTION		LOCKER
INSTRUCTOR		DATE

PROBLEMS

NOTE: Show the solution in good order below each problem. Where a chemical reaction has occurred, always write the balanced equation first, and then determine the stoichiometry based on chemical units of quantity and volume, and solve for the desired quantity. Use the proper number of significant figures and proper dimensional units.

1. Convert the following into moles.

   (a) 1000 g of H_2O _____   (d) 1000 g of NaCl   _____

   (b) 1000 g of Mg   _____   (e) 1000 g of $C_{12}H_{22}O_{11}$ _____

   (c) 1000 g of Au   _____   (f) 1000 g of H_2 gas   _____

2. Given 5.400 g Al.

   (a) How many moles are there?

   (b) How many moles of aluminum sulfate, $Al_2(SO_4)_3$, could this form by reaction with excess sulfuric acid?

   (c) How many grams of sulfur would be contained in the aluminum sulfate formed?

3. What is the weight in grams of the following?

   (a) 0.600 mole of $CuCl_2$.

   (b) 0.275 moles of silver metal.

   (c) The silver in 0.275 mole of Ag_2SO_4.

   (d) The oxygen in 0.300 mole of H_2SO_4.

4. What percentage of iron is contained in an iron ore which is 80.0% hematite, Fe_2O_3?

                                                            _____

5. The atomic weight of the hypothetical element "A" is 50.0. If 10.0 g of A reacts with exactly 30.0 g of another hypothetical element "B" to form 40.0 g of a compound whose formula is known to be A_2B_3, what is the atomic weight of the element B?

                                                            _____

6. Complete the items called for in each of the following. Indicate clearly the method of solution, including the proper use of units.

Solution	Formula Weight	Number of Moles	Formality	Number of Equiv	Normality
(a) 11.7 g of NaCl in 800 ml of solution	_____	_____	_____		
(b) 2.80 g of KOH in 400 ml of solution	_____	_____	_____	_____	_____
(c) 490 g of H_2SO_4 in 1500 ml of solution	_____	_____	_____	_____	_____

7. What amount of the solute is present in each of the following solutions?

(a) 750 ml of 2.00 F HNO_3 _____mole _____g

(b) 650 ml of 0.500 N H_2SO_4 _____equiv _____g

REPORT B	Sheet 2	NAME

8. It is desired to measure out 20.0 g of each of the solutes in the following solutions. What volume of each should be used?

(a) 3.00 *F* K_2SO_4

_______________ml

(b) 2.50 *M* C_2H_5OH

_______________ml

9. Concentrated sulfuric acid solution is 98% by weight H_2SO_4 and has a density of 1.84 g/ml. What is the formality of the concentrated sulfuric acid solution?

____

10. A 2.77 *F* solution of sodium hydroxide has a density of 1.109 g/ml. Express this concentration in each of the following ways:

(a) As percent composition (g/100 g of solution)

____

(b) As molality (moles/1000 g of solvent)

____

11. How many grams of magnesium metal, Mg(*s*), will react with 2.00 gram-formula weights of the following solutions? (Assume that all the hydrogen atoms react in each reaction.)

(a) Hydrochloric acid

____

(b) Sulfuric acid

____

12. (a) What volume of air, at 27°C and 740 mm Hg pressure, is required to burn a gallon of gasoline completely to carbon dioxide and water? (Assume that gasoline has the formula C_7H_{16}, that a gallon of gasoline weighs 2800 g, and that air is 20% oxygen.)

_____

(b) How many grams of H_2O would be produced?

_____

(c) How many grams of CO_2 would be produced?

_____

13. A popular stomach acid neutralizer consists mainly of $Mg(OH)_2$. How many grams of $Mg(OH)_2$ would be required to neutralize the daily output of gastric juice, which is about 2 liters in volume and about 0.05 F in HCl?

_____

14. A low-grade copper ore containing 1.00% cuprite, $Cu_2S(s)$, is profitably mined in open pits in Utah. This ore is processed by several metallurgical operations to form copper (II) oxide, $CuO(s)$, which is then reduced to pure copper, $Cu(s)$. The copper is then converted to copper (II) sulfate pentahydrate crystals, $CuSO_4 \cdot 5H_2O(s)$, by suitable treatment with air and sulfuric acid. If 100 kg of this low-grade ore is converted to pure copper and finally to $CuSO_4 \cdot 5H_2O$ crystals, how many grams of the crystals could be obtained, assuming 100% efficiency in each step of the process? (Solve directly for the quantity desired; avoid unnecessary intermediate steps.)

_____

THE PREPARATION OF PURE SUBSTANCES BY CHEMICAL CHANGES

PRE-STUDY

The Problem of Purification

Most ordinary materials are mixtures of two or more individual substances. Even those materials which are essentially one substance contain small amounts of other substances as impurities. An important part of the work of a chemist involves the separation of mixtures to obtain pure substances. At every stage of the purification process, it is important to test for purity by analysis so that the progress of the separation can be checked.

The methods of such separations or purifications depend on the distinctive properties of the particular substances to be isolated. For example, differences in *solubility* in water may be utilized, by dissolving and recrystallizing the sodium chloride from a mixture of salt and sand in a natural salt deposit. Differences in *boiling point* enable the oil refinery operator to separate the several fractions of gasoline, diesel oil, and heavier oil from the

original crude oil. We may apply methods in which one component is *changed chemically* into some other substance which has a different solubility or boiling point, thus permitting a separation of the desired substance as a precipitate (solid substance) or perhaps as a volatile (gaseous) substance.

Types of Chemical Change

Any chemical change involves the formation of one or more *new substances,* each with its own specific properties, and a loss or gain in *energy.*

Note the following examples of important ways to classify some common types of reactions.

1. *Elements may combine* to form compounds:[1]

$$2Cu(s) + O_2(g) \rightarrow 2CuO(s)$$

---

[1] (s) = solid, (l) = liquid, (g) = gas, (aq) = aqueous solution.

2. *Compounds may decompose* to form different compounds or elements:

$$2KClO_3(s) \rightarrow 2KCl(s) + 3O_2(g)$$

$$2HgO(s) \rightarrow 2Hg(l) + O_2(g)$$

3. *Replacement reactions* in which one free element reacts by a transfer of electrons to replace another element in a compound, and liberates this second element in the free elemental state:

$$Zn(s) + 2HCl(aq) \rightarrow ZnCl_2(aq) + H_2(g)$$

$$Cu(s) + Ag_2SO_4(aq) \rightarrow 2Ag(s) + CuSO_4(aq)$$

$$Cl_2(g) + 2KI(aq) \rightarrow I_2(s) + 2KCl(aq)$$

All of the above reactions involve changes in the oxidation state of the elements concerned, and are therefore classified as *oxidation-reduction reactions*.[2]

4. *Other types of reactions* that involve no changes in oxidation state are:
(a) *Reactions of oxides with water* to form acids or bases:

$$SO_3(g) + H_2O \rightarrow H_2SO_4(aq)$$

$$BaO(s) + H_2O \rightarrow Ba(OH)_2(aq)$$

(b) *Acid-base reactions:*

$$HCl(aq) + NaOH(aq) \rightarrow NaCl(aq) + H_2O$$

(c) *Precipitation reactions:*

$$NaCl(aq) + AgNO_3(aq) \rightarrow$$
$$AgCl(s) + NaNO_3(aq)$$

Substances in Solution

When certain classes of substances, namely the strong acids, bases, and salts, dissolve in water, their aqueous solutions actually contain separate electrically charged particles, called *ions*, rather than molecules. Thus, HCl in solution consists of the substances $H^+(aq)$ and $Cl^-(aq)$, $NaOH(aq)$ consists of $Na^+(aq)$ and $OH^-(aq)$, and $NaCl(aq)$ consists of $Na^+(aq)$ and $Cl^-(aq)$. Later, beginning with Experiment 14, we shall write the formula of such substances as separate ionic formulas, in

accord with their character as individual ions, not molecules. However, at this early stage in your training, it will be more helpful to use the molecular formulas.

Chemical Changes Occurring in this Experiment

We shall start with a sample of silver-copper alloy (such as is used in jewelry manufacture or in American coinage). By a succession of chemical and physical changes, we shall separate this sample to obtain the pure metals of silver and copper. We shall then carry out *tests* to prove their purity. The separation is to be carried out in the following steps.

1. The silver-copper alloy is dissolved by oxidation in nitric acid:

$$3Ag(s) + 4HNO_3(aq) \longrightarrow$$
silver + nitric acid   yields
$$3AgNO_3(aq) + NO(g) + 2H_2O$$
silver nitrate + nitric oxide + water

$$3Cu(s) + 8HNO_3(aq) \longrightarrow$$
copper + nitric acid   yields
$$3Cu(NO_3)_2(aq) + 2NO(g) + 4H_2O$$
cupric nitrate + nitric oxide + water

The colorless gas nitric oxide immediately combines with oxygen from the air to form the brown gas nitrogen dioxide:

$$2NO(g) + O_2(g) \rightarrow 2NO_2(g)$$

2. The solution of silver nitrate and cupric nitrate is evaporated with excess sulfuric acid until very dense white fumes of sulfur trioxide are emitted, thus insuring complete evaporation of nitric acid. (If *any* HNO_3 remains, the pure silver to be formed in step 3 will redissolve again, thus defeating the separation.) The residue now contains silver sulfate and cupric sulfate.

$$2AgNO_3(aq) + H_2SO_4(aq) \longrightarrow$$
silver nitrate + sulfuric acid   yields
$$Ag_2SO_4(s) + 2HNO_3(g)$$
silver sulfate + nitric acid

$$Cu(NO_3)_2 + H_2SO_4 \longrightarrow$$
cupric nitrate + sulfuric acid   yields
$$CuSO_4(s) + 2HNO_3(g)$$
cupric sulfate + nitric acid

---

[2] In Experiment 13 you will study an oxidation-reduction reaction and the study will be continued in detail in Experiment 32 and subsequent experiments.

3. After water is added and the mixture is heated to dissolve the salts formed, copper metal is added to precipitate the silver as a free metal by a *replacement* reaction:

$$Cu(s) + Ag_2SO_4(aq) \longrightarrow$$
copper + silver sulfate   yields

$$CuSO_4(aq) + 2Ag(s)$$
cupric sulfate +   silver

4. Finally, the cupric sulfate solution is treated with zinc metal to precipitate free copper metal by another *replacement* reaction:

$$Zn(s) + CuSO_4(aq) \longrightarrow$$
zinc + cupric sulfate yields

$$ZnSO_4(aq) + Cu(s)$$
zinc sulfate + copper

These changes may be summarized by the following flow diagram:

$$Ag\text{-}Cu \xrightarrow{HNO_3} \begin{array}{c} AgNO_3 \\ Cu(NO_3)_2 \end{array} \xrightarrow{H_2SO_4} \begin{array}{c} Ag_2SO_4 \\ CuSO_4 \end{array} \xrightarrow{Cu}$$

$$\frac{Ag}{CuSO_4} \xrightarrow{Zn} \frac{Cu}{ZnSO_4}$$

The experiment as a whole provides experience with various types of chemical change, especially replacement reactions of a more active metal for a less reactive metal from its salts.

Careful experimental technique is important for two reasons. First, the silver metal you prepare must be pure, so that you can obtain reliable quantitative data when you prepare silver chloride from this substance in Experiment 5. Second, only by careful weighing can you obtain a reliable value for the percent of silver in your alloy, and also properly evaluate and interpret the relationship between the weights of the copper and silver that interact, and the relative atomic weights of each.

EXPERIMENTAL PROCEDURE

Special Supplies: Silver-Copper alloy.[3]

Chemicals: No. 18 copper wire, 0.1 F Cu(NO$_3$)$_2$, 0.1 F AgNO$_3$, 0.1 F Zn(NO$_3$)$_2$, zinc (mossy).

---

[3]Silver-Copper alloy, of 90% Ag and 10% Cu, or of various percentages (which will provide "unknowns"), may be obtained from dealers in the noble metals—silver, gold, and platinum. The cost is only 3 to 4 cents per student. (One source is Handy and Harmon, 4140 Gibson Rd., El Monte, Calif.)

1. Pure Silver from a Silver-Copper Alloy.[4]   (a) Obtain 1 to 2 g of the alloy and about a 30-cm length of No. 18 copper wire. Shape the wire into a very loose coil 1 inch in diameter; leave 1 inch bent upright as a handle. Weigh the alloy and the wire each to a precision of ±0.01 g. Your instructor will first give you instructions for the balance to be used. Review the rules and precautions on weighing techniques presented in Experiment 1.

(b) Place the weighed alloy in a 150-ml beaker, and add 10 ml of dilute nitric acid (6 F HNO$_3$). **Caution:** *Be careful not to spill acid on your hands or clothing. If you do spill any, flush well with water. Neutralize acid on clothing with NaHCO$_3$ solution.* Arrange the apparatus as in Figure 4-1, preferably under a hood. (If a hood is not available, an inverted funnel may be clamped over the beaker and connected by a rubber tube to the water aspirator or "suction pump" to remove the acid fumes from the room.) Warm gently if necessary, but only intermittently, to maintain a moderate rate of solution of the metal. While the alloy is dissolving, review the reactions for this and the following steps in the preceding discussion. The reaction rate can be followed by observing the rate of evolution of brown NO$_2$ fumes. If these cease before complete solution of the metal, add 5 ml more of nitric acid.

(c) When the alloy is dissolved, add 25 ml of dilute sulfuric acid (3 F H$_2$SO$_4$). Heat the solution in the open beaker under the hood gently and carefully to avoid any loss by spattering, until the remaining syrupy liquid (concentrated sulfuric acid and dissolved salts) begins to emit copious, very dense, white, choking fumes of sulfur trioxide (SO$_3$). Then at once cover the beaker with a watch glass to minimize fumes in the laboratory. It is essential to evaporate until *all* nitric acid is removed; otherwise it will redissolve the metallic silver, which is to be precipitated next. *Let the beaker cool,* and then cautiously add about 125 ml of distilled water from your wash bottle. Warm the mixture (*but do not boil it*), while stirring it with a glass rod, until the white silver sulfate is dissolved. (In this acid solution, the salt is largely silver hydrogen sulfate, AgHSO$_4$.)

(d) Place the copper wire coil weighed in (a) into the solution of dissolved salts, leaving an end

---

[4]*Caution:* Wash off any spilled silver salt solutions from your hands or clothing at once; otherwise a black stain of metallic silver results. This may be removed by treating the stain first with a few drops of I$_2$ solution to form AgI, then treating it with a strong solution of sodium thiosulfate (Na$_2$S$_2$O$_3$) to remove excess I$_2$ and to dissolve the AgI(s).

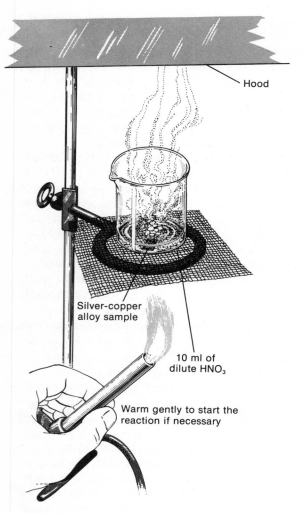

FIGURE 4-1
A procedure for dissolving the silver-copper alloy, and evaporating the solution. Note that the sample is being heated in the hood.

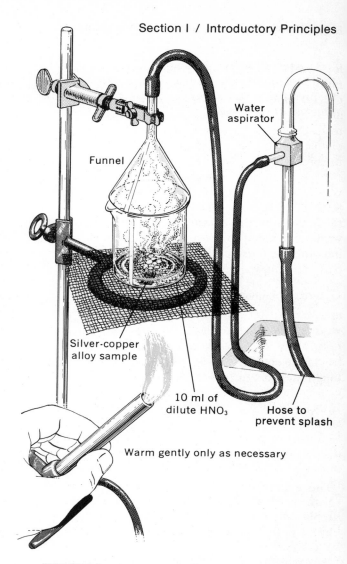

FIGURE 4-2
An alternate procedure for dissolving the silver-copper alloy and evaporating the solution. If a hood is not available, this funnel and suction arrangement is a satisfactory substitute.

of the wire above the solution as a handle. Observe the beautiful deposit of silver metal crystals. After a few minutes, periodically shake the wire gently to loosen the silver and expose fresh copper to the solution. During this reaction,[5] carry out the qualitative test reactions in (e). You will need to know them to complete the purification and separation of the silver and later of the copper.

(e) *Qualitative tests for silver and copper salts.* Five solutions—one each of silver nitrate, copper nitrate, and zinc nitrate, one of silver nitrate plus

a drop of copper nitrate, and one of copper nitrate plus a drop of silver nitrate—will be tested with two reagents: hydrochloric acid (HCl) and ammonia (NH₃) solution. In your report sheet these should be arranged in suitable columns for entry of your observations as obtained.

Place 1-ml samples of each of the five salt solutions into separate 10-cm test tubes. To each of these, add *1 drop only* of 6 F hydrochloric acid, and note the results. Prepare another set of the salt solutions, and to each add *1 ml* of 6 F ammonia. (An excess of the reagent is desired.) *Mix each solution well,* until the metal hydroxide precipitate that first forms redissolves, before observing the

---

[5]The experiment may conveniently be interrupted at this point until the next laboratory period, if necessary.

final results. The metals remain in solution as the more stable complex ions, $Ag(NH_3)_2^+$, and $Cu(NH_3)_4^{2+}$. Note the characteristic precipitates and colors, which may be useful identification tests of particular metal ions. (Zinc salts are included in your observations, since they are present in some instances where the tests are made.)

(f) To return to the reaction: when the copper wire has been in the solution for half an hour, transfer 1 ml of the solution to a small test tube, and test this for silver salts. If the test is positive, leave the wire in the solution and repeat the test at 15-minute intervals until all the silver is precipitated.[6] Compare the amount of blue copper salts now in the solution with that of the original solution. Explain. Now carefully shake loose all the silver crystals, and remove the wire from the solution. Rinse the wire thoroughly with distilled water from your wash bottle, and set it aside to dry completely. (Do not heat it, as copper oxidizes easily.) When dry, weigh it. While the wire is drying, decant as much of the clear solution as you can into a clean beaker (see Experiment 2, Figure 2-2), taking care that no silver is transferred. (If small flecks of silver tend to float, you may add a drop of diluted detergent to decrease the surface tension.) At once add the mossy zinc called for in part 2 of this procedure, so that the reaction may be completed when needed. To wash the silver, add 5 ml of distilled water, break up any large lumps of silver crystals, agitate these thoroughly, let the silver settle, and decant the wash water. Repeat this washing process, first with a 5-ml portion of 6 F ammonia solution to dissolve any impurity of AgCl, and then with 5-ml portions of distilled water, until the washings show no positive test for cupric salts when 1 ml of 6 F ammonia is added.

---

[6]The rate of reaction of silver sulfate solution with copper metal varies greatly with the concentration of acid. It is quite slow in neutral solution, requiring 5–10 hours for completion, but with the excess sulfuric acid specified, the replacement should be completed in 30–45 minutes. If it is proceeding too slowly, an additional 5 ml of sulfuric acid may be used.

(g) Heat a clean evaporating dish to dry it thoroughly. Let it cool, and weigh it carefully on the balance. Use a stirring rod and distilled water from your wash bottle to transfer all the silver to this dish, without loss, and heat it until it is thoroughly dry. Cool, and weigh it. Calculate the percentage of silver in your alloy sample.

(h) To check on the purity of your silver, dissolve a small amount of it—not more than one-twentieth of the total amount—in $\frac{1}{2}$ ml (about 10 drops) of dilute nitric acid by warming gently, and test for cupric salts as above. (Be sure to add an excess of ammonia, so the solution smells of ammonia *after mixing*.) Record your method of testing and the results, and show the test sample and record to your instructor for his approval. Save your dry silver for use in Experiment 5.

2. Pure Copper from Your Cupric Sulfate Solution. (a) Add several grams of mossy zinc to the solution from the above preparation of pure silver, if you did not already do this in part 1(f). Observe the formation of copper crystals on the surface of the zinc. What other reaction is occurring at the same time? Note the evolution of gas. What is it? (If time is available, leave this reaction mixture until the replacement is complete, as evidenced by the lack of any blue color when a small portion of the solution is tested with ammonia.) Separate the precipitated copper from the solution and from any pieces of zinc which remain. (All the zinc will eventually dissolve in the sulfuric acid solution.) Wash the copper repeatedly by decantation. The washed copper may be dried at room temperature, but it oxidizes quite readily, even without heating.

(b) To check on the purity of your copper, dissolve a small portion of it—you need not wait until it is dry—in $\frac{1}{2}$ ml (about 10 drops) of dilute nitric acid. Warm this gently until solution is complete, then dilute it with 2 ml of distilled water and test for the presence of silver salts. Record your test method and observation, and show the test sample and record to your instructor for his approval.

REPORT

4

The Preparation of
Pure Substances by
Chemical Changes

NAME

SECTION   LOCKER

INSTRUCTOR   DATE

DATA AND CALCULATIONS

1. Silver

(a–d) **Pure Silver from a Silver-Copper Alloy**

Weight of silver-copper alloy sample   _____g

Weight of copper wire before reaction   _____g

Weight of copper wire after reaction   _____g

Weight of copper which reacted with silver   _____g

Weight of evaporating dish plus pure silver   _____g

Weight of evaporating dish   _____g

Weight of pure silver obtained   _____g

Percent of silver in the alloy   _____%
Method of calculation:

(e) **Qualitative Tests for Silver and Copper Salts.** In the proper square, give the formula of any characteristic compound formed, and enter any observed result, such as "Curdy white precipitate", "No effect", the color of the solution, etc.

Test solutions	6 F HCl (1 drop)	6 F NH$_3$ (1 ml)
Pure AgNO$_3$ solution		
Pure Cu(NO$_3$)$_2$ solution		
AgNO$_3$ plus 1 drop Cu(NO$_3$)$_2$		
Cu(NO$_3$)$_2$ plus 1 drop AgNO$_3$		
Pure Zn(NO$_3$)$_2$ solution		

(f–h) **Test for Purity of Solid Silver Metal Obtained.** How completely have you succeeded in separating copper to obtain pure silver? Describe your method (reagents, observations, conclusions), starting with your solid silver.

2. Copper

(a) **Pure Copper from Cupric Sulfate Solution.** How would you expect the total weight of copper recovered to compare with the loss in weight of the copper wire? Account for the gas evolved while the zinc is replacing the copper in the acid copper sulfate solution.

(b) **Test for Purity of Solid Copper Metal Obtained.** How completely have you succeeded in separating silver to obtain pure copper? Describe your method in detail (reagents, observations, conclusions), starting with your solid copper.

APPLICATION OF PRINCIPLES

1. **Reacting Weights versus Atomic Weights.** Discover if you can any simple mathematical relationship involving the weights of silver and copper which interact, and their respective atomic weights. (Try various relationships: g × at wt, at wt/g, g/at wt. See also Study Assignment B. Relate your findings to fundamental theory. This calculation correlates with the balancing of the second equation under *Replacement reactions,* in the pre-study section.

2. **Chemical Changes Involved in this Experiment.** Write the names of any products formed in the following reactions. In the second column of blank lines, indicate cases of "No action." For those that do react, indicate those which (a) involve no oxidation-reduction ("No OR"), (b) are oxidation-reduction reactions ("OR"), and (c) are also replacement reactions ("OR-repl").

A copper wire is placed in silver sulfate solution _____ _____

A solution of silver nitrate and sulfuric acid is evaporated _____ _____

Sulfuric acid is heated to form dense white fumes _____ _____

Hydrochloric acid is added to silver nitrate _____ _____

Zinc metal is placed in dilute sulfuric acid _____ _____

Copper metal is added to dilute sulfuric acid _____ _____

Zinc metal is added to cupric sulfate solution _____ _____

Copper metal is heated in air _____ _____

Silver metal is heated in air _____ _____

Silver oxide is heated _____ _____

---

3. **The Replacement Series of the Metals.** Considering only the replacement reactions above,
list the elements copper, zinc, silver, and hydrogen in a vertical order such that each element
is above another which it replaces from solution. Comment on the meaning of such an "activity series."

_____

_____

_____

_____

QUESTIONS AND PROBLEMS

1. In this section, use any of your data, or Appendix C, Table 13.

   (a) Predict two replacement reactions that you
       have not tried, but would expect to occur.

   (b) Which of the following metals may be dis-
       solved by hydrochloric acid: gold, manganese,
       cadmium, bismuth, nickel, mercury, iron?

   (c) Suppose an aluminum wire coil, instead of copper, were used in this experiment. (i) How would this affect the
       weight of silver metal precipitated? (ii) Would cupric ion (from the solution of the alloy) remain in solution, or
       precipitate as metal along with the silver?

   (d) What weight of American coinage metal (90% silver, 10% copper) would be needed to produce, by suitable
       chemical action, 25.0 g of silver nitrate (63.5% silver)?

THE FORMULA OF A COMPOUND
FROM EXPERIMENTAL DATA

PRE-STUDY

Review the meaning of *symbols* and *formulas* in terms of *moles* and *formula weights,* as presented in Study Assignment B. Experimental data that express the relative weights (in grams) of the constituent elements in a compound *can always be expressed* in terms of *an integral number of* moles of each constituent element. This is the essence of the meaning of a formula and is a direct consequence of the atomic theory. Therefore, to find the simplest formula of a compound, we need only calculate the number of moles of each from the number of grams of each in any given amount of the compound, and find the simplest integral ratio of these.

For example, a chloride of copper is analyzed, and 5.018 g of it is found to contain 2.351 g of copper and 2.667 g of chlorine. The number of moles of each element is calculated as follows:

$$\frac{2.351 \text{ g Cu}}{63.54 \text{ g/mole}} = 0.03700 \text{ mole Cu}$$

$$\frac{2.667 \text{ g Cl}}{35.45 \text{ g/mole}} = 0.07523 \text{ mole Cl}$$

Dividing each of these numbers by the smaller one gives a ratio of 1.00 Cu to 2.033 Cl, which, within experimental error, corresponds to the simple formula $CuCl_2$.

The "simplest formula," as calculated above, gives *only* the integral proportion of the constituent atoms. For those substances in which the unit of structure is a definite molecule, the correct *molecular formula* is often some multiple of this simplest formula. Thus, the formula of hydrogen peroxide is H_2O_2, not HO, which would imply the same percent composition but only half the molecular weight. Likewise, the formula for butane is C_4H_{10}, not C_2H_5. In order to decide upon the correct molecular formula of a substance, we must have experimental evidence of the true molecular weight, as well as of the percent composition. This is considered in Experiment 12.

The Law of Definite Proportions was developed in the nineteenth century. It stated that in a given compound, the elements always combined in the same proportions by weight. Because of the precision of the measurements at that time, all compounds were found to obey this law. Thus the compound $CuCl_2$ was found to have a weight ratio of copper to chlorine that was precisely the same as the ratio of the weights of one copper atom to two chlorine atoms.

However, more precise measurements have revealed that a number of oxides and sulfides of the transition elements do not have integral ratios of the numbers of atoms in the compound, and that the ratio depends on the details of how the compound is prepared. Such compounds are called nonstoichiometric compounds. One of the challenges of this experiment will be for you to conduct the experiments as carefully as possible. Then, giving careful consideration to the precision of your experiments, determine whether the compounds you have prepared are stoichiometric or nonstoichiometric compounds.

Laboratory Technique and Measurements

The development and understanding of chemical principles depend on precise quantitative measurement of the physical relationships involved. Your chemistry course includes a number of such quantitative experiments, of which this is one. Your efficient use of laboratory time and care of the equipment correlate closely with the precision you may expect in your work. This precision is, of course, limited by the type of balance available to you. With the balances illustrated in Figures 5-1 and 5-2, weighings to ±0.001 g and ±0.0001 g re-

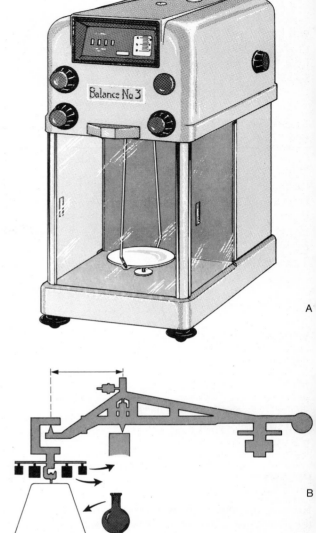

A

B

Substitution, 1-pan balance with only 2 knife edges
Constant load=constant sensitivity

FIGURE 5-2
(A) An analytical, single-pan balance for substitution weighing; (B) a schematic diagram of this instrument. In using this type of balance, one removes weights from the front portion of the beam (by turning the dial) until an amount exactly equal to the unknown mass is removed and equilibrium is restored. Since constant loading (and therefore constant sensitivity) of the beam is maintained, greater accuracy results.

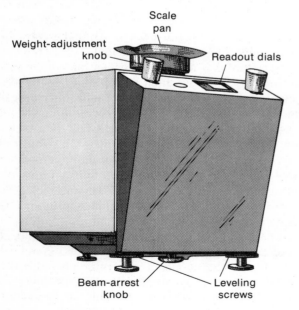

Scale pan

Weight-adjustment knob

Readout dials

Beam-arrest knob

Leveling screws

FIGURE 5-1
A top-loading single-pan balance.

spectively can be obtained. If necessary, the simpler balance (shown in Figure 1-3), which has a limiting precision of ±0.01 g, may be used to demonstrate the quantitative nature of the principles that are the basis of chemical theory. If the uncertainty in weighing a 2.00-g sample is ±0.02 g, the percentage uncertainty is 0.02 g/2.00 g × 100 + 1%. The same sample, weighed to ±0.002 g, will have an uncertainty of 0.1%. The proper use of the balance will be demonstrated by your instructor: it is essential that you have this training *before beginning the experiment.*

Experimental Precision and the Treatment of Errors

When you record experimental data and use them to make calculations, it is very important to record the values in such a way that the figures themselves convey the proper limitations on the precision of the physical measurements involved. To learn the correct manner of doing this, study Appendix B, The Treatment of Experimental Errors.

EXPERIMENTAL PROCEDURE

Chemicals: Part 1: copper wire, No. 22-26, or medium turnings, powdered sulfur; part 2: silver metal from Experiment 4 if pure, or pure silver foil; part 3: very thin lead shavings.

NOTE: Part 1, copper sulfide, can be completed in a shorter time (1–2 hours), and is somewhat less precise, owing to the method available and to possible partial oxidation. Therefore weighings to ±0.01 g are acceptable. For part 2, silver chloride, and part 3, lead chloride, balances weighing to ±0.001 g are justified and desirable, and a longer time (2–3 hours) is required for completion.

Careful quantitative determinations usually are carried out with duplicate samples, to guard against gross errors and to insure greater accuracy by averaging two results.

Your instructor will designate which of the three experiments you will perform, and whether to use duplicate or single samples; if you are to use a single sample, modify the directions accordingly.

1. A Sulfide of Copper. Support two clean crucibles on clay triangles and heat each for a moment to dry it thoroughly. Cool the crucibles, and weigh them accurately. To each crucible add 1–1.5 g of copper, wire or medium turnings, and press this well down in the crucibles. Weigh these accurately. Add about 1 g of powdered sulfur (estimated by comparison with a 1-g sample placed near the supply bottle) to each crucible. Avoid a large excess, because burning it out takes longer.

Place each crucible and contents, with lid in place, under the fume hood, and heat them—moderately at first, then more intensely. See Figure 5-3. Play the flame around the sides and lid, as well as on the bottom, until sulfur ceases to burn around the lid. This should not require more than 10–15 minutes. *Do not remove the crucible lid while the crucible is very hot,* because atmospheric oxidation will occur if you do.

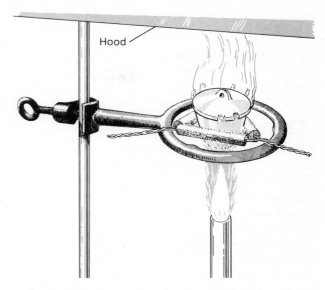

FIGURE 5-3
The formation of a sulfide of copper, and the evaporation and ignition of excess sulfur.

When each crucible has cooled, place it on a clean piece of paper, hold it firmly with your tongs, and very carefully crush the contents (without *any* loss) with the blunt end of a stirring rod so that the compound lies near the crucible bottom to facilitate thorough heating. A small amount of sulfur may be added if there is any doubt that a sufficient amount was added before. Replace the lid, and heat the crucible intensely, both top and bottom, for 5 minutes, to vaporize all sulfur not chemically combined. Again weigh each completely cooled crucible (without lid) and contents. (The crucible may be heated a second five minutes, cooled, and weighed again, as a check on the completion of the vaporization of sulfur.) From these data, calculate the combining proportions of copper and sulfur, and the formula of the compound.

Calculate the percentage of copper in your compound from your formula. Compare this percentage with that given by the formula published in the literature or from the average formula obtained by your entire laboratory section. Calculate the percent difference between these two numbers. Note that this is termed a difference and not an error. If your copper-sulfide compound turns out to be a nonstoichiometric compound, whose formula depends on the exact conditions under which it was prepared, it would be inappropriate to term this an error because the compounds may actually be different. Under these circumstances, it would be no more appropriate to calculate a percentage error in the composition of your compound and the literature value than it would be to compute a percentage error between the copper concentrations in $CuCl_2$ and $CuBr_2$.

78

Optional Experiments

NOTE: These are for students who have the time and interest to do further work. Prepare a suitable report form for the entry of all data, and the calculation of the results, *before performing each of the experiments.*

2. A Chloride of Silver. Heat, cool, and then weigh two clean evaporating dishes. (Use tongs to avoid handling the dishes.) Add to each dish approximately 0.5–1.0 g of pure dry silver saved from Experiment 4, or use pure silver foil, and weigh the dish again.

Dissolve the silver in each dish by the addition of 4 ml of dilute (6 F) HNO_3. Cover the dish with a watch glass and warm it momentarily, if necessary, only enough to maintain a modest rate of reaction. Avoid any loss of solution by spattering. When solution of the metal is complete, rinse the under side of the watch glass into the dish by spraying it with distilled water from your wash bottle (see Figure 5-4).

Add 3 ml of dilute (6 F) HCl to the solution, and mix this gently by rotation of the dish. Test the solution for completeness of precipitation with an additional drop of HCl. Evaporate the mixture very gently in a manner to avoid all spattering. It may be placed on a wire gauze 3–5 inches above a very low (1-inch) flame (as in Figure 2-3, Exp. 2); or for greater safety place the dish on a beaker of boiling water (Figure 5-5) to complete the evaporation. When the precipitate is apparently dry, again heat the dish cautiously directly on the wire gauze. Avoid spattering by wafting the flame around the edges but not directly under the dish, until it is

When the dish is nearly dry, transfer it to the top of a beaker of boiling water.

FIGURE 5-5
The evaporation of a liquid over a water bath. *Gentle* boiling of the water in the beaker is more efficient than rapid boiling.

quite hot and thoroughly dry. Do not melt the residue, but be sure all vapors of water and acid have ceased to be evolved. Cool and weigh each dish and residue. From these data, calculate the formula of silver chloride.

3. A Chloride of Lead. The same general procedure as for silver chloride may be used. Use about 0.5-g samples of lead, cut in *very* thin shavings. Lead is less reactive than silver, but will dissolve in 25–45 minutes in a mixture of 4 ml of concentrated HNO_3, and 1 ml of 6 F HCl, kept hot on the water bath (Figure 5-5). Add more HCl by drops every few minutes to maintain a fairly rapid reaction, but avoid loss of appreciable spray by the evolved gases. Much of the HCl is lost by the decomposition of the HNO_3-HCl mixture by the reaction

$$HNO_3(aq) + 3HCl(aq) \rightarrow$$
$$Cl_2(g) + NOCl(g) + 2H_2O$$

Bibliography

Bacon, E. K., *J. Chem. Educ.* **44**, 620 (1967).

Dingledy, D., and Barnard, W. M., "The Stoichiometry of Copper Sulfide Formed in an Introductory Laboratory Exercise," *J. Chem. Educ.* **44**, 242 (1967).

FIGURE 5-4
The technique of rinsing a watch glass.

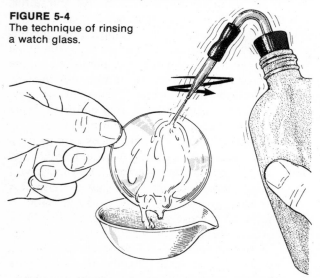

Rinse the bottom of the watch glass, back and forth, with distilled water from your wash bottle.

| | | SECTION | LOCKER |
| | | INSTRUCTOR | DATE |

DATA AND CALCULATIONS

Data[1]		1	2
Weight of crucible and copper sulfide, first heating		g	g
Weight of crucible and copper sulfide, final heating		g	g
Weight of crucible and copper		g	g
Weight of crucible, dried, at start		g	g
Atomic weight of copper[2]		g/mole	g/mole
Atomic weight of sulfur		g/mole	g/mole
Calculations[3]		1	2
Weight of copper sulfide		g	g
Weight of copper		g	g
Weight of sulfur		g	g
Number of moles of Cu in weight taken[4]		mole	mole
Number of moles of S combined with the copper		mole	mole
Ratio between moles of Cu and S (from your data)			
Formula of the compound (from your data)			
Percent of Cu in compound from literature[5] or class average		%	%
Percent of Cu in compound from your data		%	%
Percentage difference		%	%

INSTRUCTOR'S APPROVAL ———————————————

[1]Data are listed in the order most convenient for subtraction, rather than in the order obtained. Columns are provided for two samples, or for repetition of the experiment.

[2]An atomic weight table is given in Appendix C, Table 16.

[3]Indicate the method of calculation with first set of data. As soon as you have finished your calculation, report the result to your instructor.

[4]Calculate the number of moles to three significant figures, not just to "three decimal places."

[5]The letter by Bacon, referred to in the bibliography, summarizes many student results obtained in studies of this copper-sulfur compound in the course of the past 40 years.

PROBLEMS: THE UNCERTAINTY OF MEASUREMENTS

NOTE: "How shall I record my data and calculate the results to indicate properly the precision and accuracy of my measurements?" Review the discussion on the treatment of experimental errors in Appendix B.

1. State the number of significant figures in each of the following:

(a) 26.74 _____ (c) 0.0020 _____ (e) 1.030×10^{-2} _____

(b) 0.0267 _____ (d) 10.300 _____ (f) 6.02252×10^{23} _____

2. In synthesizing a sulfide of copper, 0.83 g of sulfur combined with 3.20 g of copper. Which of the following should be used in reporting the percentage of copper? (a) 79%, (b) 79.4%, (c) 79.40%, (d) 79.404%.

_____

3. The weights of three pieces of lead were designated as: 1.232 ± 0.003 g, 3.2432 g, and 2.8658 ± 0.0002 g. From the data as given, how should the total weight of the three pieces be expressed?

_____

4. A sample of silicon weighing 2.10 g is heated in air to form 4.50 g of an oxide of silicon. What is the formula of the oxide?

_____

5. A 3.000-g sample of propylene gas was found to contain 2.560 g of carbon and 0.440 g of hydrogen.

(a) What is its simplest formula?

_____

(b) The molecular weight of propylene was found to be 42.1. What is the correct molecular formula?

_____

6. A bismuth sample weighing 0.687 ± 0.003 g was converted to a chloride by reaction with HNO_3 and HCl, with careful evaporation to dryness. The weight of the bismuth chloride was 1.025 ± 0.003 g. From these data, how would you best express the percent bismuth in the compound, to indicate the precision of the weighings?

_____

7. (a) Calculate the formula of bismuth chloride based on the data of Problem 6.

_____

(b) What is the theoretical percent bismuth in bismuth chloride, based on this formula?

_____

(c) What is the percent error in the above experimental determination?

_____

ATOMIC AND
MOLECULAR STRUCTURE

EXPERIMENT 6

THE PACKING OF ATOMS
AND IONS IN CRYSTALS

PRE-STUDY

The solid state of matter nearly always consists of a regular arrangement of atoms, molecules, or ions. If we represent each building block as a point, the structure of the solid can be represented by a regularly repeating pattern called the *crystal lattice*. A crystal *structure* is an array of atoms, ions, or molecules that is centered on the lattice points of the crystal. A comparison of these two concepts is shown in Figure 6-1.

Many such crystal lattices exist. There are exactly 230 different ways in which points can be placed in a regular array in space. Such arrangements are called space groups. In order to describe these groups succinctly, the concept of the unit cell has been introduced. A unit cell is the smallest part of the crystal lattice that, when translated along the *a, b,* or *c* axes (as shown in Figure 6-3) without rotation, will generate the entire space lattice. Six unit cells are displayed in three dimensions in Figure 6-2. If dots are placed at other equivalent locations in these unit cells, such as at the centers of

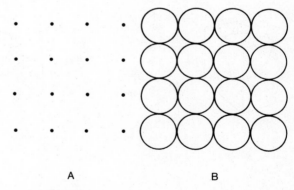

FIGURE 6-1
(A) crystal lattice. (B) A crystal structure.

faces of the unit cells, it is possible to generate 14 space lattices, called Bravais lattices after the Frenchman who described them. The unit cells and crystal lattices of the three cubic Bravais lattices are shown in Figure 6-3. Note that the unit cells of crystal structures contain fractional atoms in the following places: $\frac{1}{8}$ at the corners; $\frac{1}{4}$ at the edges;

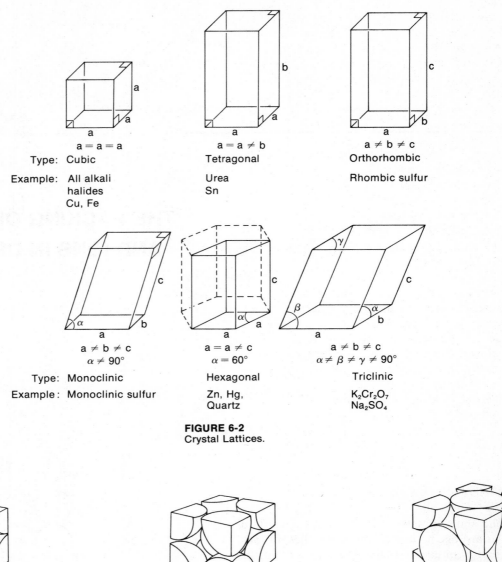

a = a = a	a = a ≠ b	a ≠ b ≠ c
Type: Cubic	Tetragonal	Orthorhombic
Example: All alkali halides Cu, Fe	Urea Sn	Rhombic sulfur

a ≠ b ≠ c, α ≠ 90°; a = a ≠ c, α = 60°; a ≠ b ≠ c, α ≠ β ≠ γ ≠ 90°

Type: Monoclinic	Hexagonal	Triclinic
Example: Monoclinic sulfur	Zn, Hg, Quartz	$K_2Cr_2O_7$ Na_2SO_4

FIGURE 6-2
Crystal Lattices.

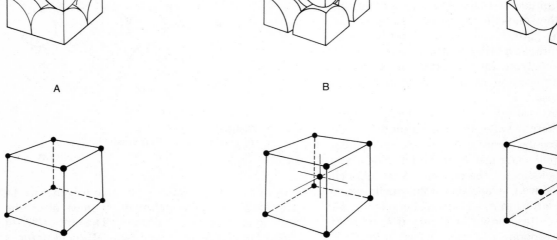

FIGURE 6-3
Unit cells: (A) a simple cubic cell; (B) a body-centered cubic cell; (C) a face-centered cubic cell.

and $\frac{1}{2}$ on the faces. The reasons are the ways in which unit cells join together: 8 join at a corner, 4 at an edge, and 2 at a face (see Figure 6-3).

In this experiment we will use styrofoam spheres as the building blocks, and we will study some of the ways they can be packed to form some typical metallic crystals. However, you must remember that the solid spheres have a rigidity that atoms and ions do not. The cloud of electrons surrounding the nucleus of an atom does not have a definite boundary, nor does its influence cease at a definite distance from the nucleus.

As you perform the experiment, bear in mind the extent of the enlargement of the radii of atoms or ions from their actual value of 1 or 2 angstrom units (10^{-8} cm) to those of the spheres used in the models. Also bear in mind the large number of particles (Avogadro's number, 6.023×10^{23}) actually involved in forming a mole of a metallic element or the formula weight of an ionic compound.

Three types of packing will be investigated: (1) body-centered cubic, (2) face-centered cubic, or cubic closest packing, and (3) hexagonal closest packing. We will observe the *coordination number* (number of nearest neighbors) and other features of the geometry of these structures.

In addition, we will investigate some of the possibilities of packing spheres of different radii into each of the three lattices and study some of the features of ionic crystals. The importance of the ratio of the ionic radii of the cations and anions as it determines the type of crystal structure and the coordination number will be observed by building the rock salt lattice, $Na^+Cl^-(s)$, the cesium chloride lattice, $Cs^+Cl^-(s)$ and the Wurtzite lattice, $Zn^{2+}S^{2-}(s)$.

Crystal Structures of Elements

Most of the metallic elements crystallize in one of the following structures.

1. *Body-centered cubic packing* (see Figure 6-4): the alkali metals Li, Na, K, Rb, and Cs, and some of the transition metals such as V, Nb, Ta, α-Cr, Mo, β-W, and α-Fe.

2. *Cubic closest packing, or face-centered cubic* (see Figure 6-5): the noble metals Cu, Ag, and Au, some of the transition metals in Group VIII, and α-Ca, Sr, and Al.

3. *Hexagonal closest packing* (see Figure 6-6): Mg, Be, and γ-Ca in Group II, and Ti and Hf in Group IVa.

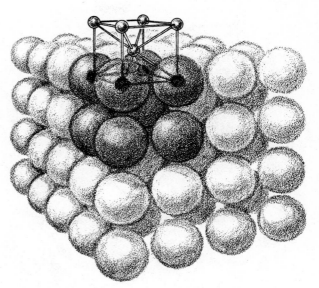

FIGURE 6-4
The body-centered cubic packing arrangement in α-iron.

FIGURE 6-5
The face-centered cubic (cubic closest) packing arrangement in copper.

A substance may crystallize in more than one form, as indicated by the Greek-letter prefixes above, but each form is stable within a definite range of pressure and temperature.

In the nonmetallic elements the building blocks of the crystals are molecules such as Cl_2, I_2, P_4, and

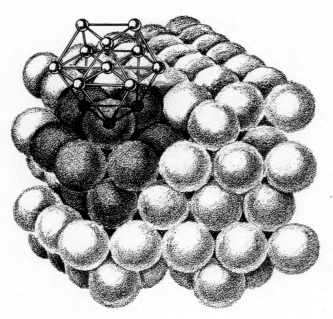

FIGURE 6-6
The hexagonal closest packing arrangement.

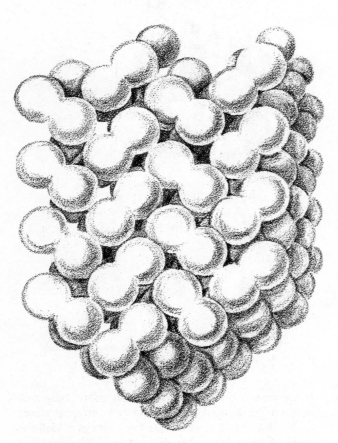

FIGURE 6-7
The molecular crystal lattice arrangement of iodine.

S_8 (see Figure 6-7). The molecules of the inert gases are monatomic and have an atomic lattice with either cubic closest packing (Ne, A, and Kr) or hexagonal closest packing (He). A less common crystal structure for the elements is the covalent type of solid. Diamond, one of the crystal forms of carbon, is the classic example. Each carbon atom is covalently bonded to four other carbon atoms yielding a very hard and high melting solid. The diamond crystal lattice is shown in Figure 6-8.

Crystal Structures of Compounds

From the standpoint of crystal structure, chemical compounds may be conveniently grouped into two broad classes: molecular crystals and ionic crystals.

In *molecular crystals* the building blocks or units occupying the points of the crystal lattice are discrete molecules, such as water (H_2O), carbon dioxide (CO_2), hydrogen chloride (HCl), or sugar ($C_{12}H_{22}O_{11}$). Within these molecules the atoms are held together by strong covalent bonds; the forces between the discrete, neutral molecules are weaker forces, called Van der Waals forces.

Ionic crystals contain positively charged particles called cations, and negatively charged particles called anions, as the building blocks. The cation may be a positively charged atom of a metal, such as Na^+, Ca^{2+}, or Ag^+, or it may be a complex cation, such as $Cu(NH_3)_4^{2+}$. The anion may be a negatively charged atom, such as Cl^-, or a complex

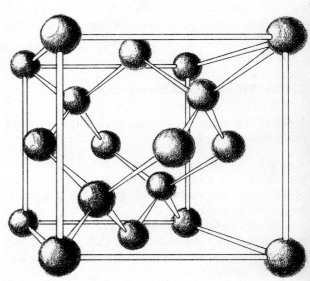

FIGURE 6-8
The crystal structure of diamond.

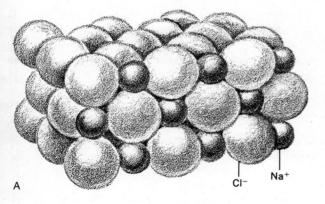

A

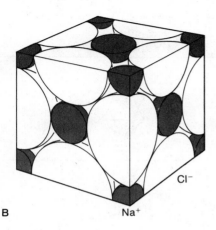

B

Cl⁻   Na⁺

FIGURE 6-9

The structure of sodium chloride. (A) The ionic crystal structure arrangement of sodium chloride. (B) The crystal structure of the sodium chloride unit cell; chloride ions are shown at each edge, another is at the center; sodium ions are at each corner and each face center; chloride ions are located on each edge and in the center of the unit cell. Observing that $\frac{1}{8}$ of each corner atom, $\frac{1}{2}$ of each face atom, and $\frac{1}{4}$ of each edge atom are in the unit cell, we see that the unit cell contains 4 NaCl units: $[8 \times \frac{1}{8} + 6 \times \frac{1}{2}]$ $(Na^+) + [12 \times \frac{1}{4} + 1](Cl^-) = 4Na^+ + 4Cl^-$. [From J. A. Campbell, *Chemical Systems, Energetics, Dynamics, Structure*, W. H. Freeman and Company. Copyright © 1970.] (C) The crystal lattice of the NaCl unit cell.

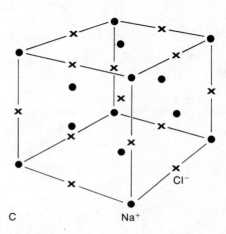

C   Cl⁻   Na⁺

anion, such as NO_3^- or $PtCl_6^{2-}$. Most inorganic salts, such as sodium chloride (NaCl), magnesium bromide ($MgBr_2$), potassium nitrate (KNO_3), and barium sulfate ($BaSO_4$), are members of this class. Figure 6-9 displays the crystal structure and the crystal lattice of the unit cell of NaCl. Contrast these figures with Figure 6-3C. The unit cell of NaCl may be viewed as two interpenetrating, face-centered sets of Na^+ and Cl^- ions. It may also be viewed as a slightly expanded cubic closest packed structure of chloride ions, with the sodium ions occupying the octahedral holes described in the following paragraph.

Holes in Crystal Lattices. Table 3 in Appendix C provides values of the ionic radii of a number of simple ions. It is apparent that the radii of anions are larger than those of cations. For this reason, a useful way to envision the structure of ionic crystals is to view them as close packed structures (face-centered cubic or hexagonal closest packed) of anions identical to the crystal structures of the

metallic elements described in the foregoing discussion (see also Figures 6-5 and 6-6). As you will note, when you begin to assemble these structures, it is impossible to pack a set of balls in such a way that there is no void space. In fact, the names cubic closest packing and hexagonal closest packing imply that the balls cannot be packed any closer, and yet each structure contains 26% empty space. These vacant places are the volumes that the cations occupy. It is interesting, challenging, and informative to examine the numbers and relative sizes of these holes in a close packed lattice.

Table 6-1 describes the geometric arrangements of cations and anions that are possible in ionic crystal lattices. The arrangement adopted by a particular ionic compound depends on the relative charges and relative sizes of the anion and cation. While you are handling the balls in this experiment, you will be able to verify columns 1 and 6.

The forces that hold an ionic crystal together are primarily electrostatic. These consist of attractive forces between the ions of opposite charge,

TABLE 6-1
Geometric arrangement of holes in crystal lattices

Number of (−) ions touching each (+) ion	Type of crystal lattice[1]	Geometric Arrangement	Radius ratio r_A/r_X	Chemical Examples	Number of holes per anion
2	ccp, hcp	Linear	0.15 or less	$F^- $―$H^+$―$F^-$	12
3	ccp, hcp	Trigonal	0.15 to 0.22	O^{2-} B^{3+} O^{2-} O^{2-}	6
4	ccp, hcp	Tetrahedral	0.22 to 0.41	O^{2-} Si^{4+} O^{2-} O^{2-} O^{2-}	2
6	ccp, hcp	Octahedral	0.41 to 0.73	Cl^- Cl^- Cl^- Na^+ Cl^- Cl^- Cl^-	1
8	bcc	Cubic	0.73 and up	Cl^- Cl^- Cl^- Cs^+ Cl^- Cl^- Cl^- Cl^-	1
12		Does not occur since spheres would have to be identical, not allowing for positive and negative ions in the same crystal			

[1]ccp—cubic closest packed; hcp—hexagonal closest packed; bcc—body centered cubic

which are always nearest neighbors in the crystal, and weaker repulsive forces between ions of like charge. The repulsive forces are weaker because the ions of like charge are further away from each other than they are from their nearest neighbors of opposite charge.

Radius Ratio. Column 4 of Table 6-1 shows how the geometric arrangement would be expected to depend on the relative size of the cation and anion in the lattice. We normalize the absolute sizes of the two sets of ions by specifying the ratio of their radii. These numbers are the ratios of the ionic radius of the cation to that of the anion. As you can see, these values are almost all less than one, again reflecting the fact that anions are generally larger than cations.

To understand the effect of the radius ratio, suppose that we have three X ions surrounding an A ion. The condition for stability is that each X ion is in contact with A so that the distance between nearest neighbors is minimized. The limiting condition arises when the X ions are also in contact with one another, so that A and X ions are just touching one another, and X ions are just touching one another also. The following relation exists between r_A and r_X, the radii of A and X respectively:

$$r_A/r_X = 0.155$$

If the radius ratio r_A/r_X falls below this value, then the X ions can no longer all touch the central A ion and this arrangement becomes unstable. For tetrahedral coordination, the limiting value of the radius ratio is 0.225, so that for values of r_A/r_X between

TABLE 6-2
The Types of Solids

Type of solid	Particles that occupy lattice points	Types of bonding	Properties	Examples
Molecular	Molecules	Van der Waals	Low melting points	H_2O, SO_2, CO_2 H_2, Cl_2
		Dipole-dipole	Soft	
		Dipole-induced dipole	Poor electrical conductors	
		London dispersion		
		Hydrogen bonding		
Ionic	Ions	Electrostatic attraction	High melting points	NaCl, KBr, CaO, KNO_3, $MgSO_4$
			Hard and brittle	
			Poor electrical conductors	
Metallic	Positive ions	Electrostatic attraction	High melting points	Na, K, Fe, Co, Ni, Cu, Ag
			Hard or soft	
			Good electrical conductors	
Covalent	Atoms	Shared electrons	High melting points	SiC, C(diamond), SiO_2
			Hard	
			Poor electrical conductors	

0.225 and 0.155 we would expect that the arrangement in which each A ion is surrounded by three X ions in a triangular configuration would be most stable. Thus each geometric arrangement has its range of radius ratios within which it would be expected to be most stable.

At first glance, you might feel that the limiting ratio calculated for each geometric arrangement should represent the *upper* limit of the range. But we must remember that the anions possess an important property that styrofoam balls do not. They are negatively charged. Thus, when they are in contact, strong and repulsive electrostatic forces are operative. These are often called Coulomb forces, after the Frenchman who first observed and measured them. Even if the cation is large enough to spread the anion lattice slightly, the crystal lattice remains stable because, although the attractive forces between cation and anion may decrease because of the increased distance between nearest neighbors, the repulsive forces between the cations and between the anions decrease also.

We may expect that when the radius ratio is near the critical limit, that a very delicate balance of these forces will come into play and determine which of the two possible geometric arrangements

will be most stable. So we may expect some deviations, as indeed we find in some of the alkali halides. But, in general, the correlation of radius ratio with coordination number is satisfactory.

Table 6-2 provides a summary of the four types of solids described in the foregoing discussion.

EXPERIMENTAL PROCEDURE

Special Supplies: A set of styrofoam spheres in size ratios 0.75 to 1 to 1.5 to 2. A convenient set is 36 2-inch, 4 1½-inch, 13 1-inch, and 13 ¾-inch spheres. The spheres can be connected with short lengths of pipe cleaners, chenille-covered wire, or toothpicks.

NOTE: Prepare your own report of this experiment as you make the observations, and as you study further the questions in Interpretation of the Experiment, following the Experimental Procedure.

1. A Study of Crystals of Metallic Elements. (a) *Body-centered Cubic Packing: Model A.* Construct the layers shown in Figure 6-10, using 2-inch spheres. Be sure to leave about ¼-inch space between the spheres. Place the single sphere in the center of the first layer, and then place the third layer in such a way that its spheres are directly over the first layer. Study the symmetry of this model and justify the name of this type of packing. Remember that in actual crystals there will be large numbers of these units with common corners. Note the number of neighbors and the coordination number that each atom has.

(b) *Face-centered Cubic Packing: Model B* (cubic closest packing). Construct the layers illustrated in Figure 6-11, using 2-inch spheres and connectors as before. Place the first layer flat on the desk and put the second layer on it in such a way that its spheres rest between the corner spheres of the first layer. Now add the third layer so that its spheres are directly over those of the first layer. Study the model carefully; note the coordination number and the relative density of the packing and compare with Model A.

(c) *Hexagonal Closest Packing: Model C.* Construct the layers illustrated in Figure 6-12, using 2-inch spheres. Place the first layer so that one of the vertices of the triangle is facing you, as shown in the figure. Then place the central ball of the second layer in the depression in the center of the

first layer. Be sure that each ball of the first layer contacts three balls of the second layer. Now place the third layer so that each of its three balls is directly over a ball of the first layer. If a pattern such as this were expanded into space until it contained 6.02×10^{23} atoms, you would have a model of a mole of metallic magnesium. Note the symmetry of this model and relate this to the name of this type of packing. Compare the coordination number and density of packing in this model with those of Model B.

(d) *Comparison of the Two Types of Closest Packing.* Place Models B and C before you. Rotate the three spheres in the top layer of Model C by 60° so that its spheres are no longer directly over those in the first layer. Rotate this model slightly and look for four spheres facing you that form a square. Now remove the top layer of Model B (face-centered cubic) and place it on the four spheres you located, so that a face-centered cube is formed, just like Model B but tilted toward you. Now observe the orientation of layers one and three in Model C to determine the slight difference that exists between hexagonal and cubic closest packing.

2. A Study of Ionic Crystals. Ionic crystals are formed if positive and negative ions are packed alternately into a lattice. The method of packing depends markedly on the relative radii of the ions. You will first verify some of the data on holes, presented in Table 6-1, and then study three different types of ionic crystals, which illustrate the effects of different radius ratios.

(a) *Holes in Close-packed Lattices.* Verify the data of column 7 in Table 6-1. For each type of hole, possible configurations of balls are suggested that may enable you to count the number of holes per anion.

Linear holes: in order to count these holes, refer to any of the close packed models already constructed, such as Models B and C.

Trigonal holes: examine the arrangement of balls in the middle diagram of Figure 6-12.

Tetrahedral holes: the arrangement shown in Figure 6-13 should be of help in making this count.

Octahedral holes: these are particularly hard to count in a large crystal lattice. The arrangement shown in Figure 6-14 is one that will enable you to count accurately.

(b) *The Sodium Chloride, NaCl(s), Lattice: Model D.* This is one of the most common ionic structures for salts that contain an equal number of positive and negative ions. Since the sodium ion,

FIGURE 6-10
Model A. Layers of atoms for the body-centered cubic arrangement.

FIGURE 6-11
Model B. Layers of atoms for the face-centered cubic arrangement.

FIGURE 6-12
Model C. Layers of atoms for the hexagonal closest packing arrangement.

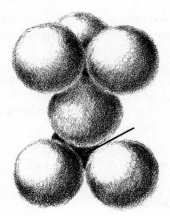

FIGURE 6-13
Tetrahedral holes in a close packed lattice.

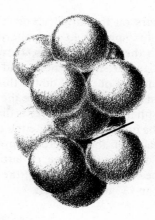

FIGURE 6-14
Octahedral holes in a close packed lattice.

Na^+, has a radius of 0.95 Å and the chloride ion, Cl^-, a radius of 1.81 Å, you may approximate the relative sizes of these ions by using 1-inch spheres for Na^+ ions and 2-inch spheres for Cl^- ions. Use Model B with its 2-inch spheres as the lattice for the Cl^- ions. Now insert the 13 1-inch spheres representing Na^+ ions into the spaces between the Cl^- ions, as shown in Figure 6-15. Arrange the three layers above one another to form the Na^+Cl^- crystal lattice. Note that this lattice is an interpenetrating set of face-centered cubes—one involving Cl^- ions, and one Na^+ ions. Note the coordination number of each Na^+ ion and of each Cl^- ion.

(c) *The Cesium Chloride, CsCl(s), Lattice: Model E.* Since the cesium ion, Cs^+, has a radius of 1.69 Å compared with 1.81 Å for the chloride ion, Cl^-, you may approximate the relative sizes of these ions by using $1\frac{1}{2}$-inch spheres for the Cs^+ ions and 2-inch spheres for the Cl^- ions. Note first that the $1\frac{1}{2}$-inch spheres will not fit into the spaces in Model B as did the 1-inch spheres used to make the Na^+Cl^- lattice. Construct the first and third layers illustrated in Figure 6-16, using 2-inch spheres for the Cl^- ions. Place 4 loose $1\frac{1}{2}$-inch spheres in the depression between the 9 spheres of the first layer. Then place the third layer directly over the first layer. Compare this packing to that of Model A. Note the coordination number and how the radius ratio changed the type of packing from that in Model D, $Na^+Cl^-(s)$.

(d) *The Zinc Sulfide, ZnS(s), Wurtzite Lattice.* There are two crystal forms of zinc sulfide—Sphalerite (or zinc blende) and Wurtzite. These two crystal forms are called *polymorphs*. Sphalerite consists of a cubic closest packed array of sulfide ions, with the zinc ions occupying alternate tetra-

hedral holes. Wurtzite has a structure similar to that of Sphalerite except the sulfide ions are in a hexagonal closest packed array.

Since the Zn^{2+} ion has a radius of 0.75 Å and the S^{2-} has a radius of 1.85 Å, we shall use $\frac{3}{4}$-inch spheres for the Zn^{2+} ion and 2-inch spheres for the S^{2-} ion to approximate the relative sizes.

Use Model C with its hexagonal closest packing to represent the lattice of the larger S^{2-} ions. Attach one of the smaller 1-inch spheres directly above each of the larger spheres in each of the three-layers of Model C, using pieces of toothpicks or of pipe-stem cleaners.

Place the large layer on the table top with the smaller spheres pointed down. Place one of the

FIGURE 6-15
Model D. Layers of atoms for the sodium chloride ionic arrangement.

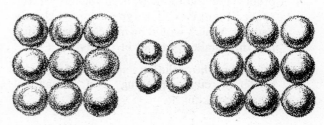

FIGURE 6-16
Model E. Layers of atoms for the cesium chloride ionic arrangement.

triangular layers on the larger layer in such a way that the smaller spheres fit into alternate depressions between the larger spheres. Now hold these two layers together, invert them, and place them on the table in front of you. Add the third triangular layer, with small spheres up, above the larger layer so that each sphere of the top layer is directly above a sphere on the bottom layer. Note the coordination number of each of the spheres representing the Zn^{2+} ions. Compare this type of lattice with the Na^+Cl^- lattice, taking into account the comparison of Models B and C made earlier.

Bibliography

Gehman, W. G., "Standard Ionic Crystal Structures," *J. Chem. Educ.* **40**, 54 (1963).

Ho, S. M., and Douglas, B. E., "A Broader View of Close Packing to Include Body-Centered and Simple Cubic Systems," *J. Chem. Educ.* **45**, 474 (1968).

Sime, R. J., "Some Models of Close Packing," *J. Chem. Educ.* **40**, 61 (1963).

INTERPRETATION OF THE EXPERIMENT

For this experiment you will be required to prepare your own report form. In your report, be sure to answer all of the questions posed throughout the directions for the experiment. In addition, respond to the several questions given below.

Exercises

1. (a) What is the coordination number of the atoms in each of the models you constructed for study of the packing of atoms in metals?

   (b) In one of the models illustrating metallic crystals, the spheres occupy about two-thirds of the space in the unit cell and in the others they fill about three-fourths of the space. Identify which type is more closely packed. Which would be more dense? Which has the smaller number of bonds?

   (c) Describe the relationship between body-centered cubic packing and hexagonal closest packing.

2. From your consideration of the models of ionic crystals constructed in 2 (a) and 2 (b) in the Experimental Procedure, what relation can you

deduce concerning the radius ratio of ions and the coordination number in the crystals?

3. (a) Compare the relationship between the sodium chloride lattice and the face-centered model of part 1(b) with the relationship between the zinc sulfide lattice and the hexagonal model of part 1(c).

   (b) How does the radius ratio Na^+/Cl^- compare with the Zn^{2+}/S^{2-} radius ratio?

Optional Exercises

1. Suppose you have a crystal composed of elements X and Y in which the radius ratio is approximately $\frac{1}{2}$, so that the sodium chloride lattice is possible. Each ion is about the same size as Na^+ and Cl^-, respectively, but each ion is doubly charged X^{2+} and Y^{2-}. Would $XY(s)$ have a higher or lower melting point than $NaCl(s)$? Suggest a specific compound that would meet the criteria of $XY(s)$ listed above. Look up its melting point and compare it with that of $NaCl(s)$. Data for both this question and the next one can be obtained from the *Handbook of Chemistry and Physics,* published annually by the Chemical Rubber Company, Cleveland.

2. Suppose you have a crystal composed of elements A and B in which each ion has a charge of 1, A^+ and B^-, like Na^+ and Cl^-, but in which the radii of A and B are proportionally larger but still 1 to 2. Would $AB(s)$ have a higher or lower melting point than $NaCl(s)$? Suggest a specific compound that would meet the criteria of $AB(s)$ given above. Look up its melting point and compare it with that of $NaCl(s)$.

3. Calculate the ratio of the radius of the cation to that of the anion, so that the cation will just fit in an octahedral hole, shown in the figure below.

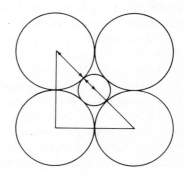

THE DETERMINATION OF AVOGADRO'S NUMBER

PRE-STUDY

Matter consists of atoms. These atoms generally are joined together to form molecules. Chemists and those people who think about chemistry are required to interpret the structure of matter and interactions between various forms of matter on an atomic or a molecular basis. The problem that arises in the introductory laboratory is that you will never see a molecule, much less an atom. You will examine the properties of very large numbers of atoms.

The analytical balance is probably the most sophisticated instrument you will use. The prospect of weighing an atom or a molecule, or even two or three hundred thousand of them, is unlikely. The balance weighs to 0.0001 g. If the substance you wished to examine had a molecular weight of 600, the smallest quantity of it that you could weigh would be 10^{17} molecules! Clearly a connection must be found between the microscopic world of the atom and the macroscopic world of human observation. The connection is made by means of Avogadro's number, which we have already discussed briefly

in Study Assignment B. It is given the symbol N_A. Avogadro's number is approximately equal to 6.02×10^{23} particles/mole. (It is one of the very few numbers you will encounter in introductory chemistry that is worth remembering.)

Avogadro's number is defined as the number of atoms contained in exactly 12 g of the $^{12}_6C$ isotope. Note that it is a *defined number,* meaning that it is arbitrary. It could have been defined as the number of atoms in 12 pounds of $^{12}_6C$ in which case it would have been 454 times larger. It could have been (and until a few years ago was) defined on the basis of another isotope or mixture of isotopes.

The importance of Avogadro's number is sensed most clearly as we attempt to carry out and understand chemical reactions. If we wish to make H_2O from hydrogen (H_2) and oxygen (O_2), we can write the reaction

$$H_2(g) + \tfrac{1}{2}O_2(g) \rightarrow H_2O(l) \qquad (1)$$

However, we cannot arrange to isolate these individual molecules experimentally, nor can we work with half a molecule. If we wish to write an equation suitable at the molecular level we can write

2 molecules H_2 + 1 molecule O_2 →

$$2 \text{ molecules } H_2O \quad (2)$$

However, there is still no way in which we can actually gather up 2 molecules of hydrogen and 1 molecule of oxygen and arrange for them to react. We must work not with thousands, or even trillions of molecules, but with numbers approximating 10^{19}, if we are to have any hope even of measuring them with simple chemical instruments.

It is crucially important that you realize that the following equation is entirely wrong:

$$2 \text{ g } H_2 + 1 \text{ g } O_2 \neq 2 \text{ g } H_2O \quad (3)$$

Integral ratios of the *numbers* of molecules react with each other, but integral ratios of *weights* do not necessarily do so.

How, then, does one measure the quantities of reactants and products in a chemical reaction? We cannot operate on a molecular scale, but neither can we make calculations or carry out reactions on a gram scale if that scale simply measures the weights of the molecules. The problem is resolved by means of Avogadro's number and the concepts of the mole and molecular weights.

If equation (2) is correct, we can multiply both sides by any number and the mathematical identity is retained. In particular, we can multiply both sides by Avogadro's number.

$2(6.02 \times 10^{23})$ molecules H_2

$+ 1(6.02 \times 10^{23})$ molecules O_2 →

$$2(6.02 \times 10^{23}) \text{ molecules } H_2O \quad (4)$$

A mole of a substance is defined as the weight in grams of Avogadro's number of the particles of which that substance is composed. Because the assignment of the base of the atomic weight scale is $^{12}_{6}C$ and because the definition of Avogadro's number has the same base, the mass of one mole of any substance is equal to the sum of the atomic weights of the atoms in the molecule or ion. Thus we can write the two equivalent equations:

$$2 \text{ moles } H_2 + 1 \text{ mole } O_2 → 2 \text{ moles } H_2O \quad (5)$$

and

$2(2 \times 1.0080)$ g H_2 + (2×15.999) g O_2 →

$$36.030 \text{ g } H_2O \quad (6)$$

We finally have a way to relate macroscopic quantities that we can measure to the stoichiometry that occurs at the molecular level.

It is important in scientific work to have accurate values for significant constants. A variety of experimental methods have been devised to measure Avogadro's number. A French scientist, Jean Perrin, determined the first value in 1908. He measured the vertical distribution in a gravitational field of gamboge (resin) particles suspended in an organic liquid and obtained values of 6.0 $\times 10^{23}$ and 5.4 $\times 10^{23}$. (Henry, 1966; Slabaugh, 1965). More refined values have been obtained by accurate measurements of crystals by x-ray diffraction (division of the volume of 1 mole by the volume of 1 molecule yields N_A).

All of the above methods require sophisticated and expensive equipment and considerable care in experimental technique and treatment of data. The payoff is the most accurate value of Avogadro's number that we have to date: $N_A = 6.02209 \times 10^{23}$ particles/mole.

The purpose of this experiment is to illustrate how a rather simple method can be used to obtain a crude value of N_A. This value will be obtained after about one hour of experimentation and two hours of calculation. The most sophisticated items of equipment needed are a dish, an eye dropper, and some soaplike solution.

Concept of the Experiment

Matter exists in three states. The fact that a gas can be condensed to a liquid and a liquid frozen to a solid indicates that there are attractive forces between all molecules. We can schematically represent these forces at the surface of a liquid by the arrows in Figure 7-1. In the interior of the liquid, the forces exerted on a given molecule are

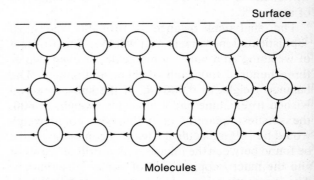

FIGURE 7-1
The molecular structure of a liquid near its surface.

uniform in all directions. At the surface, however, it is clear that there is a net force attracting each surface molecule inward. These molecules have higher energies than interior molecules, thus giving rise to the force known as surface tension. It is because of this force that liquid droplets are spherical. A spherical shape presents the least surface area for a given volume.

If our liquid is water, the surface tension is especially strong because particularly strong bonds, called hydrogen bonds, exist between the water molecules. These bonds arise whenever a hydrogen atom attached to a highly electronegative atom, such as oxygen, has access to an unbonded pair of electrons, such as those of another oxygen atom.

Another property displayed by water is that it is polar. Polar molecules possess a separation of charge. In ionic compounds, such as NaCl, there is a separation of a full unit charge, $Na^+ \cdots Cl^-$. Polar compounds display a partial separation of charge denoted by a delta, δ. An arrow is used to display this charge separation. The dipole moment of a polar molecule is equal to the partial charge times the distance separating the charges, and it is represented by an arrow pointing toward the negative end of the molecule. Figure 7-2 displays the polar nature of the water molecule.

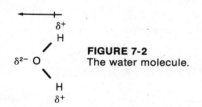

FIGURE 7-2
The water molecule.

Polar molecules attract each other. The negative end of the dipole of one molecule is attracted to the positive end of the dipole of another molecule. For this reason, water dissolves formic acid, H—COOH, which has a dipole moment, but it does not dissolve butane, $CH_3CH_2CH_2CH_3$, which has a nearly uniform charge distribution. If a molecule possessing the properties of both of these molecules is brought up to the surface of water, the polar part of the molecule will be attracted to the surface and the nonpolar portion will be repelled. If the nonpolar part is much larger than the polar portion, the molecule will not dissolve in water but will simply stick to its surface. Consequently it will lower the energy of the surface water molecules and of the adhering molecules.

The molecule we will use in this experiment, stearic acid, behaves in just this way. Stearic acid

has a polar end consisting of a carboxyl group, —COOH and a large nonpolar "tail" consisting of 16 methylene groups, —CH_2—, terminating in a methyl group, —CH_3. Figure 7-3 is a reasonably accurate representation of this molecule.

The addition of a limited number of stearic acid molecules to a water surface results in a monolayer being formed, as illustrated in Figure 7-4. However, after the surface is covered with a monolayer of stearic acid molecules, the addition of more molecules causes the stearic acid molecules to cluster in globular aggregates. The polar heads attract each other, leving the globular mass with the hydrocarbon tails sticking out. Figure 7-5 schematically illustrates such an aggregate.

The properties of the water surface and the stearic acid molecule permit us to perform what amounts to a titration of the water surface. We can add stearic acid molecules to the water surface until a monolayer covers the entire surface. Further addition will cause a globular lens to form on

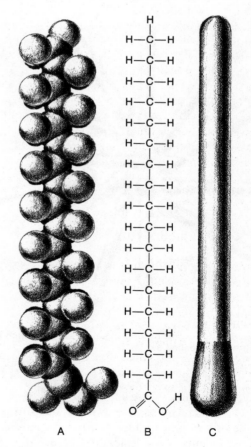

FIGURE 7-3
The stearic acid molecule: (A) space-filling model; (B) structural formula; (C) schematic representation.

New surface

Monolayer

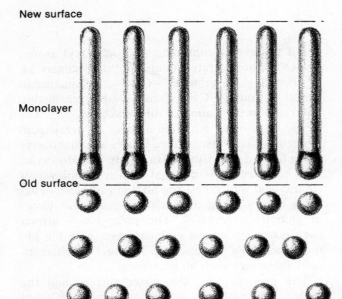

Old surface

FIGURE 7-4
Monolayer of long-chain molecules lying on top of a water surface.

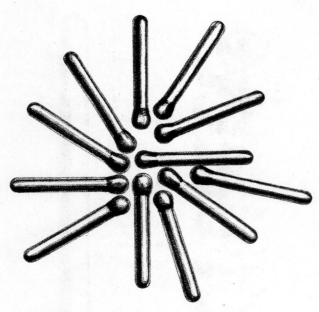

FIGURE 7-5
Globular aggregate of long-chain molecules. Such an aggregate is called a micelle.

the liquid surface. If we know the area of the surface of water and have a way of measuring the volume of substance added to form the monolayer, we can calculate the thickness, t, of the layer. This thickness equals approximately the length of the stearic acid molecule.

As we have seen above, this molecule consists of 18 carbon atoms linked together. If we make the simple assumption that these carbon atoms are linearly linked together and we assume that the ends of the molecule are equivalent to two carbon atoms, the diameter of a carbon atom is given by $t/20$. Of course, knowledge of the diameter of a sphere allows us to calculate its volume and we have exactly half of the data needed to compute Avogadro's number.

The other half of the data requires no experimental work on your part. The other data needed consist of the volume of a mole of these balls, which have the diameter of a carbon atom. One way to approximate this value is to consider the three-dimensional array of carbon atoms in diamond, which is pure carbon. The density of diamond is 3.51 g/ml. You may recall that you can calculate the molar volume of an element by dividing its atomic weight by its density. Verify this idea by examining the dimensions of this quotient. (As indicated by calculation number 5 and question 2 in the report form, the molar volume computed by this procedure is too large leading to a high value for N_A.)

Avogadro's number results from one final step.

$$N_A = \frac{\text{molar volume (ml/mole)}}{\text{atomic volume (ml/atom)}}$$

$$= \text{number of atoms/mole} \qquad (7)$$

EXPERIMENTAL PROCEDURE

Special Supplies: 14-cm watch glass, 10-ml graduated cylinder, medicine dropper.

Chemicals: Benzene and a solution of stearic acid in benzene. Your instructor will give you the weight/volume concentration of this solution.

1. Preparation of Equipment. Stearic acid is a solid. It is conveniently measured and applied to the water surface by dropping a solution of stearic acid in benzene onto the water. The benzene is insoluble in water and, because it has a high vapor pressure, it rapidly evaporates, leaving the stearic acid spread on the water surface.

Because stearic acid is one of the fatty acids present in soaps, your equipment must be scrupulously clean. Any dirt, soap, or grease present will dissolve the stearic acid and lead to meaningless results.

Obtain a 14-cm watch glass and wash it thoroughly with detergent. Rinse the detergent off completely by placing it under a full stream of cold water for two minutes. Repeat this procedure after each experiment. Set the glass aside and do not touch the inside of it with anything. Be especially careful to keep your fingers (which will be slightly greasy) off the glass.

Your instructor will demonstrate how to pull the tip of your medicine dropper out to a fine point. Alternatively, a piece of glass tubing can be pulled out, the end flared, and a rubber bulb attached. Watch your fingers. Molten glass is *really* hot and *stays* hot for a long period of time.

After your dropper has cooled, calibrate it by filling it with benzene. While holding the dropper vertically, count the number of drops that must be expressed into the graduated cylinder to equal 1.00 ml. The best way to do this is to fill the cylinder accurately to one of the milliliter marks and then add your drops from the dropper until the meniscus reaches the next mark. Between 100 and 150 drops should be required. If fewer than 100 drops equal 1 ml, ask your instructor for help in pulling out another capillary tip. Repeat your calibration until your results agree within one or two drops.

2. Measurement of Volume of Stearic Acid Solution to Cover the Surface. Fill the clean watch glass to the brim with distilled water. Carefully measure the diameter of the water surface with a meter stick. Rinse your calibrated dropper several times with stearic acid solution. Then add this solution

drop by drop to the water surface, counting the drops. Wait about 10 seconds between drops. The solution will spread across the entire surface initially and will continue to do so until a complete monolayer of stearic acid has been produced. As this point is approached, the spreading will become slower and slower until, finally, a drop will not spread out but will instead form a thick, lens-shaped layer. If this lens persists for about 30 seconds, you may safely assume that you have added one drop more than is required to form a complete monolayer.

Thoroughly clean the watch glass, rinse out the dropper with stearic acid solution several times, and repeat the experiment. Repeat until two results agree within two or three drops.

Bibliography

Henry, P. S., "Evaluation of Avogadro's Number," [by the method of Perrin], *J. Chem. Educ.* **43**, 251 (1966).

King, L. C., and Neilsen, E. K., "Estimation of Avogadro's Number," *J. Chem. Educ.* **35**, 198 (1958).

Robinson, A. L., "Metrology: A More Accurate Value for Avogadro's Number," *Science* **185**, 1037 (1974).

Slabaugh, W. H., "Determination of Avogadro's Number by Perrin's Law," *J. Chem. Educ.* **42**, 471 (1965).

Slabaugh, W. H., "Avogadro's Number by Four Methods," *J. Chem Educ.* **46**, 40 (1969).

REPORT

7

**The Determination of
Avogadro's Number**

NAME

SECTION   LOCKER

INSTRUCTOR   DATE

DATA AND CALCULATIONS

NOTE: A number of calculations are required in this report. Show the calculations for only one of your trials, but give the results for both trials, those from trial 1 in the left blank, and those for trial 2 in the right.

1. Calibration of the Dropper

(a) Note the number of drops required to equal 1.00 ml.

_____   _____

(b) Calculate the fraction of a ml per drop.

_____   _____

2. Experimental Results

(a) Note the diameter of the water surface.

_____cm   _____cm

(b) Note the number of drops required to cover the surface.

_____   _____

3. Calculation of the Length of the Stearic Acid Molecule

(a) Give the volume of solution required to form a monolayer: use the data from parts 1(b) and 2(b).

_____ml   _____ml

(b) Give the weight of stearic acid in that volume (concentration in g/ml will be given to you).

_____g   _____g

(c) Give the volume of stearic acid, V: use the density, ρ, of solid stearic acid, which is $\rho = 0.847$ g/ml.

_____ml   _____ml

(d) Give the area of the monolayer ($A = \pi r^2$, $r =$ radius of water surface).

_____cm^2 _____cm^2

(e) Give the thickness of the monolayer ($t = V/A$).

_____cm   _____cm

4. Calculation of the Volume of a Carbon Atom

The main portion of the stearic acid molecule consists of 18 carbon atoms linked together. In addition, there is an extra hydrogen atom at one end and an oxygen atom and hydroxyl group at the other. We can roughly account for the presence of these extra atoms by assuming that they are the equivalent of 2 more carbon atoms. Thus we can picture the stearic acid molecule as being 20 balls of equal diameter linked together.

The question now is, how are they linked together?

(a) Give the volume of a carbon atom based upon a linear array.

Calculate the diameter of a carbon atom, assuming that the 20 balls are joined in a straight line.

_____cm   _____cm

Calculate the volume ($V = 4/3\pi r^3$) of the carbon atom, using this diameter.

_____cm³   _____cm³

(b) Volume of a carbon atom based upon the actual bond angle.

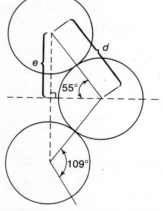

The actual bond angle between these carbon atoms is 109°. Thus the chain is better represented by the figure to the left. You can calculate the diameter, d, of the carbon atom by reference to the triangle of this figure. The side of the triangle labeled e is just $\frac{1}{20}$ of the length of the molecule.

Value of e _____cm   _____cm

Value of d _____cm   _____cm

A portion of a carbon-carbon chain.

Calculate the volume of a carbon atom using this diameter.

_____cm³   _____cm³

5. Calculation of the Volume of a Mole of Carbon Atoms

(a) Calculate the molar volume of carbon, using the density given earlier and the atomic weight of carbon.

_____cm³

(b) Comment on the validity of this calculation as a measure of the volume of a mole of carbon atoms.

6. Calculation of Avogadro's Number

(a) Calculate Avogadro's number from the appropriate ratio of volumes.

_____particles/mole   _____particles/mole

(b) Your *best value* for N_A is:

_____particles/mole

QUESTIONS

1. Calculate the percent error between your value and the accepted literature value.

2. Diamond has a tetrahedral bonding arrangement, which is equivalent to a body-centered cubic structure with half of the corners unoccupied. If your geometry is good enough, try to show what fraction of the crystal lattice is occupied by carbon atoms. (This is not a simple problem. See Figure 6-8 for the diamond structure.) The answer is $3\sqrt{3}\pi/32 \times 100\%$ or 51%. What difference does this make in your estimate of N_A?

3. Criticize the several assumptions made throughout this experiment. Comment on which assumption or which experimental error you think was primarily responsible for your percentage error.

EMISSION SPECTRA AND ATOMIC STRUCTURE

PRE-STUDY

Some of the most convincing evidence on atomic structure and the quantization of the energy levels of electrons has been obtained from the absorption and emission spectra of the elements. In this experiment, you will construct a spectroscope from a cardboard box, a 10-cent plastic replica diffraction grating, and a razor blade. This simple device will permit you to measure with a precision of 5% the wavelengths of the lines emitted by several elements when their atoms have been excited by an external source. After calibrating the spectroscope with the lines of mercury (which we will regard as a known element), you will measure the wavelengths of the lines of hydrogen and some metallic elements. The data for hydrogen will be used to construct quantitatively the energy level diagram for hydrogen.

Waves and Diffraction. Light is electromagnetic radiation. This means that it possesses electric and magnetic properties. These properties vary sinusoidally and in phase with each other, as shown in Figure 8-1.

Because only the electric part of the wave interacts with the electrons in atoms, Figure 8-1 is often simplified to the representation of Figure 8-2. The wavelength of this wave, λ, is defined as the distance between any two repeating portions of the wave. The wavelength of light visible to our eyes is 4×10^{-5} cm (blue) to 7.5×10^{-5} cm (red). Be-

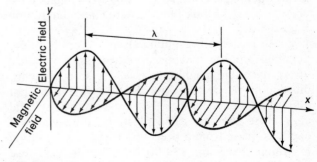

FIGURE 8-1
The electric and magnetic field components of a light wave.

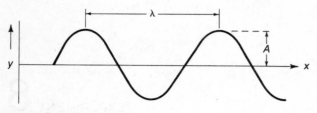

FIGURE 8-2
The amplitude (A) and wavelength (λ) of a light wave.

cause these numbers are quite small, the units of wavelengths are often given in angstroms (Å), where $1 \text{ Å} = 10^{-10}$ m, or in nanometers (nm), where $1 \text{ nm} = 10^{-9}$ m. Demonstrate for yourself that the limits of the visible spectrum in these units are 4000 Å to 7500 Å or 400 nm to 750 nm.

Another important property of waves is frequency, ν. The frequency of a wave is the number of wavelengths that pass a given point in unit time. The units are generally number per second.

For a given wave, wavelength and frequency are not independent of each other. The higher the frequency, the smaller the wavelength. The reason for this relationship is that all light waves travel at the same velocity. This velocity is the speed of light, c, and is equal to 3.00×10^{10} cm/sec. Some reflection on these three quantities (an analogy with a train in which $\lambda = $ length of one car is helpful) leads to the correct mathematical expression:

$$c(\text{cm/sec}) = \lambda(\text{cm}) \cdot \nu(\text{sec}^{-1}) \qquad (1)$$

One of the unusual properties of waves is diffraction. All waves display this phenomenon, whether they are ocean waves, light waves, or sound waves. Perhaps you have seen ocean waves striking a small opening or several openings in a breakwater and observed the interesting patterns formed by the waves after they have passed through these openings. One pattern that you might have observed would look like the schematic diagram in Figure 8-3.

These patterns are obtained only when both the openings and the spacing, d, between the openings are of the same magnitude as λ. Thus, in order for diffraction to occur with light waves, the light must pass between slits that are very narrow and very close together. The plastic grating you will use was made from a precision ruled metal grating in which there are 13,800 lines per inch. These gratings are of the transmission type. This means that light does not pass through the grating where

the lines have been drawn but rather through the spaces between the lines.

The quantitative description of the behavior of waves on passing through slits of the appropriate size is the famous Bragg equation:

$$n\lambda = d \sin \theta \qquad (2)$$

The quantity n is a small integer having values 0, 1, 2, 3, ... The value d is the distance between the lines of the grating. The meaning of the angle θ is displayed in Figure 8-4.

The usefulness of the Bragg equation, as applied to a diffraction grating, is that it shows that the longer the wavelength is, the greater the angle of

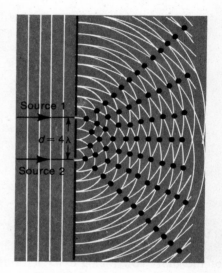

FIGURE 8-3
Diffraction of ocean waves by openings in a barrier.

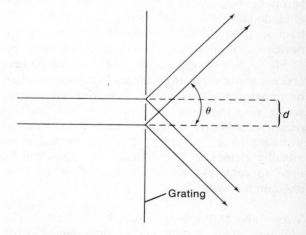

FIGURE 8-4
Diffraction of light by a transmission grating.

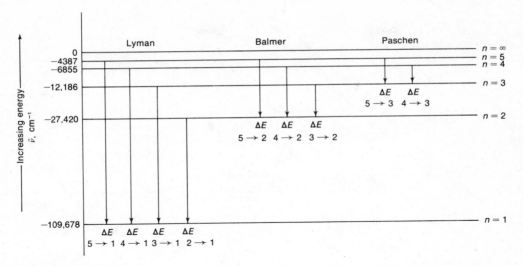

FIGURE 8-5
The energy levels of the hydrogen atom. The numbers on the left are the energy levels expressed in wave numbers. The integers on the right are the principal quantum numbers. The vertical arrows correspond to the energy differences ΔE, between levels described by the indicated quantum numbers. The energy differences represent observable lines in the hydrogen spectrum (corresponding to the spectral transition—that is, the transition from one energy state to another state). Each of the three groups of spectral lines shown here is named for its discoverer.

deviation, θ, will be. In all measurements made in this experiment, you will be observing the first order or $n = 1$ diffraction patterns. This means that, if light of differing wavelengths strikes the grating, it will be bent to varying angles and in fact separated into its component wavelengths. White light consists of light of all wavelengths. The diffraction pattern of an ordinary tungsten light therefore is just a lovely rainbow. One of the first observations you should make with your spectroscope is of the light from a fluorescent lamp. In addition to the rainbow pattern observed with the tungsten lamp, you will find a bright green and a bright violet line. This is your first demonstration of the existence of line spectra of elements.

Energy Levels and Line Spectra of Elements. Our present theory of atomic structure states that electrons in an atom can possess only discrete energy values. The term quantized is used to describe discrete values. Every element has a characteristic set of energy levels. The energy level diagram for hydrogen is given in Figure 8-5. The energy levels are characterized by an integer n,[1] which is

called the principal quantum number. At room temperature, most of the hydrogen atoms have energy corresponding to the $n = 1$ level, which is called the ground state.

If energy is supplied to a hydrogen atom in the $n = 1$ state by an electrical discharge or by heat, some of the atoms will absorb this energy and enter the $n = 2$, $n = 3$, or higher levels. These atoms having extra energy are called excited atoms. They can lose some of this extra energy in discrete amounts and drop back down to the $n = 3$ or $n = 2$ or the ground state, as indicated in Figure 8-5.

The crucial question now is, how do atoms lose energy? Frequently they do so by emitting light of discrete wavelengths. It is important to note that all light is quantized in units called photons, which can be described as both particles and waves (or, indeed, as neither): in truth, they are particle-waves. Photons possess an energy proportional to the frequency of the light wave. The proportionality constant is Planck's constant, h, which has a value of 6.62×10^{-27} erg sec. Thus the electron of the hydrogen atom can lose energy by emitting a photon that has an energy corresponding to the transition from n_2 to n_1, where n_2 and n_1 represent the quantum numbers of two different energy levels, and we have not specified their numerical values.

If a transition takes place between two different

---

[1]The quantum number, n, should not be confused with the symbol n used to denote the diffraction order in Equation (2), the Bragg equation.

FIGURE 8-6
The construction of the spectroscope.

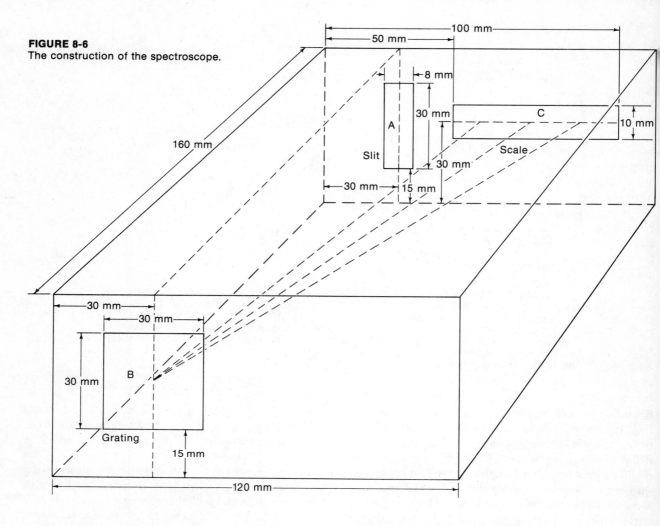

energy levels, n_2 and n_1, the energy difference between the n_2 and n_1 levels is related to the wavelength of the photon by the equation

$$\Delta E = E_{n_2} - E_{n_1} = h\nu = \frac{hc}{\lambda} \qquad (3)$$

The last identity is derived from equation (1), which expresses the relationship between the velocity of light, frequency and wavelength. The energy, ΔE, can be expressed in ergs or wavenumbers, $\bar{\nu}$. Wavenumber is defined as the reciprocal of the wavelength, where λ is given in cm. The units of $\bar{\nu}$ are cm^{-1}. This definition is useful because energy then is directly proportional to $\bar{\nu}$.

$$\Delta E = hc\bar{\nu} \qquad (4)$$

Sometimes, as in Figure 8-5, the product hc is divided into ΔE and the units of energy are given simply in cm^{-1}.

EXPERIMENTAL PROCEDURE

Special Supplies: A cardboard box with lid,[2] 120 mm × 160 mm × 60 mm, spray can of flat black paint, single-edged razor blade, double-edged razor blade, small piece of translucent, millimeter graph paper, metric ruler, plastic replica diffraction grating, tape, variety of light sources (Hg, H_2 and others).

Chemicals: Several salts such as $Ca(NO_3)_2$, $Ba(NO_3)_2$, $RbNO_3$.

1. Construction of the Spectroscope. The spectroscope consists of the cardboard box with three holes cut in it to provide for the razor blade slit, the grating, and the scale. Note the dimensions given in Figure 8-6, mark the positions of the holes on the outside of the box, and cut out the three holes. Cut the holes through the box and lid simultaneously, using the single-edged razor blade.

---

[2]A No. 8 two-piece, set-up mailing box, 6 × 4.5 × 2.25 inches, is about the right size.

Spray the interior of the box with flat black paint. This reduces internal reflections within the box and makes the line spectra easier to observe.

Carefully (razor blades are quite sharp) break the double-edged razor blade lengthwise in two. Tape the sharp edges facing each other to the outside of the box at A with their edges 0.2–0.3 mm apart. (No diffraction is obtained from this slit because this opening is many times the λ of visible light.) Make sure that the edges of the slit are parallel to each other and perpendicular to the flat sides of the box.

Observe the grating carefully. You will probably be able to observe some striations on the plastic surface. If you can see these reflections, mount the grating at B with the lines perpendicular to the top and bottom of the box. If you cannot observe these striations, close the box and hold the box with the slit pointing toward a fluorescent light. Hold the grating next to the hole at B and observe the diffraction pattern. That orientation of the grating which produces the diffraction pattern on the right and left of the slit is the correct one. Tape the grating to the inside of the box at B, with the grating in the correct orientation. Measure the distance from the slit to the grating.

Tape the millimeter graph paper over the slot at C. If too much light comes through the graph paper, making it difficult to see the lines, mask off all but 2–3 mm of the slot with opaque tape or paper. Measure the distance from the center of the slit to the first thick line on the graph paper.

After all the components have been taped in place, securely tape the box and lid together.

2. Calibration of the Spectroscope.

All of the lamps you will study are rather bright. Do not look directly at them for any longer than is necessary. In addition, several of the lamps emit appreciable amounts of UV radiation. Wear your glasses at all times while observing the lamps through your spectroscope, and observe the hydrogen and mercury lamps only when the UV filter is in place.

The following precautions and directions will enable you to measure reproducibly the positions of the lines on the scale to about 0.2 cm.

(a) The lamp should be far enough away from the spectroscope that the direction of propagation of the light is well-defined, or collimated (Figure 8-7). If you do not observe this precaution, you will find that, whenever you move your eye, the spectral lines will wander, rather than remaining in stationary relationship to the scale.

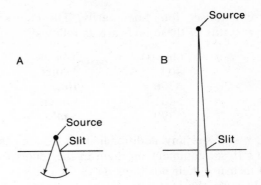

FIGURE 8-7
(A) Poorly collimated light; (B) well collimated light.

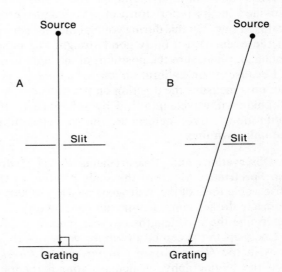

FIGURE 8-8
Incident angle of light in relation to the grating: (A) correct; (B) incorrect.

(b) The lamp should not be too far away from the spectroscope, or the intensity of light falling on the slit will be too low for you to observe the spectrum clearly.

(c) The light must be incident to the grating at 90° (Figure 8-8), if the angle of diffraction is to satisfy the equation given previously.

We will use the emission lines of mercury as a "known" to calibrate the spectroscope. Look at the spectrum of a mercury arc lamp.[3] You should

---
[3]Table 4, Appendix C, shows the spectrum of a fluorescent lamp, which displays the prominent lines of the mercury spectrum in the visible region.

be able to see four lines easily. The colors and wavelengths of these lines are as follows:

WAVELENGTH, Å	COLOR
4047	Violet
4358	Blue
5461	Green
5790	Yellow

Do you see any additional lines besides these four? If so, comment on them in the report form, and measure their positions on the scale.

Measure the positions of the four prominent lines on your millimeter graph paper scale to the nearest $\frac{1}{10}$ or $\frac{2}{10}$ of a millimeter. Record the average of several observations.

Draw a calibration graph on the graph paper provided in the report form of wavelength versus scale reading. Do this during your laboratory period. In the event that a fairly good straight line is not obtained, remeasure the positions of the lines again.

Because your calibration curve is linear, you can now measure the position on the scale of a line of unknown wavelength and by reference to the calibration curve, determine the wavelength of the unknown line.

3. Observations and Measurements of the Hydrogen Spectrum. Measure the scale position of the observable lines of the hydrogen emission spectrum from a hydrogen lamp. From your calibration graph, determine the wavelengths of these lines.

Look up the accepted literature values of these lines in the *Handbook of Chemistry and Physics* (see the bibliography). Calculate your percentage error.

Convert your values of wavelengths for the hydrogen spectrum to wavenumbers ($\bar{\nu} = 1/\lambda$). Compare these numbers with the known values of the wavenumbers corresponding to the energy transitions displayed in Figure 8-5. (Remember that the wavenumbers of the spectral lines correspond to *differences* in the energy levels.)

Draw that portion of the hydrogen energy level diagram that you have measured, using wavenumbers as energy units. Make your figure to scale on the graph paper provided.

Calculate the Rydberg constant. This constant, R_H, is the number that relates the energy of a given transition to the principal quantum numbers involved.

$$\bar{\nu}(\text{cm}^{-1}) = R_H \left[\frac{1}{n_2^2} - \frac{1}{n_1^2} \right] \qquad (5)$$

For each of the three $\bar{\nu}$ values you have measured, calculate the corresponding value of R_H. (The values of n_2 and n_1 may be obtained from Figure 8-5.)

4. Other Uses of the Spectroscope. You will be provided with several salts. Observe the emission spectra of the cations of these salts with your spectroscope while a partner holds some of this salt on a metal spatula in the flame of a bunsen burner.

Observe the spectrum of the light from a fluorescent lamp with your spectroscope, and comment on your observation.

Bibliography

Edwards, R. K., Brandt, W. W., and Companion, A. A., "A Simple and Inexpensive Student Spectroscope," *J. Chem. Educ.*, **39**, 147 (1962).

Harris, S. P., *J. Chem. Educ* **39**, 319 (1962).

Handbook of Chemistry and Physics. Published annually by the Chemical Rubber Company, Cleveland.

Emission Spectra and Atomic Structure

NAME

SECTION   LOCKER

INSTRUCTOR   DATE

OBSERVATIONS AND DATA

1. Construction of the Spectroscope

(a) Give the distance from slit to grating. _____cm

(b) Give the distance from center of slit to first major scale mark. _____cm

2. Calibration of the Spectroscope

(a) Measure the positions of the four prominent lines in the spectrum of mercury.

Color	Known wavelength (Å)	Position on spectroscope scale (cm)
Violet	4047	_____
Blue	4358	_____
Green	5461	_____
Yellow	5790	_____

(b) Comment on other lines observed:

(c) Plot wavelength (in angstroms) versus the scale reading (in cm) on the graph below.

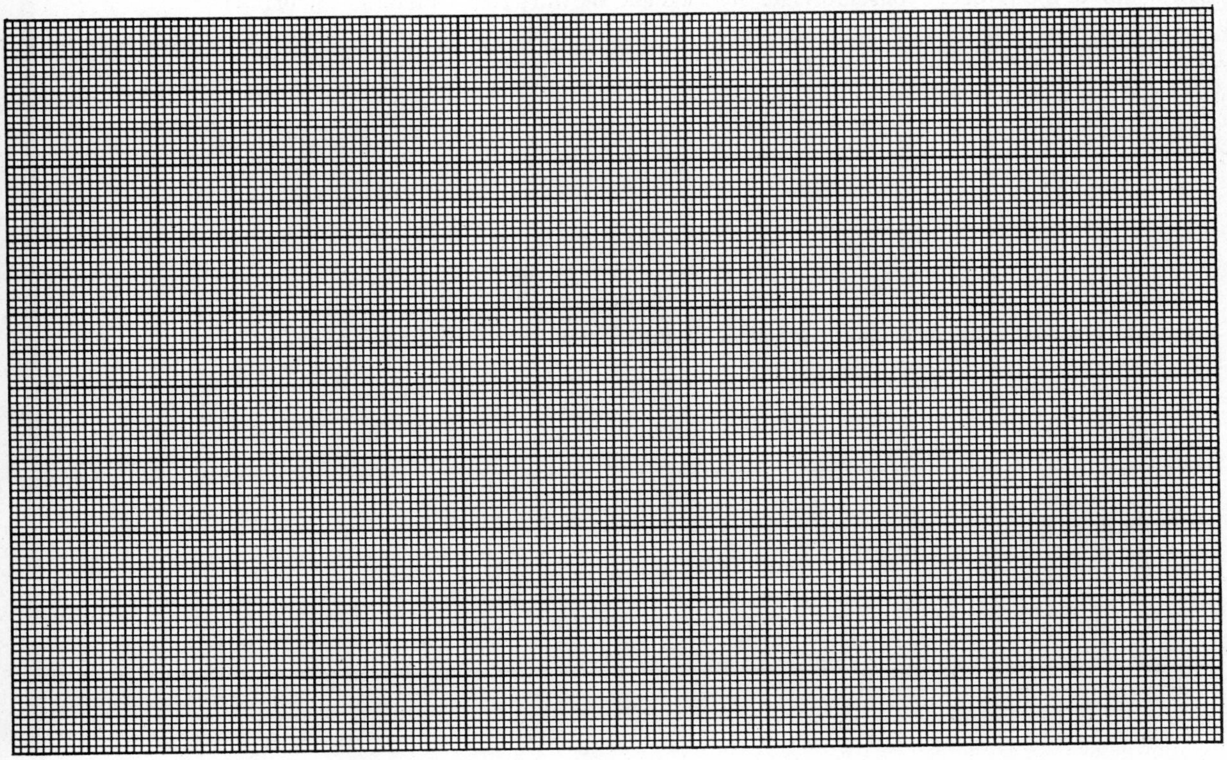

3. Observations and Measurements of the Hydrogen Spectrum

(a) You should observe at least three lines in the hydrogen spectrum. If you can see an additional line enter the data for it in the table below.

Color	Scale position (cm)	Wavelength (Å) from calibration graph	Known wavelength (Å)	Percentage error
_____	_____	_____	_____	_____
_____	_____	_____	_____	_____
_____	_____	_____	_____	_____
_____	_____	_____	_____	_____

(b) Show how you convert one of the above λ values in Å to the corresponding $\bar{\nu}$ in cm^{-1}.

(c) Convert all of your wavelengths to the corresponding wavenumbers.

Your wavelength (Å)	Wavenumber (cm^{-1})	Literature value for wavenumber (cm^{-1})
_____	_____	_____
_____	_____	_____
_____	_____	_____
_____	_____	_____

(d) Calculation of the Rydberg constant. Give one example showing your method of calculation.

(e) Look up the actual value of R_H and compute your percentage error for each line.

Your wavelength (Å)	Experimental value of R_H	Percentage error
_____	_____	_____
_____	_____	_____
_____	_____	_____
_____	_____	_____

4. Other Uses of the Spectroscope

(a) Record your observations of the cation emission spectra.

Salt	Color of flame	Scale reading of observed lines	Wavelengths of these lines

(b) Describe the spectrum of the light from a fluorescent lamp, and try to explain what you observe.

QUESTIONS

1. Use the Bragg equation to predict the value on the scale of the spectroscope where 5000 Å should fall. This calculation will require the data entered in part 1 of your report form and an understanding of the meaning of the sine of an angle. Compare this scale value with that read from your calibration graph at 5000 Å. Calculate the percent difference in these numbers. This percentage is a rough measure of the accuracy of your instrument.

2. Figure 8-6 shows the light coming from the scale. Does it really? (Hint: You may want to consult an elementary physics book to answer this question. Look up the term *virtual image*.)

3. How could you use the observation of the emission spectra of cations, made in part 4 of your report? Suggest a test to prove that you observed the cation spectra and not the anion spectra.

PROPERTIES OF GASES

ANALYSIS OF SUBSTANCES
IN THE ATMOSPHERE

PRE-STUDY

What is in air—that ordinary substance on which we, like all other living creatures, are so dependent for our very existence? This question has become an important contemporary concern because our automobiles and factories, too, gulp it in huge quantities and exhale deadly poisons. Consequently, it has become increasingly apparent, that the time may come when we will be unable to live in the befouled environments of some of our major cities.

What makes air good, and what makes it bad? Data on the chemical composition of clean, dry air are reproduced in Table 9-1.

This is the air we and rabbits and birds can breathe all of our lives and never suffer from any lung damage or respiratory ailment. The problem is that many of us breathe in air that contains other substances or some of the same substances but in higher concentrations. Some of these contaminants are the following.

Several sulfur oxides, SO_2 and SO_3, collectively called SO_x are produced by burning coal, an important source of energy for power plants and one destined to become more important in America in the future. A variety of nitrogen oxides, collectively represented as NO_x, are produced primarily by the internal combustion engine. Other unpleasant contaminants are these: carbon monoxide, CO; unsaturated hydrocarbons,

$$\underset{R}{\overset{R}{\diagup}} C = C \underset{R}{\overset{R}{\diagdown}}$$

where R is a hydrogen atom or organic radical; ozone, O_3; and small particulate matter such as dust and smoke.

TABLE 9-1
Composition of clean, dry air near sea level[1]

Compound	Formula	Percent by volume	Compound	Formula	Percent by volume
Nitrogen	N_2	78.09	Nitrous oxide	N_2O	0.000025
Oxygen	O_2	20.94	Hydrogen	H_2	0.00005
Argon	Ar	0.93	Methane	CH_4	0.00015
Carbon dioxide	CO_2	0.0318	Nitrogen dioxide	NO_2	0.0000001
Neon	Ne	0.0018	Ozone	O_3	0.000002
Helium	He	0.00052	Sulfur dioxide	SO_2	0.00000002
Krypton	Kr	0.0001	Carbon monoxide	CO	0.00001
Xenon	Xe	0.000008	Ammonia	NH_3	0.000001

[1]Source: The American Chemical Society, *Cleaning our Environment—The Chemical Basis for Action*. Washington, D.C., 1969.

The sulfur oxides and particulates combine to form the thick type of smog that killed 20 people in Donora, Pennsylvania on October 26, 1948, and 4000 people in London on December 5, 1952. All of the above contaminants except SO_x contribute to the formation of photochemical smog. These ingredients, in the presence of a temperature inversion and sufficient sunlight, generate PAN, or peroxyacyl nitrate. This component has been identified as the agent responsible for the dramatic decrease in the productivity of orange trees and grape vines in southern California, and for the death of millions of Ponderosa pine trees in the same region.

It would be interesting if we could analyze for some of these contaminants in this experiment. Unfortunately, they are present in such trace amounts that sophisticated instruments are required to detect them and measure their concentrations. We will have to be satisfied with the qualitative demonstration of the presence of three of the components of air and a quantitative determination of a fourth. None of these four molecules (H_2O, CO_2, N_2, and O_2) are considered to be contaminants. While you are taking these measurements however, you might want to think about the other molecules that are considered to be contaminants, and perhaps investigate the ways in which they are detected and means by which you yourself might help reduce their production.

Experimental Method

Oxygen, O_2, will be detected by reaction with pyrogallol. This substance reacts with gaseous oxygen and removes it from the gas phase. The reaction is quantitative if the volume percentage of oxygen in the gas is 25% or less, if the temperature is above 20°C, and if the solution is thoroughly shaken with the gas. The structure of pyrogallol is

The molecule is stable in water but reacts slowly with O_2 in pure water. It reacts rapidly with O_2 in an alkaline solution but is not stable very long in a basic solution. The problem is solved by mixing the pyrogallol solution with NaOH immediately before its reaction with the air (see the Experimental Procedure, which follows).

EXPERIMENTAL PROCEDURE

Chemicals: 0.1 F Ba(OH)$_2$,[1] CaCl$_2$ (anhyd., 4-mesh), Fe (filings), Mg (ribbon), 30% pyrogallol solution (fresh).

1. Some Qualitative Observations. (a) *Water Vapor.* Place several granules of anhydrous calcium chloride on a watch glass and leave it exposed to the air. At the close of the period, or during the next laboratory period, observe any change that has occurred. (The reasons for the absorption of

[1]The "*F*" preceding a formula indicates a solution of a specified concentration. Thus, a 0.1 F solution contains 0.1 of the gram formula weight per liter of solution. Units of concentration are discussed in Study Assignment B.

water by very soluble substances like calcium chloride will be considered when you study solutions. In extremely dry weather, no absorbed water at all may be observed.)

(b) *Carbon Dioxide*. Place about 15 ml of 0.1 F Ba(OH)$_2$ in a 500-ml flask. Stopper the flask and shake it for a moment. Observe the change. (A barium hydroxide solution provides a more sensitive test for carbon dioxide than does the more commonly used limewater, or calcium hydroxide solution, which you used in Experiment 2. The reaction is similar—a precipitate of barium carbonate, BaCO$_3$, is formed.)

As a comparison, check on the relative amount of carbon dioxide in your breath by inhaling deeply, then exhaling, through a piece of glass or rubber tubing, into another 500-ml flask containing 15 ml of 0.1 F Ba(OH)$_2$. Close and shake the flask. Compare the two amounts of precipitate.

(c) *Nitrogen*. Place about 15 cm of magnesium ribbon, packed rather compactly, into a crucible, and cover it with the crucible lid. Heat intensely with the Bunsen burner for about 10 minutes. Cool the crucible, and replace the lid with a small watch glass, which has a small piece of moist red litmus paper attached to the bottom side. Lift the watch glass just enough to moisten the white residue with a few drops of water. Warm the crucible slightly. Note any change in the litmus.

The reactions are as follows. Along with the reaction of magnesium to form magnesium oxide, some of the magnesium reacts, at high temperature, with nitrogen in the air to form magnesium nitride:[2]

$$3Mg(s) + N_2(g) \rightarrow Mg_3N_2(s)$$

This, by reaction with water, forms ammonia (NH$_3$) and the slightly soluble base, magnesium hydroxide [Mg(OH)$_2$]. Ammonia gas dissolves in water to form a basic solution.

$$Mg_3N_2(s) + 6H_2O \rightarrow 3Mg(OH)_2(s) + 2NH_3(g)$$

$$NH_3(g) + H_2O \rightarrow NH_3(aq) \rightleftharpoons NH_4^+ + OH^-$$

You may detect the odor of ammonia gas above the moist, white, solid magnesium hydroxide.

2. Quantitative Determination of Oxygen in the Air. Prepare the apparatus sketched in Figure 9-1, using a 250-ml Erlenmeyer flask. Remove the stopper and fill the flask and the 10-cm test tube (still in the

[2](s) = solid, (g) = gas, (l) = liquid, (aq) = aqueous solution.

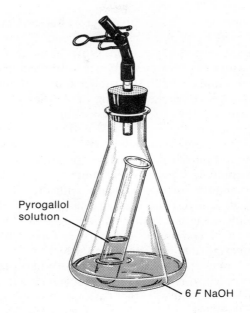

Pyrogallol solution

6 F NaOH

Equalize the levels inside and outside the flask and replace the clamp.

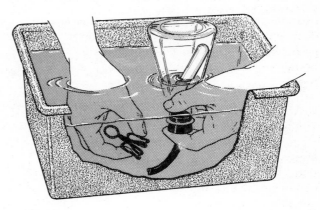

FIGURE 9-1
Apparatus for the absorption of oxygen from the air by pyrogallol solution.

flask) with tap water. Temporarily disconnect the rubber tube and clamp, and firmly reinsert the stopper and glass tube into the flask, making certain no air bubbles are trapped. Then remove the stopper and glass tube, and accurately measure the volume of the water with your graduated cylinder, to obtain the capacity of the apparatus. Record this volume and all subsequent data, at once, in your report sheet.

Empty the apparatus, but do not dry it. Measure about 20 ml of 6 F NaOH into the flask, and 5–6 ml of a solution of 30 weight % pyrogallol[3] (which is also called pyrogallic acid) into the test tube. Record the exact volumes used. Now firmly close the flask with the stopper and glass tube (there must be no leak), and reattach the rubber tube without the clamp. Avoid undue handling of the flask, as this will warm it above room temperature. Finally, tightly clamp the rubber tube just above the glass tube. Invert the apparatus repeatedly, and shake it moderately, at intervals, to mix the contents. After about 15 minutes, during which time the oxygen should completely react chemically with the alkaline pyrogallol, invert the flask in a large vessel of water, such as a pneumatic trough, so that the rubber tube is completely submerged. Open the clamp to permit the entry of water, which will replace the volume of oxygen absorbed. When the flask has cooled to room temperature, equalize the levels inside and outside the flask by raising or lowering the flask in the trough. Then close the

clamp. (Again avoid handling that part of the flask which is exposed to the remaining air within.) This procedure equalizes the pressure so that all measurements are made under the same conditions. Place the flask upright on the desk top.

At this point, test the character of the residual gas by thrusting a burning splint into the flask just as you remove the stopper. Results? With your graduated cylinder, carefully measure the volume of the brown liquid in the flask. From these data calculate the volume of oxygen absorbed and the original volume of air, taking account of the volumes of all liquids used. Calculate the volume percentage of oxygen in the air. If time is available repeat the experiment.

Bibliography

Cleaning Our Environment—The Chemical Basis for Action, American Chemical Society, Washington, D.C. (1969), p. 24.

Stephens, E. R., "Chemistry of Atmospheric Oxidants," Journal of the Air Pollution Control Association 19, 182 (1969).

---

[3]A 30 weight % solution means 30 g of solute dissolved in 70 g of water. This should be freshly prepared for class use, or you may use 2.0 g of pyrogallic acid dissolved in 5 ml of water.

REPORT

9

**Analysis of Substances
in the Atmosphere**

NAME

SECTION   LOCKER

INSTRUCTOR   DATE

OBSERVATIONS AND DATA

1. Some Qualitative Observations

Describe briefly the qualitative evidence that you obtained for the presence or absence of the following in the air:
(a) Water vapor

(b) Carbon dioxide

(c) Nitrogen

2. Quantitative Determination of Oxygen in the Air

	1	2
Volume of water required to fill flask with test tube in it	_____ml	_____ml
Volume of 6 F NaOH	_____ml	_____ml
Volume of pyrogallol solution	_____ml	_____ml
Volume of all solutions in flask (after absorption)	_____ml	_____ml
Volume of oxygen absorbed	_____ml	_____ml
Volume of air originally in the stoppered flask	_____ml	_____ml
Volume percentage of oxygen in the air	_____%	_____%

Method of calculation:

QUESTIONS

1. How would the results be affected if: (1) All the measurements in part 2 were carried out at a higher, but constant, temperature? (2) The temperature of the initial measurements of gas volume differed from that of the final measurements?

2. The amount of water vapor in the air is variable, up to a maximum of 2.0 volume percentage (at room temperature, 20°C). If dry air contains 20.99 volume percentage oxygen, what is the volume percentage of oxygen in this same air when it is saturated with water vapor? (Hint: 20.99 parts of oxygen in 100 parts of dry air corresponds to 20.99 parts oxygen in 102 parts of moist air.)

_____Vol. %

3. Calculate your percentage error for the analysis for oxygen, assuming (since the vessels were wet) that the air you analyzed was saturated with water vapor. (See the discussion on the treatment of experimental errors in Appendix B.)

_____%

4. In the analysis for oxygen, the carbon dioxide in the air, 0.03%, is absorbed, as well as the oxygen, by the sodium hydroxide in the solution. How much of an error in the calculation of the percentage of oxygen is caused by neglecting this factor?

_____%

5. Summarize any physical and chemical properties of oxygen and of nitrogen that you observed, or that it was necessary to assume, in performing these experiments.

6. List as many reasons as you can in support of the statement that the air is a mixture, rather than a compound, of its principal constituents.

PRESSURE-VOLUME-TEMPERATURE
RELATIONSHIPS IN GASES

PRE-STUDY

The Kinetic Molecular
Theory of Gases

Review in your text the fundamental concepts of the kinetic molecular theory and the explanation of the properties of the gaseous state of matter. We may summarize the essential concepts as follows:

1. Gaseous matter consists of minute, discrete particles, called *molecules,* which ordinarily are relatively far apart. (The mean free path of a gas is the average distance a gas molecule travels before it collides with another molecule. An approximate value for the mean free path of a gas at 1 atm is 10^{-5} cm, about 300 times the average diameter of a gaseous molecule.)

2. These molecules are in *rapid translational motion,* moving in straight lines until they collide with and rebound from other molecules. They also possess rotational and vibrational motion. See Figure 10-1.

3. Such molecular collisions are *perfectly elastic*; that is, there is no loss in total energy resulting from the collisions of the moving molecules.

4. The sum effect of the thrust of these countless individual molecular collisions against the walls of the container is the *gas pressure.*

5. *Temperature* is a manifestation of molecular motion itself, and depends on the energy due to the molecular motion, namely $\frac{1}{2}mv^2$, where m is the mass and v is the velocity of the molecules. Thus the minimum temperature possible is that temperature at which all translational and rotational molecular motion ceases. It is called *absolute zero.*

6. *Attractive forces* between molecules are relatively small, a phenomenon that permits a substance to exist in the gaseous state. In the idealized

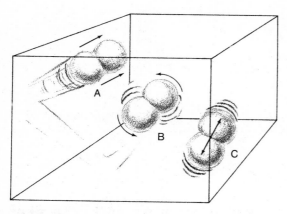

FIGURE 10-1
Types of molecular motion: (A) translational
motion, the movement of the center of mass of the
molecule; (B) rotational motion, in which the
molecule rotates about its center of gravity;
(C) vibrational motion, in which the atoms move
alternately toward and away from their center
of gravity.

or *perfect gas* (see Study Assignment B), there
would be no attractive forces at all.

7. The volume occupied by the molecules is
negligible compared to the total volume of the
container.

We shall observe experimentally the ways in
which the pressure, temperature, and volume of a
sample of gas are interdependent, and express the
physical laws thus demonstrated by mathematical
equations and by graphs of the data obtained.

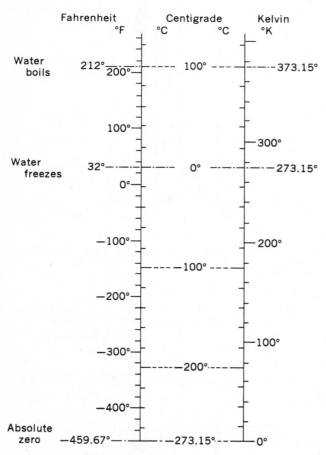

FIGURE 10-2
A comparison of temperature scales.

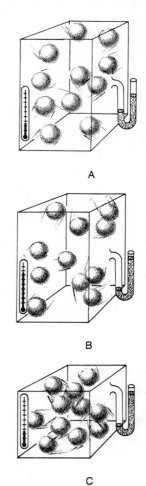

FIGURE 10-3
Kinetic theory interpretation of the effect of
temperature and volume on gas pressure. A
quantity of gas in a given container (A) will exert
a greater pressure under the following conditions:
if it is heated to a higher temperature (B), so that
the rate of momentum transfer by molecular
collisions with the wall is increased because of
the increased molecular velocity; or if it is
crowded into a smaller volume (C), so that the
number of collisions per square centimeter
against the walls of the container is increased.

EXPERIMENTAL PROCEDURE

Special Supplies: Demonstration pressure-volume apparatus (Fig. 10-4), metric rules, ice, thermometer, large beakers, pneumatic trough.

NOTE: Because of the expense of obtaining mercury and the possible danger inherent in handling large quantities of it, it is recommended that part 1 be performed by the instructor. Part 2 can be carried out by individuals or pairs of students.

---

1. Pressure-Volume Relationships at Constant Temperature. Study the apparatus illustrated by Figure 10-4. A definite number of molecules of gas are contained in the enclosed volume of the gas buret above the mercury column. This volume changes in accord with the changes in the total pressure applied to the system by raising or lowering the leveling bulb. Obviously, when the mercury levels are equal, the pressure of the confined gas is equal to the external atmospheric pressure. Since pressure is measured in terms of the height of a mercury column, the gas pressure at other volumes (when the mercury levels are not equal) may be calculated by adding or subtracting the measured difference in mercury levels for each corresponding volume.

   (a) *Obtaining the Data.* Adjust the leveling bulb to equalize the mercury levels in the bulb and in the gas buret. Read the volume as precisely as possible. Read the laboratory barometer in units of torr. (The pressure unit, mm Hg, is now called a torr.) Record these and subsequent corresponding values of the volume and the pressure of the gas, in tabular form, in your experiment report.

   Adjust the leveling bulb at two or three positions (up to about 400 mm difference in mercury levels) above the mercury level in the gas measuring tube. For each position carefully read the volume of the gas and the height of both liquid levels from the meter stick. Repeat for two or three positions below the mercury level in the tube. Calculate the corresponding gas pressure for each reading by adding (or subtracting) the difference in mercury heights and the barometric pressure.

   (b) *Calculating and Interpreting the Data.* Examine the corresponding relative values of volume and pressure obtained for this sample of gas. What mathematical operation (addition, subtraction, multiplication, division) would be the best method for checking on any fundamental relationship governing these relative volumes and pressures? Carry out this calculation for each set of data obtained, and compare the results. State both in a mathematical equation and in words the physical law (known

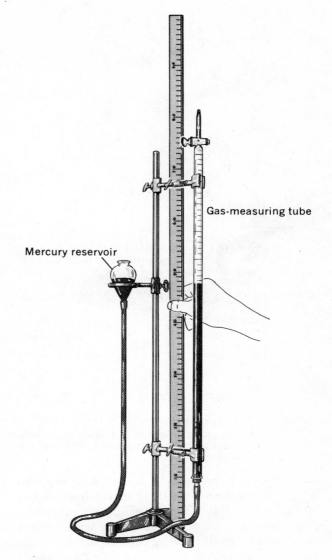

FIGURE 10-4
Apparatus for the determination of the relation between the pressure and the volume of a gas at constant temperature.

as Boyle's Law) that expresses this behavior. Plot your data on the report form, following the directions given there.

2. Volume-Temperature Relationships at Constant Pressure. The apparatus used is illustrated by Figure 10-5. A flask of dry air, heated to a definite temperature in the water bath, will be inverted under cold water in a pneumatic trough without loss of any of the contained air. From the contraction in volume, as measured by the water that

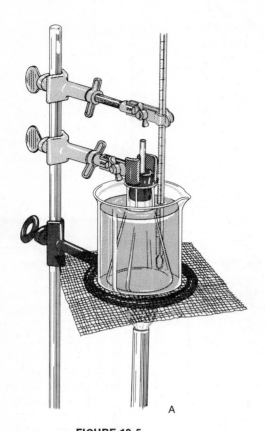

Don't warm flask with your hand.

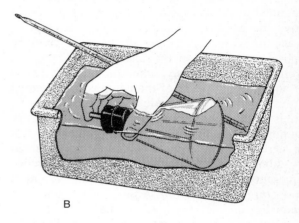

A   B

FIGURE 10-5
Apparatus for the determination of the relation between the volume and the temperature of
a gas at constant pressure: (A) the high-temperature bath; (B) the low-temperature bath.

enters the flask, you will be able to determine the law governing the relationship between the volumes of the air and the corresponding temperatures. Four determinations[1] are desirable, with the hot bath varying from the boiling point of water to about 30°C, and, in each determination, the flask cooled to 0°C in an ice bath.

(a) *Obtaining the Data.* Heat about 700 ml of water in a 1-liter beaker, as shown in Figure 10-5A. While this is heating, fit a 250-ml Erlenmeyer flask with a one-hole stopper into which a 7-cm length of glass tubing has been inserted.[2] *Dry the flask thoroughly before use* by warming it over a flame and then passing air first through a $CaCl_2$ tube (to

dry it) and then through a tube leading to the bottom of the flask. (If compressed air is not available, a water-vacuum aspirator may be used, with reverse connections to draw dry air through the warm flask.)

Fit the rubber stopper and glass tube firmly into the flask of dry air. Immerse the flask in the beaker of water, supporting the flask loosely with a buret clamp to keep it almost completely submerged, in order to maintain a uniform temperature. Heat the water to boiling. When the thermometer reading in the hot water bath remains constant for 2 minutes, read the temperature. Then hold your finger *firmly* over the end of the tube, remove the flask, and quickly invert it in a pneumatic trough of cold water, which is maintained at 0°C by chipped ice. See Figure 10-5B. Remove your finger, keeping the flask completely submerged (with the glass tube down so no air is lost) for 2 minutes to attain temperature equilibrium. Now equalize the pressure inside the flask with that of the atmosphere by quickly raising the flask (keep the glass tube

---

[1]These can all be carried out by a student during a laboratory period, but if time is short, a student may work at only one or two hot temperatures, and the results of different class members may be pooled in a common graph of the data. If students do pool results, each one should clearly label his own data.

[2]If you are doing all four determinations, time will be saved if you prepare four such flasks, or at least two, and dry a flask for the next determination while the preceding one is being completed.

submerged) so that the water levels inside and outside are equal, and at once (the air in the flask must not be permitted to warm) close the end of the tube with the finger, withdraw the flask, and place it upright on the desk top. With a graduated cylinder, measure the volume of water in the flask. Measure the volume of the flask completely filled with water to the bottom of the stopper. Enter these volumes, and the high and low temperatures, in your experiment report. Read the barometer, and record the atmospheric pressure.

Repeat the entire experiment three times, maintaining the high temperature for 2 minutes each time at about 75°, 50°, and 30°C respectively. The flask must be thoroughly dry at the start each time. Read the exact temperature at the time you transfer the flask to the cold bath for each determination. Record all data.

(b) *Treatment of Data.* The temperature data that have been obtained when the flask is closed have no bearing on the variation of volume with temperature. Although the temperature is varied in the four experiments, the volume of the hot air always remains equal to the volume of the flask. What, then, is varying?

You have probably studied the ideal gas equation

$$PV = nRT \qquad (1)$$

where P refers to the pressure, V to the volume of the flask, n to the number of moles of gas within the flask, R to the gas constant, and T to the temperature of the flask. The units of P and V must be consistent with those of the gas constant. The temperature T must always be expressed in units of absolute temperature, K. The pressure, P, is constant in all parts of this experiment, equaling the atmospheric pressure. At the elevated temperature, V is also constant. The gas constant R is of course a constant. This means that, as the temperature T of the flask at the elevated temperature varies, n, the number of moles of gas inside the flask, must vary. As you place your finger firmly over the glass tube after temperature equilibrium is obtained at the high temperature, you do indeed trap differing numbers of moles of gas. Thus we confront the peculiar situation of attempting to establish the relationship between V and T when all that we have accomplished at this time is to obtain some data relating n and T.

The problem is resolved in the next step and the subsequent calculations. What is accomplished by this procedure is essentially to calculate V/n as a function of T. A slight rearrangement of equation (1) yields

$$\frac{V}{n} = \left(\frac{R}{P}\right) \times T \qquad (2)$$

Thus a plot of V/n versus T should be a straight line of slope R/P. Although V, the volume of the gas at the high temperature, does not change during the experiment, n does, and therefore V/n also varies with the temperature. Thus our problem now consists of measuring n.

This measurement is done in the following way. Because we are interested in how V varies with T, and not in the proportionality constant, we do not compute n, but rather a quantity that is directly proportional to it. In all four determinations, you will measure the volume of the air that is trapped in the flask at the elevated temperature by immersing it in a constant temperature bath at 0°C and measuring its volume at this fixed temperature. This 0°C volume is directly proportional to n. Thus if we divide the high-temperature volume (a constant) by the low-temperature volume (a variable) we have calculated a quantity proportional to V/n. A plot of this quantity versus T will reveal how the volume of a gas varies with temperature.

A small correction to the volume measured at 0°C is required because the air at that point becomes saturated with water vapor as water enters the flask. Dalton's Law of Partial Pressures (which is discussed more fully in Experiment 11) states that the total pressure of a mixture of gases is the sum of the partial pressures of the individual gases in the mixture. Thus it is necessary to multiply the volume of the moist air by the ratio of the corrected P to the total P.

The instructions that follow, and the format of the report form, should help you to record your data and to complete the necessary calculations discussed above. Your data consist of the following.

1. The temperature of the hot bath.
2. The temperature of the cold bath.
3. The volume of hot, dry air (volume of the flask up to the stopper).
4. The volume of water entering the flask at atmospheric pressure.

From these data, calculate the following items.

5. The volume of cold, moist air (subtract item 4 from item 3).
6. The corresponding volume of cold, *dry* air.

Since the vapor pressure of water at 0°C is about 5 mm, this corrected dry volume[3] will be

$$\frac{P_{atm}(Torr) - 5\ Torr}{P_{atm}(Torr)}$$

7. The volume of hot, dry air *per ml* of cold, dry air. (To compare data at different temperatures, *you must compare the same amount of gas,* that is, the same number of molecules.) This volume will be the ratio of the volume of hot, dry air (item 3) to the volume of cold, dry air (item 6).

Next, carefully graph your data for the four experimental runs at different temperatures by plotting the calculated volumes (item 7) on the vertical axis against the corresponding temperatures on the horizontal axis (scale from −300°C to +100°C).[4]

Include also the point (1 ml, 0°C) on your graph. Draw the straight-line graph that fits your experimental points best, and extrapolate the data by extending this line (dotted) to its intersection with the zero-volume axis. Read the corresponding temperature. What name and what specific significance applies to this point? How do your data compare with the accepted value for this point?

8. Now convert all your temperature readings for the four runs to this Kelvin scale.

9. By study of your graph, discover and calculate the relationship between the temperature (expressed on this new scale) and the corresponding volume for each run. State this law (Gay-Lussac's or Charles' Law) as a mathematical equation.

---

[3]You will understand this small correction more fully after studying Dalton's Law of Partial Pressures in Experiment 11.

[4]A more accurate value of the zero-volume temperature can be obtained by plotting your data on a scale from 0°C to 100°C and calculating absolute zero by algebraic extrapolation. (0°K = −intercept/slope. See Appendix A for definitions of slope and intercept.)

DATA

1. Pressure-Volume Relationships at Constant Temperature

(a) Data. Calculate the pressure on the gas in torr by adding (or subtracting if appropriate) column 4 to the barometric pressure to obtain column 5.

Reading number	Height of reservoir (mm)	Height of mercury in measuring tube (mm)	Difference in heights (torr)	P (torr)	V (ml)	$P \times V$ (torr × ml)
1						
2						
3						
4						
5						
6						
7						

(b) Data analysis. Plot the data of columns 5 and 6 on the graph below. Plot the volume as the ordinate (vertical axis) and pressure as the abscissa (horizontal axis). Choose appropriate scales to utilize as much of the graph paper as possible; circle the data points. Draw a smooth curve connecting these circles, but do not draw through the circles.

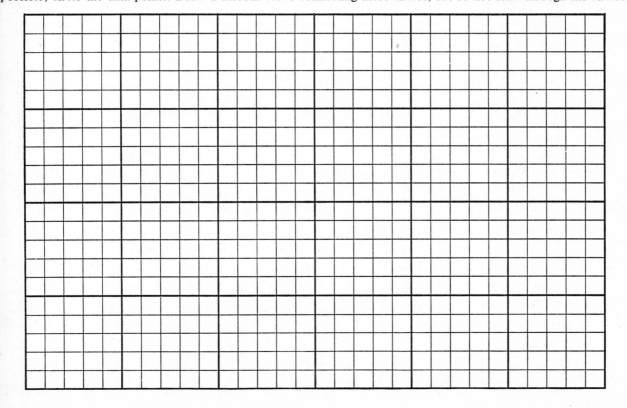

From your interpretation of the data and the graph, write the mathematical equation that expresses the fundamental relationship you have shown to be true.

State this physical law in words.

---

2. Volume-Temperature Relationships at Constant Pressure

(a) Data

Barometric pressure, P_____

	1	2	3	4
(1) Temperature of hot bath				
(2) Temperature of cold bath				
(3) Volume of hot, dry air (flask capacity to stopper)				
(4) Volume of H_2O entering flask				

(b) Treatment of Data

(5) Volume of cold, moist air (items 3-4)

(6) Volume of cold, dry air $\left[\dfrac{P - 5 \text{ Torr}}{P} \times V \right]$

(7) Volume of hot, dry air per ml of cold, dry air $\left[\dfrac{\text{Vol (3)}}{\text{Vol (6)}} \right]$

(*Enter next items after plotting data as directed.*)

(8) Temperature of hot air, K

(9) Ratio $\dfrac{\text{Vol (7)}}{K}$

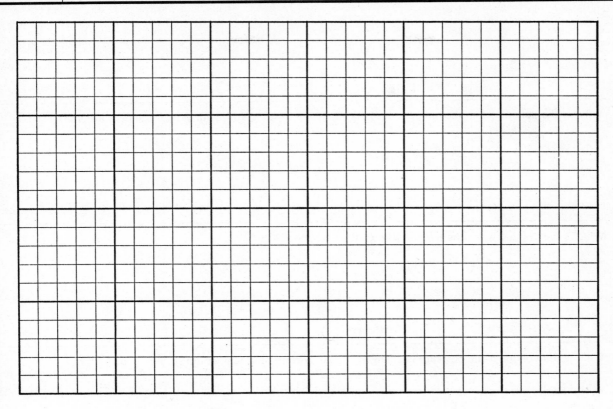

From a consideration of the calculated data, and from the graph, derive the fundamental relationship between the volume of a gas and its temperature, as a mathematical equation.

State this law in words.

PROBLEMS

For each of the following, assume 260 ml of nitrogen gas at a pressure of 740 Torr and at a temperature of 20°C. *Show your method in each problem.*

1. What is the volume when the pressure is increased to 760 Torr?

2. What is the pressure when the volume is increased to 900 ml?

_____

3. What is the volume when the temperature is increased to 50°C? (Note your proper temperature units.)

_____

4. What is the volume when the temperature is decreased to −30°C?

_____

5. What is the volume if the pressure is changed to 600 Torr and the temperature is raised to 27°C?

_____

6. What is the pressure if the volume is increased to 390 ml and the temperature is decreased to 0°C?

_____

APPLICATION OF PRINCIPLES

Explain why the following statements are true, basing your explanations on the fundamental concepts of the kinetic molecular theory as given in the pre-study section.

1. A gas always completely fills any container in which it is placed, regardless of the volume of the container.

2. If the volume of a gas sample is decreased to one-third, the pressure of the sample will be increased three-fold.

3. If 1.0 liter of N_2 gas at 250 Torr pressure is mixed with 1.0 liter of O_2 gas at 600 Torr pressure in a 1.0-liter container, each gas "fills" the container completely, and the final pressure is 850 Torr.

4. If a gas contained in a vessel of constant volume is heated, the pressure increases.

QUESTIONS

1. Will Boyle's Law be obeyed more closely by a gas at low pressure or at high pressure? Explain your answer.

2. At which temperature will the behavior of a substance correspond more closely to that of a *perfect* gas, at a temperature only slightly above its boiling point, or at a considerably higher temperature? Explain your answer.

THE MOLAR VOLUME OF OXYGEN

PRE-STUDY

Avogadro's Law and the Molar Volume

This experiment deals with one of the most important quantitative relationships involving gases — a relationship which depends on the following two facts.

1. One mole of any substance contains the same number of molecules as 1 mole of any other substance. This is implied from the definition of a mole.

2. Equal volumes of all gases, under identical conditions of temperature and pressure, contain equal numbers of molecules. This follows from "the kinetic theory of gases," and was first stated about 1811 by *Amedeo Avogadro,* an Italian chemist.

It follows, then, that under identical conditions of temperature and pressure, 1 mole of any and all gases occupies equal volumes — this is called *the molar volume of a gas.*

The Molar Volume at Standard Conditions. The Perfect Gas

To evaluate the molar volume in metric units, recall that the symbol O stands for 1 mole — namely, 15.9994 g — of oxygen, and that 1 mole of oxygen gas, O_2, therefore weighs 31.9988 g. In this experiment we shall measure the weight and corresponding volume of a sample of oxygen gas and from this data, calculate the volume at standard conditions; then we shall calculate the volume of 32.0 g (1 mole) of oxygen gas.

All real gases deviate more or less from the ideal behavior described above because their molecules do have some slight attraction for one another and do occupy some slight volume themselves. *A perfect gas* is that idealized gas in which the molecules would have no attractive forces whatever and likewise would be mere "points" without significant size. The standard molar volume for a perfect gas has been calculated from measurements on real gases at very low pressures to be 22.4136 liters. Most common gases, unless they have a high molecular weight or are measured quite near their boiling point, have molar volumes that do not deivate more than about 1% from this volume.

Calculation of Gas Volumes

The quantity of a gas sample can be measured more easily by volume than by weight. In measuring the volume of a gas, it is also necessary to measure its temperature and pressure. Why? In Experiment 10 we discovered and verified the separate laws relating pressure to volume, and relating either pressure or volume to absolute temperature:

$$PV = k_1, \quad \text{and } P = k_2T, \quad \text{or} \quad V = k_3T$$

For a given gas sample, these two laws may be combined into one equation, which shows the way in which all three variables—pressure, volume, and absolute temperature—are interdependent:

$$PV = kT, \quad \text{or} \quad \frac{PV}{T} = k \qquad (1)$$

Since *any* two corresponding sets of PV/T measurements will be equal to k and to each other for a given amount of gas, we may write

$$\frac{P_1V_1}{T_1} = \frac{P_2V_2}{T_2} \qquad (2)$$

This may be transposed to give

$$V_1 = V_2 \times \frac{T_1}{T_2} \times \frac{P_2}{P_1} \qquad (3)$$

Note in equations (2) and (3) that if the temperature is constant ($T_1 = T_2$), the inverse proportionality of pressure and volume (Boyle's law) is expressed. Likewise, for constant pressures ($P_1 = P_2$), the direct proportionality of volume and absolute temperature (Charles' law) is expressed. If any five of the quantities in equation (2) are known, the

sixth can of course be calculated by simple algebraic means.

Many students and teachers prefer to reason out the pressure-volume-temperature relationships, rather than blindly to "follow a formula," and to apply corrective factors for pressure (and for temperature) according to whether the pressure change (and the temperature change) will cause an increase (or a decrease) in the volume. Either approach yields the same result.

Example 1. 300 ml of CO_2 gas, measured at 30°C and 780 Torr pressure, will have its volume decreased by decreasing temperature (to 0°C) and increased by decreasing pressure (to 760 Torr), in accordance with the equation

$$V_1 = 300 \text{ ml} \times \frac{273 \text{ } K}{303 \text{ } K} \times \frac{780 \text{ Torr}}{760 \text{ Torr}} = 277 \text{ ml}$$

The General Gas Law Equation

For specific amounts of the same gas, or for any different gases, equation (1) may be restated in its most general form:

$$PV = nRT \qquad (4)$$

Here, n is the number of moles of gas, and R is a proportionality constant, called the "gas constant," which has the same value for all gases under all conditions, namely, 0.0821 liter atmospheres per mole per degree. In all calculations in which this constant is employed, pressure must be expressed in atmospheres, volume in liters, and temperature in absolute or Kelvin degrees.

As an example of the application of this general gas law where both volume and weight of a gas sample are involved, consider the following. What weight of chlorine gas, Cl_2, would be contained in a 5.00-liter flask at 20°C and at 600 Torr (600/760 atm) pressure? Substituting in equation (4), transposed to give, n, the number of moles, we have

$$PV = nRT, \quad \text{or} \quad n = \frac{PV}{RT}$$

$$n = \frac{600/760 \text{ atm} \times 5.00 \text{ liters}}{0.0821 \frac{\text{liter atm}}{\text{mol deg}} \times 294 \text{ } K} = 0.165 \text{ mole}$$

and, in grams,

$$0.164 \text{ mole} \times 70.9 \frac{\text{g}}{\text{mole}} \text{ } Cl_2 = 11.6 \text{ g } Cl_2$$

Aqueous Vapor Pressure.
Dalton's Law of Partial Pressures

When any gas in a closed container is collected over liquid water, or exposed to it, the water evaporates until a saturated vapor results—that is, until the opposing rates of evaporation and condensation of water molecules at the liquid surface reach a "balance." These gaseous water molecules contribute to the total gas pressure against the walls of the container. Thus, of all the gas molecules, if 3% are water molecules and 97% are oxygen molecules, then 3% of the total pressure is due to water vapor

and 97% of the total pressure is due to oxygen. *Each gas exerts its own pressure regardless of the presence of other gases.* This is Dalton's Law of Partial Pressures. Stated as an equation:

$$P_{\text{total}} = P_{H_2O} + P_{O_2}$$

or, if it is transposed,

$$P_{O_2} = P_{\text{total}} - P_{H_2O}$$

To illustrate Dalton's law, Figure 11-1 shows a mixture of oxygen molecules and water vapor molecules. In (b), the water molecules have been removed, but all the oxygen molecules are still present, in the same volume. The pressure has been reduced by an amount equal to the vapor pressure of the water.

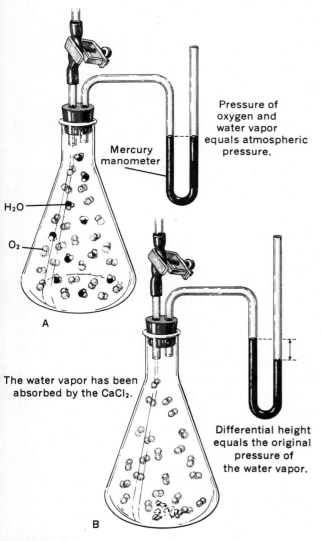

Pressure of oxygen and water vapor equals atmospheric pressure.

Mercury manometer

H_2O

O_2

A

The water vapor has been absorbed by the $CaCl_2$.

Differential height equals the original pressure of the water vapor.

B

FIGURE 11-1
The application of Dalton's law of partial pressures when water vapor is removed from a gas mixture. Note that, in this figure, the gas molecules are greatly exaggerated in size.

EXPERIMENTAL PROCEDURE

Special Supplies: Analytical weights, 600-ml beaker, 100°C thermometer, 500-ml graduated cylinder.

Chemicals: Prepared unknown mixture of $KClO_3$, KCl, and MnO_2; or $KClO_3$, MnO_2.

Prepare the apparatus as illustrated in Figure 11-2. Note the details carefully. Test tube A must be clean and *thoroughly dry.* Fill the flask B with water, and have a small amount of water in the beaker D. Fill the rubber tube C with water by blowing into the rubber tube E momentarily. Syphon water back and forth through tube C by raising and lowering the beaker, to expel all air bubbles. Finally, with the flask nearly filled, but not quite to the top (water must not enter the short glass tube connected to the rubber tube E), close the clamp which has been placed on tube C near the flask.

Obtain an unknown mixture of $KClO_3$, KCl and MnO_2 from the instructor and at once record its code number on your report sheet.[1] By completely

---

[1]If you are not using an unknown mixture, and are determining only the molar volume of O_2, about 5 g $KClO_3$ may be dried by just melting it in test tube A (disconnected and held by a test tube holder, and heated throughout its entire length). Let this cool, then add a very small amount (10–15 mg) of dry MnO_2. Weigh the test tube and connecting stopper and glass tube with the contents. *Warning:* $KClO_3$ can cause a dangerous explosion when heated with certain reducing agents. *Be sure you read the label correctly.* As a safety precaution, test a small amount of the $KClO_3$ (the size of a pea), mixed with a very small amount of the MnO_2 you will use, by heating it in a test tube to observe if it decomposes safely without obvious combustion.

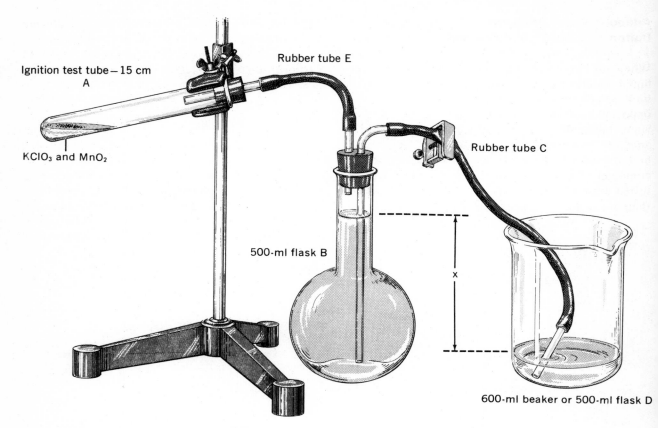

Ignition test tube—15 cm
A

Rubber tube E

KClO₃ and MnO₂

Rubber tube C

500-ml flask B

x

600-ml beaker or 500-ml flask D

FIGURE 11-2
Apparatus for determining the molar volume of oxygen.

decomposing a weighed sample of this and measuring the volume of the oxygen liberated at a known temperature and pressure, the molar volume of oxygen, and also the percent of KClO₃ in the sample, may be determined. First weigh the empty test tube A, together with its connecting stopper and glass tube, as precisely as your balance permits. Then add 1.5–4.0 g of the KClO₃ mixture to the test tube. (Your instructor will suggest the approximate amount for you to use, because you must not use an amount that will generate oxygen in excess of the capacity of the flask B.) Again weigh the test tube assembly.

Connect the test tube and stopper to tube E, and make sure that your apparatus is air tight by opening the clamp on tube C, and noting that water does not flow out of the flask, even when the water levels are quite uneven. Have your apparatus approved by the instructor. Equalize the pressures inside and outside the flask by raising the beaker until both water levels are the same. Water can syphon in or

out of the flask until the internal and external pressures are equal. Close the clamp and discard the water in the beaker. Drain the beaker carefully and completely, but do not dry it. Replace the tube in the beaker, and open the clamp. During the heating that follows, see that the end of tube C is kept under the water in the beaker.

Heat the KClO₃ mixture gently at first from the top down, then more strongly, to maintain a moderate rate of evolution of oxygen gas. If white vapors appear in the tube, decrease the heating until they disappear. Be careful to insure that the mixture is not heated so violently that some of the KClO₃ mixture is expelled from the bottom of the tube. If this material enters tube E and blocks it, the apparatus will be forced apart at one of the connections with unfortunate consequences for your data and for yourself. Continue the heating as long as any oxygen gas is liberated, until there is no further transfer of water into the flask. (If you have prepared your own mixture, using excess pure $KClO_3$ and MnO_2, *stop the heating* in time so that the water level does not go lower than 1–2 cm above the end of the tube in the flask.) Do not remove the tube from the beaker until the apparatus has cooled completely to room temperature, and the internal and external pressures have been equalized. *Also do not handle the flask* in such a way that the gas contained in it will be warmed by the hands. Adjust the levels in the flask and beaker so they are equal, and then close the clamp on tube C. Obain the final weight of the test tube, its contents, and the glass tube attached to it.

Take the temperature of the oxygen by placing a thermometer directly in the gas. Measure the volume of oxygen by carefully measuring the water in the beaker, using a 500-ml graduated cylinder. The aqueous vapor pressure may be obtained from Table 5 in Appendix C. Obtain the barometer reading for the day.

Repeat your experiment if time permits. From your data, calculate the volume, at standard conditions, of 32.0 g (1 mole) of oxygen gas. If you used an unknown mixture of KCl and $KClO_3$, calculate also the percentage of $KClO_3$.

The Molar Volume of Oxygen

NAME	
SECTION	LOCKER
INSTRUCTOR	DATE

DATA AND CALCULATIONS

Unknown sample number[1]_____   *Instructor's approval of apparatus*_____

Data	1	2
Weight of tube and contents before heating	g	g
Weight of tube and residue after heating	g	g
Weight of empty tube[1]	g	g
Temperature of the oxygen	°C	°C
Volume of oxygen collected	ml	ml
Barometer reading	Torr	Torr
Aqueous vapor pressure at temperature of gas	Torr	Torr

Calculations		1	2
Weight of oxygen		g	g
Temperature, absolute		K	K
Barometric pressure		Torr	Torr
Pressure of oxygen alone, in the flask		Torr	Torr
Volume of your oxygen at standard conditions		ml	ml
The molar volume of oxygen		l	l
Percentage error		%	%
Moles of oxygen[1]		moles	moles
Moles of $KClO_3$ decomposed[1]		moles	moles
Weight of $KClO_3$ in sample[1]		g	g
Percent of $KClO_3$ in sample[1]		%	%

---

[1]Leave these data and calculation items blank if you are determining the molar volume of oxygen only, and do not have an unknown sample for analysis.

EXERCISES AND PROBLEMS

1. Why must the end of the delivery tube C (Figure 11-2) remain under water while the apparatus is cooling, and why must you wait until the test tube is cool before adjusting the water levels and measuring the volume of water in the beaker?

_____

2. What objection is there to weighing an object, such as the tube of $KClO_3$, on the balance while the tube is still warm?

_____

3. Calculate the percent of uncertainty of your data that is due to the balance used, if you can weigh the $KClO_3$ tube to a precision of 0.003 g (two weighings, total uncertainty 0.006 g), and if the loss in weight (weight of oxygen) was 0.925 g. Label and give units.

_____

4. A pilot balloon containing 300 liters of helium gas at 730 Torr and 27°C rises to a height of 10,000 ft, where the pressure is 500 Torr and the temperature is −33°C. To what volume will the helium expand?

_____

5. Dry air is 78.0% N_2 and 21.0% O_2. What is the partial pressure of each of these gases when the atmospheric pressure is 745 Torr?

N_2————————————

O_2————————————

6. A mixture of $KClO_3$ and KCl, weighing 62.5 g, was heated enough to drive off all the oxygen, yielding 7.64 liters of $O_2(g)$ at standard conditions.

   (a) Write the equation for the reaction.

   _____

   (b) How many moles of O_2 were liberated?

   _____

   (c) How many moles of $KClO_3$ were present?

   _____

   (d) How many grams of $KClO_3$ were present?

   _____

   (e) What was the percent of $KClO_3$ in the mixture?

   _____

7. All real gases deviate to some extent from the behavior of perfect gases. At standard conditions, the density of $O_2(g)$ is 0.0014290 g/ml, that of $H_2(g)$ is 0.00008988 g/ml, and that of $CO_2(g)$ is 0.0019769 g/ml.

   (a) Using these values and the exact atomic weights, calculate the molar volume of each of these, in milliliters, to five significant figures.

   O_2_____

   H_2_____

   CO_2_____

   (b) Correlate the values of these three gases with the molecular volume of a perfect gas.

   Which gas deviates least?   _____

   Which gas deviates most?   _____

   Can you suggest reasons for this behavior?

THE MOLECULAR WEIGHT OF A GAS

PRE-STUDY

Review the relationships between Avogadro's law, the molar volume, and the general gas law equation as developed in the last experiment. From these considerations we may state that *the molecular weight of any gas is the weight in grams of 22.4 liters of that gas at standard conditions*. Likewise, the molecular weight of a volatile liquid or solid, whose vapor may be readily studied at known conditions and then converted to standard conditions, can be determined by this same method.

In this experiment we shall make direct measurements of the weight and corresponding volume of a given gas at the laboratory temperature and pressure. We shall first obtain the weight of the gas container, a 200- to 300-ml flask, filled with dry air. From the known density of air and the volume of the flask, the weight of the contained air, and hence the weight of the empty (evacuated) flask, can then be calculated. The difference between this weight, and the weight when the flask is filled with the gas under consideration, will of course give the weight of the volume of the gas contained in the flask.

Various gases, such as carbon dioxide (CO_2), sulfur dioxide (SO_2), ammonia (NH_3), nitrogen (N_2), nitrous oxide (N_2O), methane (CH_4), or acetylene (C_2H_2), may be used in the experiment. However, unless an adequate number of well-ventilated hoods are available, the selection of gases should be restricted to the less noxious ones.

When tanks of the compressed gas are available, they are convenient; but you will learn more chemistry if you prepare your own sample of gas. Before you begin, you should study, in a general chemistry text, the properties and method of preparation of the gas to be used. In the following discussion on experimental procedure, directions will be given for the preparation of carbon dioxide. Sulfur dioxide may be prepared in a similar apparatus by substituting sodium bisulfite ($NaHSO_3$) in place of the marble chips, and warming the mixture slightly to drive the more soluble sulfur dioxide gas out of solution. The gas SO_2 is a particularly hazardous one and should be handled with considerable caution.

EXPERIMENTAL PROCEDURE

Special Supplies: Analytical weights, thistle tube, calcium chloride drying tube, 110°C thermometer.

Chemicals: $CaCO_3(s)$, calcium carbonate (marble chips) for CO_2, $NaHSO_3(s)$ for SO_2, 6 F HCl.

The apparatus to be used is sketched in Figure 12-1. First clean and thoroughly dry a 125-ml flask (C), fitted with a one-hole rubber stopper and a glass tube extending very nearly to the bottom of the flask. This tube is connected by a section of rubber tubing to a drying tube B filled with 4-mesh calcium chloride, which is protected at each end by a loose plug of cotton. Leave the stopper in C loose in order to permit gas to escape, and connect the drying tube to the compressed air supply. Pass a very gentle stream of dried air through C for at least five minutes. If C has been warmed previously, and the air passed through while it is cooling, more complete drying will be assured. (If compressed air is not available, a two-hole stopper, with a short exit tube, may be used in C, and this connected to a water aspirator, so that air will be sucked through B and C. If you do use this type of stopper, you should plug the second hole when making all weighings, to prevent undue diffusion of CO_2.)

In the meantime, prepare the carbon dioxide generator A. Use a 200-ml Erlenmeyer flask that has a thistle tube[1] reaching almost to the bottom and that has a short, right-angle exit tube, to be connected to the calcium chloride tube when ready to generate the gas. Place about 30 g of marble chips ($CaCO_3$) in the generator, and add about 10 ml of water to cover the end of the thistle tube. Do not add acid until you are ready to generate the carbon dioxide.

Disconnect the rubber tube from the flask C, and weigh this flask, including the stopper and glass tube, to an accuracy of at least 0.001 g. After this

---

[1]If thistle tubes are not available, you may substitute a long-stemmed funnel; or you may simply use a one-hole stopper for the right-angle tubing, remove the stopper to add the acid, and quickly restopper the flask.

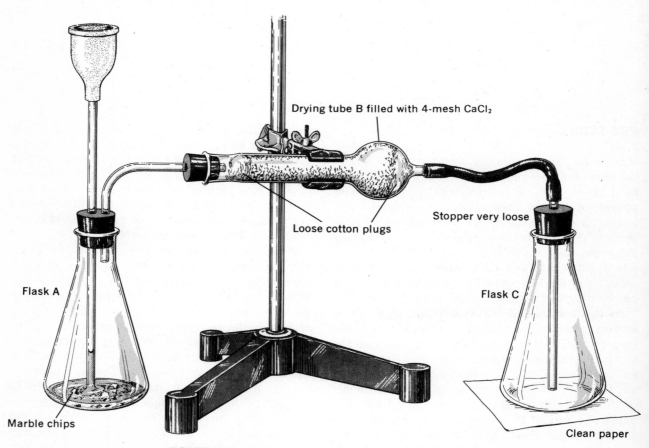

Drying tube B filled with 4-mesh $CaCl_2$

Loose cotton plugs

Stopper very loose

Flask A

Flask C

Marble chips

Clean paper

FIGURE 12-1
Apparatus for determining the molecular weight of a gas.

weighing, avoid all unnecessary handling of the flask. Hold it by the rim, and reattach it to the drying tube and CO_2 generator, placing it on a clean dry square of paper to avoid contamination by the table. Be sure the stopper is left loose so the displaced air and excess gas can escape between it and the flask. When ready add dilute hydrochloric acid (HCl) through the thistle tube, a little at a time as needed, so as to maintain a gentle, but fairly rapid, evolution of $CO_2(g)$. (You may momentarily tighten the stopper in C and note the rate at which liquid backs up in the thistle tube, in order to estimate the rate at which gas is being generated.)

Let the generator run for at least 20 minutes to displace all the air in C by CO_2 gas. Finally, detach the rubber tube from C and push in the stopper firmly. Without undue handling, at once weigh the flask and contents. It is necessary to weigh at once to avoid undue diffusion of air into C through the glass tube. Again connect C to the drying tube, gently release the stopper in C, and pass CO_2 through it for an additional 10–15 minutes. Weigh the flask C as before. The two weights should agree if the flask was filled with CO_2 the first time.

Take the temperature of the gas with a thermometer placed in the gas shortly after the last weighing, and obtain the barometer reading for the day. To measure the exact volume of the flask to the bottom of the stopper, fill the flask with water, replace the stopper and glass tube, and wipe off excess water. Weigh the full flask on the platform balance to the nearest gram. (Because a weight as heavy as this flask might injure an analytical balance, the more rugged platform balance is used. Note also that great accuracy is not required for this weight: why not?) How does the weight of water enable you to calculate its volume? Look up in a reference source[2] the density of dry air at your laboratory conditions. Enter all data and calculations on your report sheet.

From the volume of CO_2, calculated to standard conditions, and the weight of this CO_2, calculate its molecular weight (the weight in grams of the molar volume).

---

[2]See, for example, *Lange's Handbook of Chemistry*, McGraw-Hill, New York, 11th ed., 1973; or *Handbook of Chemistry and Physics*, published annually by the Chemical Rubber Company, Cleveland.

The Molecular Weight of a Gas

NAME		
SECTION		LOCKER
INSTRUCTOR		DATE

DATA AND CALCULATIONS

Data	1	2
Weight of flask and stopper, filled with CO_2	g	g
Weight of flask and stopper, filled with water	g	g
Weight of flask and stopper, filled with air	g	g
Temperature of flask	°C	°C
Barometer reading	Torr	Torr
Density of dry air at flask temperature and pressure Reference:	$\frac{g}{ml}$	$\frac{g}{ml}$

Calculations	1	2
Temperature, absolute	K	K
Barometer reading	Torr	Torr
Volume of flask[1]	ml	ml
Volume of carbon dioxide at standard conditions	ml	ml
Weight of air in flask at the start	g	g
Weight of empty flask and stopper	g	g
Weight of carbon dioxide contained in the flask	g	g
Molecular weight of CO_2 (from your data)	$\frac{g}{mole}$	$\frac{g}{mole}$
Percentage error	%	%

[1]Why can we neglect the weight of the air in the flask when obtaining the weight of the water (from which we calculate the volume of the flask), but cannot neglect it when calculating the weight of the carbon dioxide? Explain. If you are unable to answer this question, discuss it with your instructor before leaving the laboratory.

PROBLEMS

1. Write the equation for the reaction taking place in the preparation of carbon dioxide:

_____

2. How many moles are there in the 30.0 g of $CaCO_3$ used?

_____mole

_____

3. How many liters of carbon dioxide could be generated at standard conditions by this weight of $CaCO_3$?

_____liters

_____

4. What volume would this amount of carbon dioxide occupy at ordinary laboratory conditions? (Assume 25°C and 710 Torr.)

_____liters

_____

5. What is the weight of a liter of each of the following gases at standard conditions? (Use no other data than atomic weights and the molar volume.)

Fluorine, $F_2(g)$ _____g

Propane, $C_3H_8(g)$ _____g

Ammonia, $NH_3(g)$ _____g

_____

6. 360 g of water is vaporized at 100°C and 735 Torr pressure. What is the volume of the steam formed? (*Hint:* First convert the weight to moles.)

_____liters

_____

7. 1.60 g of an impure sodium acetate sample was mixed with excess $NaOH(s)$ and heated in a test tube to decompose it completely according to the equation

$$NaC_2H_3O_2(s) + NaOH(s) \rightarrow Na_2CO_3(s) + CH_4(g)$$

The evolved methane gas, $CH_4(g)$, was collected over water at 27°C and 747 torr pressure, resulting in a volume of 320 ml of gas collected. Calculate the percent of pure anhydrous sodium acetate, NaAc, in the sample. (*Suggestion:* Save arithmetic by using $PV = nRT$.)

_____%

_____

THE REACTIVITY OF METALS WITH HCL

PRE-STUDY

The Concept of Equivalent Weight

The term equivalent weight, as the adjective "equivalent" implies, is used to designate the relative amounts of substances that are chemically equivalent—that is, of those substances which just react with or replace one another in chemical reactions. In some experiments this term is defined as the weight of metal that reacts with $\frac{1}{4}$ mole (8.000 g) of oxygen (O_2) gas. In this experiment, we shall consider the equivalent weight of a metal as that weight of it which reacts with hydrogen ion to produce 1 mole of hydrogen atoms or $\frac{1}{2}$ mole of hydrogen (H_2) gas (1.008 g). We will produce this type of reaction by dissolving the weighed metal in acid and measuring the volume of evolved hydrogen gas. A typical example is

$$Ca(s) + 2HCl(aq) \rightarrow CaCl_2(aq) + H_2(g)$$

Note that for each mole of calcium that is dissolved to Ca^{2+} ion (valence +2), 1 mole of hydrogen gas is formed. The equivalent weight of calcium in this reaction is therefore half its atomic weight. The equivalent weight is always either the atomic weight (for monovalent elements) or a simple fraction of the atomic weight, depending on the change in ionic valence of the metal produced. A general defining equation is therefore

$$\text{Equivalent weight} = \frac{\text{atomic weight}}{\text{change in ionic valence}} \quad (1)$$

Some elements, such as iron, can have more than one ionic valence in their compounds, and therefore can have *more than one possible equivalent weight*. Consider the following possible reactions:

$$Fe(s) + 2HCl(aq) \rightarrow FeCl_2(aq) + H_2(g) \quad (2)$$

$$2Fe(s) + 3Cl_2(g) \rightarrow 2FeCl_3(aq) \quad (3)$$

$$2FeCl_2(aq) + Cl_2(g) \rightarrow 2FeCl_3(aq) \quad (4)$$

In the first reaction, the change in ionic valence is +2, in the second it is +3, and in the third it is +1. The corresponding equivalent weights of iron, respectively, are 27.92 ($\frac{1}{2}$ the atomic weight), 18.62 ($\frac{1}{3}$ the atomic weight), and 55.85 (the atomic weight). Thus it is always necessary to specify the applicable reaction whenever an equivalent weight is given.

An alternative view of the definition of equivalent weight is that it is that weight of substance which releases or accepts one mole of electrons. The defining equation then becomes

Equivalent weight =

$$\frac{\text{atomic weight}}{\text{moles of electrons transferred}} \quad (5)$$

Thus, in equation (2), iron changes from an oxidation state (see Study Assignment A) of 0 to oxidation state +2. It loses 2 electrons per mole Fe in so doing. The equivalent weight of Fe *in this reaction* therefore is

$$\frac{55.85}{2} = 27.92$$

This is the same result obtained by means of defining equation (1). Verify that the equivalent weights for reactions (3) and (4) are 18.62 and 55.85, using defining equation (5).

Equation (5) emphasizes that all of the reactions studied in this experiment are oxidation-reduction reactions, reactions in which a transfer of electrons has caused changes in oxidation states. This concept and these reactions are studied in greater detail in Section VIII.

The use of the term equivalent weight is not restricted to elements — either metals or nonmetals. In calculations, it is often convenient to designate the equivalent weights of various types of compounds in such a way that *one equivalent always reacts with one equivalent*. The concept of equivalent weight, and the way in which it can be employed to simplify the relationships and calculations for various types of reactions, will be considered in detail where applicable — particularly in Experiment 23, in which acid-base equivalents are treated.

EXPERIMENTAL PROCEDURE

Special Supplies: Thermometer, pieces of metals (cut to size or issued as unknowns), 50-ml buret (for Method I only).

NOTE: The analysis samples for this experiment may be issued (1) as unknown metals, to calculate the equivalent weight, (2) as preweighed samples,[1] to calculate and report the sample weight from the known equivalent weight, or (3) as Al-Zn alloys of different compositions, to calculate the percent composition from the known equivalent weights. Prepare the legends in your report form accordingly, *before beginning the experiment.*

Your instructor will designate which of the preceding analyses and calculations you will do, and also which of the following alternate procedures you will follow. Method I is faster, but limits you to smaller samples, to a fixed acid concentration, and to room temperature. Method II permits larger samples and the control of both temperature and acid concentration during the reaction.

Obtain two samples of the metal to be used. Weigh these precisely on the analytical balance, at the same time noting that the weights do not exceed the maximum permitted, in order not to generate more hydrogen than the capacity of your apparatus in either Method I or Method II. (For all laboratories except those at high elevation, maximum weights are: for Method I, with 50-ml buret, 0.12 g Zn, 0.032 g Al, 0.042 g Mg, 0.10 g Mn; for Method II, with 500-ml flask, 1.10 g Zn, 0.40 g Mg, 0.30 g Al, 0.90 g Mn, 0.90 g Fe, 1.90 g Sn, 1.90 g Cd).

Method 1

Compress the weighed samples into compact bundles, and wrap each sample in all directions with about 20 cm of fine copper wire, forming a small basket or cage, leaving 5 cm of the wire straight as a handle. This confines the particles as the metal dissolves, and also speeds the reaction.[2]

Obtain and clean a 50-ml buret. Next measure the uncalibrated volume of the buret between the stopcock and the 50-ml graduation. Measure by filling the buret with water and draining the buret

---

[1] *To the instructor:* Samples may be preweighed by a stock assistant on a rapid 1-pan balance, or if in wire or ribbon form cut to exact length to give a known weight, and individually coded. The student may then report the corrected volume of hydrogen as a preliminary check on his work, and finally the calculated equivalent weight, or the weight of the sample from the known equivalent weight.

[2] As the more active metal dissolves, it gives up electrons which move easily to the less reactive copper, where they react with hydrogen ions of the acid to form hydrogen gas. Note that the bubbles of gas form on the copper wire, and thus keep a larger surface of active metal exposed to the acid.

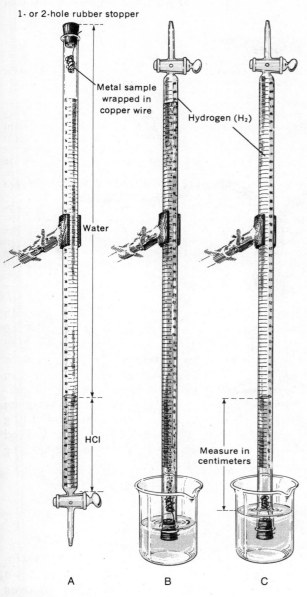

1- or 2-hole rubber stopper

Metal sample
wrapped in
copper wire

Hydrogen (H₂)

Water

HCl

Measure in
centimeters

A   B   C

FIGURE 13-1
The appropriate volume of concentrated (12 *F*) HCl is
added to the buret, and water is then layered on top of it
until the buret is completely filled, as in (A). Inversion of
the buret in a beaker of water begins the reaction (B),
which continues until all of the metal is gone and the
buret is nearly full of hydrogen gas, as shown in (C).

through the stopcock until the liquid level falls
exactly to the 50-ml mark. Then use a 10-ml graduated cylinder to measure the volume delivered
when the water level is lowered to the top of the
stopcock. Then pour into the buret the required
amount of concentrated hydrochloric acid. Use a
funnel for this addition, being careful not to allow
the acid to touch your skin or clothing. Because of
differences in the activity of the metals used, it is

necessary to vary the amount of acid. For magnesium, use about 3 ml; for aluminum or zinc, about
20 ml; for manganese, about 7 ml. Fill the buret
completely with water, slowly and carefully to
avoid undue mixing of the acid. Insert the metal
sample about 4 cm into the buret, and clamp it
there by the copper wire handle, using a 1- or 2-hole
rubber stopper. Make certain no air is entrapped in
the buret. Cover the stopper hole(s) with your finger, and invert the buret (Figure 13-1) in a 400-ml
beaker partly filled with water. The acid, being
more dense, quickly sinks and diffuses down the
buret and reacts with the metal. As the H₂ is generated, it collects at the top of the buret expelling
the HCl-water solution out the hole in the stopper
at the bottom. (*Caution·* If the reaction is too rapid
and the metal too close to the end of the buret,
small bubbles of hydrogen may escape from the
buret along with the acid solution as it is expelled.
If so, repeat the experiment, using less acid.)

After complete solution of the metal, let the apparatus cool to room temperature, since heat is
generated by the reaction. Free any hydrogen
bubbles adhering to the sides of the vessel or copper
wire by tapping the apparatus. Measure the volume
of gas liberated[3] and — without changing the position of the buret — measure the difference in height
of the two water levels with a metric rule, and calculate the equivalent pressure in millimeters of
mercury (torr). Take the temperature of the gas by
holding a thermometer in contact with the side of
the buret. Raise the buret up out of the HCl solution in the beaker and allow the remainder of the
solution to drain down out of the buret. Allow the
H₂ to escape into the atmosphere, flush the HCl
solution down the drain with plenty of water, and
discard the Cu wire in the waste basket.

Obtain the barometer reading for the day. Repeat the determination with your second sample.

Method II

Set up the apparatus as sketched in Figure 13-2,
utilizing a *500-ml flask* to contain the evolved hydrogen. The exit tube C from the test tube, connected to the flask E, must not extend below either
rubber stopper (so that gas will not be trapped).
The longer glass tube in the test tube should extend
nearly to the bottom, and should be constricted to

---

[3]If a 50-ml buret is used, the volume of gas liberated equals
(50 minus the final buret reading) plus the volume of the uncalibrated portion of the buret.

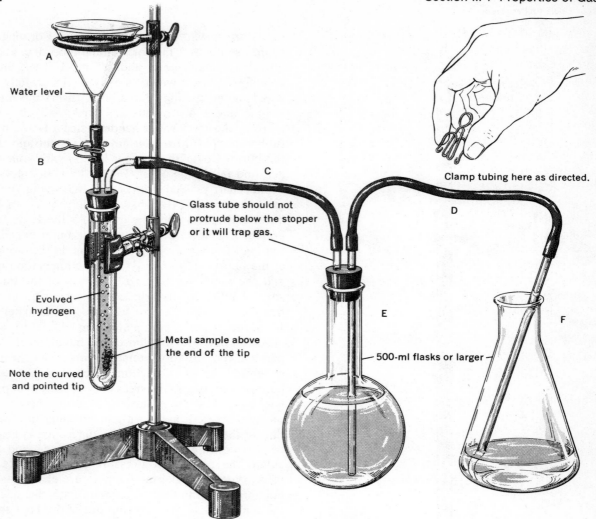

A

Water level

B

C

Glass tube should not
protrude below the stopper
or it will trap gas.

Clamp tubing here as directed.

D

Evolved
hydrogen

E

500-ml flasks or larger

Metal sample above
the end of the tip

F

Note the curved
and pointed tip

FIGURE 13-2
An alternate apparatus for the determination of the equivalent weight of a metal.

a small capillary and bent as illustrated. The flask
E is filled with water, and the flask F is partially
filled. The siphon tube D, extending to the bottom
of both flasks, is also filled with water.

Place the first carefully weighed sample of metal
in the 15-cm test tube, as indicated. A fine copper
wire, wrapped about the sample in all directions,
like a cage, as in Method I, may be of some help in
increasing the rate of reaction and in confining small
bits of metal as it dissolves, although this is not
entirely necessary. With the stopper in flask E
loosened to permit air to escape, pour water into
the funnel to fill completely the test tube and tubes
B and C, and close clamp B. If flask E is not completely filled with water, raise flask F so that water
siphons back into it, then push in the stopper in
E tightly.

When all air bubbles have been thus removed
from the apparatus, release clamps B and D just
enough to permit the water level in the funnel to
fall just to the stem top, *but no further;* then close
the clamp B. Now empty and drain flask F, but do
not dry it, leaving tube D filled with water and
clamp D open. When all is ready, add exactly 25.0
ml of concentrated hydrochloric acid into the funnel. Release clamp B momentarily to permit a
little acid to flow into the test tube and react with
the metal at a moderate rate.

A volume of water equal to the volume of hydrogen generated will siphon from E to F. You can
control the rate of reaction by regulating the amount
of acid you add, and by heating the tube gently
(for less reactive metals) by a 1-inch Bunsen flame

applied intermittently as needed. When all the metal has dissolved and the reaction mixture has cooled to room temperature, release clamp B to permit the acid level to fall to the funnel stem, *but no further*. Carefully measure 25.0 ml of water, add this to the funnel, and again release the clamp to let the level fall exactly to the funnel stem; then close the clamp tightly. All gas should now be displaced from the test tube and connecting tubes into flask E. If not, repeat the addition of a measured amount of water. Adjust the levels in flasks E and F by raising or lowering one of them until they are even (avoid warming the gas in E with your hands on the flask), then close clamp D tightly. Obtain the temperature of the hydrogen by removing the stopper enough to place a thermometer in flask E. Measure the volume of water in flask F by pouring it into a 500-ml graduated cylinder. This volume, minus the volumes of acid and water added to the funnel, will be the volume of the hydrogen generated. Obtain the barometer reading. Repeat the determination with your second sample.

Calculation of the Experiment

Recall that the standard molar volume of any gas is 22.4 liters. The volume of $\frac{1}{2}$ mole of hydrogen gas (1.008 g) at standard conditions is thus 11.2 liters, or 11,200 ml. *The equivalent weight of your metal sample therefore is that weight which will produce 11,200 ml of hydrogen gas at standard conditions from an acid.* Also note that you must consider the difference in liquid levels (Method I) and the vapor pressure of water in arriving at the correct hydrogen gas pressure.

Calculate and report your sample according to the unknown in your particular analysis: (1) the equivalent weight of the metal, (2) the weight of a preweighed sample, or (3) the percent composition of an Al-Zn alloy. For this latter calculation, note that

(a) wt sample = wt Al + wt Zn

and

(b) vol H_2 (at STP) =

$$\left(\frac{\text{wt Al}}{\text{equiv wt Al}} + \frac{\text{wt Zn}}{\text{equiv wt Zn}} \right) 11{,}200 \, \frac{\text{ml}}{\text{equiv}}$$

Report also the percent error in your analysis.

Bibliography

Masterton, W. L. "Analysis of an Aluminum-Zinc Alloy," *J. Chem. Educ.* **38**, 558 (1961).

The Reactivity of
Metals with HCL

NAME

SECTION   LOCKER

INSTRUCTOR   DATE

DATA AND CALCULATIONS

*Unknown sample number*_____

NOTE: Before beginning the experiment, complete the legends in the blank spaces provided for the entry of the required calculated values, to show whether you are reporting (1) the equivalent weight of the sample, (2) the weight of a preweighed sample, or (3) the percent composition of an Al-Zn alloy.

Data	1	2
Weight of sample (except for preweighed sample)	g	g
Volume of uncalibrated portion of buret	ml	ml
Final buret reading	ml	ml
Volume of hydrogen	ml	ml
Temperature of hydrogen	°C	°C
Barometric reading	Torr	Torr
Difference in water levels inside and outside tube (Method I)	mm H_2O	mm H_2O
Aqueous vapor pressure at temperature of hydrogen	Torr	Torr

Calculations		1	2
Barometric pressure		Torr	Torr
Mercury equivalent of difference of water levels (Method I)		Torr	Torr
Pressure of H_2, after correction for difference in H_2O levels and for vapor pressure		Torr	Torr
Temperature, absolute		K	K
Volume, dry H_2 at standard conditions		ml	ml
Percentage error		%	%

PROBLEMS

1. The atomic weight of tin is 118.69. What is its equivalent weight in the following reactions:

   (a) When it reacts with chlorine to form $SnCl_4$? _____

   (b) When it reacts with hydrochloric acid to form $SnCl_2$? _____

2. What volume of hydrogen gas, at standard conditions, will be liberated by the reaction of sulfuric acid with the following substances? Relate your results to the position of the elements in the periodic table and to their oxidation numbers in compounds.

   (a) 0.0230 g Na?

   _____

   (b) 0.0243 g Mg?

   _____

   (c) 0.0270 g Al?

   _____

3. A 1.70-g sample of a certain metal reacts with acid to liberate 380 ml of hydrogen gas, when collected over water at 20°C and 720 torr pressure. What is the equivalent weight of the metal?

   _____

4. Calculate the maximum weight of calcium metal that should be used in an equivalent weight determination, with a 100-ml gas measuring tube for collection. Assume 80.0 ml of hydrogen gas at standard conditions for purposes of calculation.

   _____

DESCRIPTIVE INORGANIC CHEMISTRY

WRITING EQUATIONS FOR IONIC REACTIONS

THE PRINCIPLES OF IONIC REACTIONS

Application of the principles of this Study Assignment will enable you to write ionic equations for a large number of reactions, such as those you will study in Experiment 14. Ions in solution will combine to form the following:

weakly ionized molecules, such as H_2O, H_2CO_3, $HC_2H_3O_2$, $HgCl_2$;[1]

slightly soluble compounds, such as $AgCl(s)$, $BaSO_4(s)$, $Fe(OH)_3(s)$;

complex ions, such as $Ag(NH_3)_2{}^+$, $Cu(NH_3)_4{}^{2+}$, $Zn(OH)_3{}^-$, $Cd(NH_3)_4{}^{2+}$.

By expanding your knowledge of strong and weak acids, bases, and salts (Table 9, Appendix C and of slightly soluble compounds (Table 12, Appendix C), you will have at your fingertips the information to write hundreds of equations.

---

[1]Although mercuric chloride is a salt, it is only slightly dissociated in aqueous solution so that a $HgCl_2$ solution contains mostly molecular $HgCl_2$ rather than Hg^{2+} and Cl^- ions.

Three examples will illustrate the reasoning underlying ionic reactions.

1. NaOH and Fe(NO₃)₃ solutions are mixed, giving a red-brown precipitate. Both substances are readily soluble (Appendix C, Table 11) and highly ionized; therefore, before mixing, Na^+, OH^-, Fe^{3+}, and $NO_3{}^-$ are present. Na^+ and $NO_3{}^-$ are called "spectator" ions and do not react. $NaNO_3$ is readily soluble (Appendix C, Table 11, Rule 1) and nearly all soluble salts are completely ionized. Fe^{3+} and OH^-, however, cannot remain in solution, because $Fe(OH)_3$ is a virtually insoluble red-brown precipitate (Appendix C, Table 11, Rule 6). The total ionic equation is therefore

$$Fe^{3+} + 3NO_3{}^- + 3Na^+ + 3OH^- \rightarrow$$
$$Fe(OH)_3(s) + 3Na^+ + 3NO_3{}^-$$

Na^+ and $NO_3{}^-$ which do not react, may be eliminated to give the net ionic equation

$$Fe^{3+} + 3OH^- \rightarrow Fe(OH)_3(s)$$

2. NH_3 and $Fe(NO_3)_3$ solutions are mixed, giving a red-brown precipitate. The only difference here from example (1) is that NH_3 is a fairly weak base, consisting largely of NH_3 molecules and some NH_4^+ and OH^- ions. The precipitation of $Fe(OH)_3$ reduces the OH^- concentration, causing more NH_3 to react with H_2O and therefore increasing the NH_4^+ concentration.[2] The overall effect is expressed by the total ionic equation

$$Fe^{3+} + 3NO_3^- + 3NH_3 + 3H_2O \rightarrow$$
$$Fe(OH)_3(s) + 3NH_4^+ + 3NO_3^-$$

Omitting the unreactive NO_3^-, the net ionic equation is

$$Fe^{3+} + 3NH_3 + 3H_2O \rightarrow Fe(OH)_3(s) + 3NH_4^+$$

[2]This involves the principles of chemical equilibrium, discussed in Study Assignment E.

3. $Cu(NO_3)_2$ and an aqueous solution of NH_3 are mixed, giving first a greenish gelatinous precipitate, owing to the formation of $Cu(OH)_2$:
First, NH_3 reacts with water to give OH^-

$$NH_3 + HOH \rightarrow NH_4^+ + OH^-$$

Then OH^- reacts with Cu^{2+} to give $Cu(OH)_2(s)$.

$$Cu^{2+} + 2NO_3^- + 2OH^- \rightarrow Cu(OH)_2(s) + 2NO_3^-$$

Omitting the ions which do not participate, we give the net ionic reaction as

$$Cu^{2+} + 2NH_3 + 2H_2O \rightarrow Cu(OH)_2(s) + 2NH_4^+$$

If more NH_3 solution is added, the precipitate dissolves and forms an intensely colored deep blue complex, $Cu(NH_3)_4^{2+}$

$$Cu(OH)_2(s) + 4NH_3 \rightarrow Cu(NH_3)_4^{2+} + 2OH^-$$

REPORT	Writing Equations	NAME	
C	for Ionic Reactions	SECTION	LOCKER
		INSTRUCTOR	DATE

EXERCISES

NOTE: These exercises provide the practice which is absolutely essential to attain proficiency in writing net ionic equations—a proficiency which every chemist must develop. Consult Tables 9, 11, and 12, Appendix C, as needed.

1. **Recognition of Ionic or Molecular Species Present.** For each substance (the molecular formula is given), write the formula(s) of the *principal species* (molecules or ions) present in major amount in the aqueous solution it the substance is soluble; if it is only slightly soluble, use the molecular formula, followed by (s).

Examples: (1) $BaSO_4$: $BaSO_4(s)$; (2) $HC_2H_3O_2$: $HC_2H_3O_2$; (3) NaCl: Na^+, Cl^-.

(a) $NH_4C_2H_3O_2$_____

(b) $Ba(OH)_2$_____

(c) $Cr(OH)_3$_____

(d) CuS_____

(e) H_2SO_3_____

(f) $HgCl_2$_____

(g) $Al_2(SO_4)_3$_____

(h) HBr_____

(i) C_2H_5OH_____

2. **Recognition of Common Properties of Substances.** In each of the following groups, identify, by underlining, the two substances whose aqueous solutions have common properties to a marked degree.

(a) $Mg(OH)_2$, KOH, C_2H_5OH, $Ba(OH)_2$.

(b) $NH_4C_2H_3O_2$, H_2SO_4, $HC_2H_3O_2$, $NaC_2H_3O_2$.

(c) $BaSO_4$, H_2SO_4, $BaCl_2$, HNO_3.

(d) $(NH_4)_2SO_4$, NH_4Cl, AgCl, NH_3.

(e) $HgCl_2$, AgCl, KCl, HCl.

3. **Interpretation of Reactions by Net Ionic Equations.** Aqueous solutions of the following substances, or their mixtures with water if they are only slightly soluble, are mixed. Write first the *total ionic equation* to gain an understanding of all principal species present, both before and after reaction. Then condense this to the *net ionic equation*. If you predict no appreciable reaction, indicate this and state why.

(a) Ammonium sulfate and potassium hydroxide _____

(b) Barium chloride and sodium carbonate _____

(c) Nitric acid and aluminum hydroxide _____

(d) Calcium chloride and sodium nitrate _____

(e) Hydrochloric acid and magnesium acetate _____

162

(f) Zinc nitrate and
    excess sodium hydroxide

(g) Cadmium chloride and
    excess ammonia solution

(h) Ammonia solution
    and acetic acid

(i) Silver chloride
    and nitric acid

(j) Ammonia solution
    and sulfuric acid

(k) Ammonia solution
    and magnesium chloride

(l) Sodium hydrogen carbonate
    and sulfuric acid

4. **Prediction of Results.** Separate solutions of 0.01 F $Ba(C_2H_3O_2)_2$ and 0.01 F H_2SO_4 are tested for electrical conductivity. Equal volumes of these solutions are then mixed and the conductivity tested again. Predict the result, and justify your answer with an appropriate equation.

THE NOMENCLATURE OF INORGANIC COMPOUNDS

NOTE: Study Assignment A, The Language of Chemistry, introduced to you some elementary rules for naming chemical compounds. Since the chemistry of the nonmetals involves the study of compounds in a great variety of oxidation states, a more thorough treatment of nomenclature is given here.

The standards of nomenclature are established by the International Union of Pure and Applied Chemistry (IUPAC). Since this international body must coordinate different languages and conflicting views, it is not surprising that the rules reflect a compromise: as a result, American chemists occasionally employ usages not officially sanctioned by the IUPAC rules. In this Study Assignment we have followed the IUPAC rules[1] except those conflicting with current American usage. Most of the conflicts are minor, and the beginning student of chemistry is not likely to encounter any serious conflicts in applying the rules listed here.

[1]For the most recent IUPAC rules on the nomenclature of inorganic chemistry, see "Nomenclature of Inorganic Chemistry, Definitive Rules 1970," *Pure and Applied Chemistry* **29**, 1 (1971). Other IUPAC publications on nomenclature have been listed in the *Journal of Chemical Education* **50**, 341 (1973).

OXIDATION NUMBER

Oxidation number (or oxidation state) is an empirical concept defined by formal rules; it is not synonymous with the number of bonds to an atom, or with the actual electronic charge (electron density) on an atom, except in the simplest cases. The oxidation number of an element in any chemical entity is the charge that would be present on an atom of the element if the electrons in each bond to that atom were assigned to the more electronegative atom. Some examples of the oxidation numbers of elements composing various compounds follow.

COMPOUND	OXIDATION NUMBERS	
MnO_4^-	$Mn = +7,$	$O = -2$
ClO^-	$Cl = +1,$	$O = -2$
NH_4^+	$N = -3,$	$H = +1$
NF_4^+	$N = +5,$	$F = -1$
AlH_4^-	$Al = +3,$	$H = -1$

Note that hydrogen is assigned +1 in combination with nonmetals and −1 in combination with metals (metal hydrides). By convention oxygen is assigned oxidation number -2 except in peroxides or in combination with the more electronegative element, fluorine:

$$O_2F_2 \qquad O = +1, \quad F = -1$$

In the elementary state, the atoms have oxidation numbers of zero, and a bond between atoms of the same element makes no contribution to the oxidation number. Thus,

$$P_2H_4 \qquad P = -2, \quad H = +1$$

but

$$P_4 \qquad P = 0$$

Difficulties in assigning oxidation numbers may arise if the elements in a compound have similar electronegativities, such as those in NCl_3 and S_4N_4.

BINARY COMPOUNDS

Many chemical compounds are essentially binary in nature and can be regarded as combinations of ions or radicals;[2] others may be treated as such for the purposes of nomenclature. The characteristic ending -ide is used for all binary compounds whether they are covalent or ionic.

Covalent Compounds. In the chemical formulas of covalent compounds, the less electronegative element is written first, followed by the more electronegative element. In the English language, the same pattern is used in naming the compounds— one names the less electronegative element first, and then the stem of the name of the more electronegative element plus the ending -ide. The Greek prefixes *mono-, di-, tri-, tetra-, penta-*, and so on, are used to indicate the number of atoms of each element:

CO_2	carbon *dioxide*
PCl_3	phosphorus *trichloride*
PCl_5	phosphorus *pentachloride*
N_2O_4	*dinitrogen tetraoxide*
Cl_2O	*dichlorine monoxide*
$HCl(g)$	hydrogen chloride
$H_2S(g)$	hydrogen sulfide

Cations. Monoatomic cations are given the same names as their corresponding elements with no change, unless the *-ous-ic* system (to be discussed later) is used. The NH_4^+ ion, *ammonium* ion, is named as if it were a metal ion because of the salt-like properties of the ammonium salts. Examples of monoatomic cations include:

Li^+	lithium ion	Mg^{2+}	magnesium ion
Na^+	sodium ion	Al^{3+}	aluminum[3] ion

Monoatomic Anions. The names for monoatomic anions consist of the name of the element (sometimes abbreviated) and the suffix -ide. Thus

H^-	hydride	Te^{2-}	telluride
F^-	fluoride	N^{3-}	nitride
Cl^-	chloride	P^{3-}	phosphide
Br^-	bromide	As^{3-}	arsenide
I^-	iodide	Sb^{3-}	antimonide
O^{2-}	oxide	C^{4-}	carbide
S^{2-}	sulfide	Si^{4-}	silicide
Se^-	selenide	B^{3-}	boride

Polyatomic Anions. Certain very stable groups of atoms are named as if they were like ions formed from nonmetal elements, and they have the binary ending -ide:

OH^-	hydroxide	N_3^-	azide
O_2^{2-}	peroxide	CN^-	cyanide
I_3^-	triiodide	NH_2^-	amide

Salts. Salts are ionic compounds formed by reaction of an acid and a base. (Note that this is a generic term, not to be confused with our everyday usage of the term for table salt, $NaCl$). In simple salts consisting of a monoatomic cation and a monoatomic anion, one names the metal first, and then the stem of the anion plus the -ide ending:

$NaCl$	sodium chlor*ide*
CaO	calcium ox*ide*
K_2S	potassium sulf*ide*
$MgBr_2$	magnesium brom*ide*
LiH	lithium hydr*ide*
BaF_2	barium fluor*ide*

Note that the prefixes *mono-, di-, tri-*, and so on are not used with metals in Groups I_A and II_A, which do not show variable valence.

---

[2] A radical is a group of atoms that occurs repeatedly in a number of different compounds. The group may be neutral or charged

---

[3] This is the American spelling. In most other languages, including British standard English, it is spelled alumin*ium*.

Metals of Variable Valence. When the metals are those of variable valence, it is necessary to distinguish the various salts by one of the following systems:

1. The *-ous-ic* system is applied to elements that exhibit only two stable oxidation states.[4] The ending *-ous* is used to denote the lower oxidation state of the metal, and the ending *-ic* the higher oxidation state:

$FeCl_2$	ferr*ous* chloride
$FeCl_3$	ferr*ic* chloride
Hg_2O	mercur*ous* oxide
HgO	mercur*ic* oxide
$SnCl_2$	stann*ous* chloride
$SnCl_4$	stann*ic* chloride

2. In the *Stock* system, the oxidation state of the metal is designated by a Roman numeral placed in parentheses just after the name of the element. This system is especially useful if the metal has more than two oxidation states and the *-ous-ic* system is inadequate.

$FeBr_2$	iron(II) bromide
$FeBr_3$	iron(III) bromide
VCl_2	vanadium(II) chloride
VCl_3	vanadium(III) chloride
VCl_4	vanadium(IV) chloride
VCl_5	vanadium(V) chloride

3. In the *Ewens-Bassett* system, the charge of an ion is indicated by an Arabic numeral and the sign of the charge, in parentheses, immediately after the name of the ion:

$FeCl_2$	iron(2+) chloride
Hg_2Cl_2	dimercury(2+) chloride
$TlCl_3$	thallium(3+) chloride

Peroxides. The O_2^{2-} ion contains two oxygen atoms with a single covalent bond between them and is given the distinguishing name "peroxide". Care must be taken to distinguish between peroxides, such as BaO_2 (barium peroxide), H_2O_2 (hydrogen peroxide), Na_2O_2 (sodium peroxide), and normal dioxides, such as MnO_2 (manganese dioxide), TiO_2 (titanium dioxide), and SiO_2 (silicon dioxide).

Trivial Names. Some common binary compounds are designated by trivial names that have been assigned arbitrarily. Examples are H_2O (water),

----

[4]The use of the *-ous-ic* system is discouraged by the IUPAC rules, but reluctantly allowed because it is so firmly entrenched.

NH_3 (ammonia), PH_3 (phosphine), AsH_3 (arsine), and many carbon compounds, such as CH_4 (methane).

These names are so universally used that they are allowed by the IUPAC rules of nomenclature. In addition, because aqueous solutions of HF, HCl, HBr, and HI are acidic, they are often called by the following trivial names:

HF(*aq*)	hydrofluoric acid
HCl(*aq*)	hydrochloric acid
HBr(*aq*)	hydrobromic acid
HI(*aq*)	hydriodic acid

TERNARY COMPOUNDS

Oxoacids and Salts. The oxides of the nonmetals react with water to form hydroxides that are acidic. In some cases there may be a series of oxoacids, each one containing the nonmetal in a different oxidation state. In order that these acids and their respective salts can be distinguished from one another, characteristic prefixes and suffixes are used, as illustrated below for the oxoacids containing chlorine and their salts:

OXIDATION STATE OF Cl	ACID	NAME
+1	HClO	*hypo*chlor*ous* acid
+3	$HClO_2$	chlor*ous* acid
+5	$HClO_3$	chlor*ic* acid
+7	$HClO_4$	*per*chlor*ic* acid

	SALT	NAME
+1	NaClO	sodium *hypo*chlor*ite*
+3	$NaClO_2$	sodium chlor*ite*
+5	$NaClO_3$	sodium chlor*ate*
+7	$NaClO_4$	sodium *per*chlor*ate*

This nomenclature for the oxoacids and their salts is highly traditional, and its roots can be traced back to Lavoisier (1743–1794). Unfortunately, the passage of time has revealed the great diversity of the oxoacids. Thus the prevailing system that has evolved has many limitations, not the least of which is the major feat of memory required of the student in order to distinguish the various oxoacids.

The prefix *hypo-*, Greek for "under," is used to denote the lowest oxidation state of the nonmetal, with the characteristic ending *-ous*. The prefix *per-*, from the Greek *hyper*, meaning "above," is used to denote the highest oxidation state of the nonmetal. Note that for acids whose names end in *-ous*, the name of the corresponding salt ends in

-ite; and for acids whose names end in -ic, the name of the salt ends in -ate.

Recognizing the -ic Acid.

Except in the Group VII oxoacids, the oxoacid with the nonmetal in its highest oxidation state is given the -ic ending. The following are examples of -ic acids, and their anions, listed in order of their periodic table groups.

III	IV	V	VI	VII
H_3BO_3	H_2CO_3	HNO_3	H_2SO_4	$HClO_3$
	H_2SiO_3	H_3PO_4	H_2SeO_4	$HBrO_3$
		H_3AsO_4	H_6TeO_6	HIO_3

	ACID		ANION
H_3BO_3	boric acid	BO_3^{3-}	borate
H_2CO_3	carbonic acid	CO_3^{2-}	carbonate
H_2SiO_3	silicic acid	SiO_3^{2-}	silicate
HNO_3	nitric acid	NO_3^-	nitrate
H_3PO_4	phosphoric acid	PO_4^{3-}	phosphate
H_3AsO_4	arsenic acid	AsO_4^{3-}	arsenate
H_2SO_4	sulfuric acid	SO_4^{2-}	sulfate
H_2SeO_4	selenic acid	SeO_4^{2-}	selenate
H_6TeO_6	telluric acid	TeO_6^{6-}	tellurate
$HClO_3$	chloric acid	ClO_3^-	chlorate
$HBrO_3$	bromic acid	BrO_3^-	bromate
HIO_3	iodic acid	IO_3^-	iodate

Note that in Groups III, IV, V, and VI, the non-metal has the oxidation number corresponding to the group number in each of the oxoacids with the -ic ending. It is necessary to use the per- prefix for the Group VII oxoacids in their highest oxidation states because names must be provided for Group VII oxoacids with four different oxidation states.

The acids ending in -ous usually have an oxidation state of two less than the -ic acid. Shown below are the names of some of the -ous acids and their corresponding anions:

	ACID		ANION
HNO_2	nitrous acid	NO_2^-	nitrite
H_3AsO_3	arsenious acid	AsO_3^{3-}	arsenite
H_2SO_3	sulfurous acid	SO_3^{2-}	sulfite
$HBrO_2$	bromous acid	BrO_2^-	bromite

Ortho, Meta, Di-Acids.

The prefixes ortho- and meta- have been used to distinguish acids differing in the "content of water." The prefix ortho- applies to the common acid (from the Greek word meaning "regular"). The acid with one molecule of water less than the ortho acid is a meta acid. The acid whose formula may be derived by removing one molecule of water from two molecules of the ortho acid has the prefix di-.[5] These prefixes may be applied to either -ous or -ic acids.

Some anhydrides react with water to produce acids (ortho), which may be dehydrated to form di or meta acids, such as in the following reactions:

$P_4O_{10} + 6H_2O \rightarrow 4H_3PO_4$	orthophosphoric acid
$2H_3PO_4 \rightarrow H_2O + H_4P_2O_7$	diphosphoric acid
	(pyrophosphoric acid)
$H_3PO_4 \rightarrow H_2O + HPO_3$	metaphosphoric acid
H_3BO_3	orthoboric acid
HBO_2	metaboric acid
H_3PO_3	orthophosphorous acid
HPO_2	metaphosphorous acid
H_4SiO_4	orthosilicic acid
H_2SiO_3	metasilicic acid
H_2SO_4	(ortho)sulfuric acid
$H_2S_2O_7$	disulfuric acid
	(pyrosulfuric)

Salts of Polyprotic Acids.

When salts of polyprotic acids, such as H_2S, H_2SO_4, or H_3PO_4 are formed, one or more of the hydrogen ions may be replaced by metal ions. Several systems of nomenclature have been used in the past to differentiate between these salts. In the examples below, the first name given is the preferred name specified by IUPAC rules.

NaH_2PO_4	sodium dihydrogen phosphate[6]
	monosodium phosphate
Na_2HPO_4	disodium hydrogen phosphate
	disodium phosphate
Na_3PO_4	trisodium phosphate
NaHS	sodium hydrogen sulfide
	sodium bisulfide
Na_2S	sodium sulfide[7]
$NaHSO_4$	sodium hydrogen sulfate,
	sodium bisulfate
Na_2SO_4	sodium sulfate[7]
$NaHCO_3$	sodium hydrogen carbonate,
	sodium bicarbonate

---

[5]The prefix pyro- was formerly used.

[6]IUPAC rules recommend combining hydrogen and the name of the anion into one word, e.g., sodium dihydrogenphosphate. In American usage they are separated into two words.

[7]The di- and tri- prefixes are often omitted for Group I or Group II ions that have only one stable oxidation state.

COORDINATION COMPOUNDS

The phenomenon of complex formation is a very general one. Ions dissolved in a polar solvent attract a cluster of solvent molecules about them. Here the force of attraction is primarily that between a charged ion and the dipole created by a nonuniform electron density distribution in the polar solvent molecule. Other bonding forces may be more nearly electrostatic—such as those occurring in the formation of a complex, such as $FeF_6{}^{3-}$, in which negatively charged fluoride ions are attracted to a positively charged Fe^{3+} ion; or those occurring in a more covalent type of bonding in which a metal atom (or ion) with vacant orbitals shares the electrons donated by another atom (generally a nonmetal). The transition metal ions, because of their partially filled d electron orbitals, form numerous coordination compounds. If the product of the interaction is an ion in solution, e.g., $Co(NH_3)_6{}^{3+}$, it is generally called a *complex ion*. The neutral salt $[Co(NH_3)_6]Cl_3$, is generally called a coordination compound, a complex compound, or simply a complex.

Before we proceed to the nomenclature of coordination compounds, it will be helpful to define a number of terms that are commonly employed in the language of coordination chemistry.

1. *Ligands:* The groups attached to the central metal atom or ion in a complex are called ligands. Ligands may be monoatomic or polyatomic ions or neutral molecules.

2. *Donor atom:* Within the ligand, the atom attached directly to the metal is called the donor atom. In the complex ion $[Ag(NH_3)_2]^+$, nitrogen is the donor atom of the ammonia ligand.

3. *Coordination number:* The coordination number of the central atom in a complex is the number of atoms that are directly linked to the central atom. The attached atoms may be charged, uncharged, or part of an ion or molecule. Crystallographers define the coordination number of an atom or ion in a lattice as the number of that atom's or ion's nearest neighbors. In the complex $[Cu(NH_3)_4]^{2+}$, the coordination number of copper is four, whereas in $[Fe(CN)_6]^{3-}$, the coordination number of Fe(III) is six. Note that the coordination number is not the same as the oxidation number (or oxidation state) of a metal atom or ion.

4. *Polydentate ligand or chelating agent:* When a molecule attaches itself to more than one coordination site of a given central metal ion to form a closed ring, it is called a polydentate (or multi-dentate) ligand or chelating agent. Ligands may be bidentate, tridentate, quadridentate, and so on, depending on the number of donor atoms in the ligand. Bidentate ligands, such as ethylenediamine $(H_2NCH_2CH_2NH_2)$, are very common. Polydentate ligands can surround a metal ion in a pincer-like arrangement like the claw of a crab; such ligands are called *chelating* agents (from the Greek word for "the claw of a crab") and the complexes are called *metal chelates*.

NOMENCLATURE OF COORDINATION COMPOUNDS

The great number of complicated and structurally intricate complexes that arise requires a very systematic means of nomenclature. The system now in use is based largely on that which was originally devised by Alfred Werner and has been refined throughout the years. The basic IUPAC rules are summarized as follows:

1. In a coordination compound that is a salt, the cation is named first and then the anion, in accordance with usual rules of nomenclature.

2. In naming the complex, whether the complex is a cation, anion, or neutral molecule, the ligands are named before the central metal atom or ion.

3. The names of anionic ligands, whether inorganic or organic, end in *-o*. In general, if the anion name ends in *-ide, -ite,* or *-ate,* the final *-e* is replaced by *-o*, resulting in one of the endings *-ido, -ito,* or *-ato.* The names of some anionic ligands are given below.

SYMBOL	ION	LIGAND
F^-	fluoride	fluoro
Cl^-	chloride	chloro
Br^-	bromide	bromo
I^-	iodide	iodo
O^{2-}	oxide	oxo
$NH_2{}^-$	amide	amido
$NO_3{}^-$	nitrate	nitrato
ONO^- (O donor)	nitrite	nitrito
$NO_2{}^-$ (N donor)	nitrite	nitro
$CO_3{}^{2-}$	carbonate	carbonato
CN^-	cyanide	cyano

Neutral ligands have the same names as the corresponding molecule, except for the water ligand (which is called "aquo") and the ammonia ligand (called "ammine"). Positive ligands (which are rare) have names ending in *-ium.*

4. In the naming of coordination compounds, the ligands are named first in alphabetical order, followed by the name of the central metal atom. The multiplying prefixes *di-*, *tri-*, *tetra-*, and so on are not treated as part of the name of the ligand in determining the order of naming ligands. Thus "diammine" would be written before "bromo". Prefixes *bis-*, *tris-*, *tetrakis-*, and so on are used before more complex ligand names such as ethylenediamine (en) and ethylenediaminetetraacetic acid (EDTA). For examples, see the list of complexes at the conclusion of this study assignment.

5. The oxidation number of the central metal atom or ion is indicated in parentheses after the name of the metal. When the complex is an anion, the ending *-ate* is attached to the element name (or the stem) of the central metal atom. Some examples follow:

ELEMENT NAME	ELEMENT NAME IN AN ANIONIC COMPLEX
cobalt	cobaltate
chromium	chromate
lanthanum	lanthanate
manganese	manganate
molybdenum	molybdate
mercury	mercurate
platinum	platinate
zinc	zincate

The elements whose symbols are derived from their Latin names use the Latin stem name with the ending *-ate*, as follows:

ELEMENT NAME	LATIN NAME	ELEMENT NAME IN AN ANIONIC COMPLEX
copper	*cuprum*	cuprate
gold	*aurum*	aurate
iron	*ferrum*	ferrate
lead	*plumbum*	plumbate
nickel	*"niccolum"*	niccolate (sometimes designated as nickelate in American usage)
silver	*argentum*	argentate
tin	*stannum*	stannate
tungsten	*wolfram*	wolframate (or tungstate)

6. In writing the *formulas* of coordination compounds, place the symbol for the central atom first, and then add those for the ligands; the formula of the whole complex is enclosed in square brackets.

The preceding rules are illustrated below by a list of complexes with names and formulas.

FORMULA	NAME
$[Ag(NH_3)_2]^+$	diamminesilver(I) ion
$[Co(NH_3)_6]Br_3$	hexaamminecobalt(III) bromide
$[Cr(H_2O)_4Cl_2]NO_3$	tetraaquodichlorochromium(III) nitrate
$K_3[Fe(CN)_6]$	potassium hexacyanoferrate(III)
$[Co(en)_3]_2(SO_4)_3$	tris(ethylenediamine)cobalt(III) sulfate
$Na_2[CrOF_4]$	sodium tetrafluorooxochromate(IV)

REPORT D	The Nomenclature of Inorganic Compounds	NAME	
		SECTION	LOCKER
		INSTRUCTOR	DATE

EXERCISES

1. Name the following.

$HBr(g)$ ______________________ $HBr(aq)$ ______________________

$SnCl_2$ (two ways) ______________________ $SnCl_4$ (two ways) ______________________

______________________ ______________________

SiF_4 ______________________ SO_3 ______________________

MnO_2 ______________________ BaO_2 ______________________

$Ba(OH)_2$ ______________________ $Ca(CN)_2$ ______________________

2. Write the formulas for the following compounds.

Cuprous oxide ______________________ Potassium peroxide ______________________

Cupric oxide ______________________ Dichlorine heptoxide ______________________

Uranium(VI) fluoride ______________________ Nitrogen trichloride ______________________

Uranium(IV) fluoride ______________________ Magnesium iodide ______________________

Sodium oxide ______________________ Ammonium sulfide ______________________

3. Complete the following table.

	Name as Acid	Formula for Sodium Salt	Name of Salt
HF			
HNO_3			
HNO_2			
HBrO			
$HBrO_2$			
$HBrO_3$			
$HBrO_4$			
H_3AsO_4		Na_3AsO_4	
$HAsO_3$			
H_3AsO_3		NaH_2AsO_3	

4. Name the following salts.

$AgHSO_4$ _____ $NaHS$ _____

$NaIO$ _____ $KMnO_4$ _____

$Mg_2P_2O_7$ _____ $BaSO_3$ _____

K_2HPO_4 _____ $Ca(ClO_2)_2$ _____

$Fe(NO_3)_3$ _____ $FeSO_4$ _____

P_4O_6 _____ P_4O_{10} _____

NH_4ClO_4 _____ $HClO_3$ _____

K_2CrO_4 _____ $K_2Cr_2O_7$ _____

$NaSO_3$ _____ Na_2SO_4 _____

5. Write chemical formulas for the following complex ions.

tetrachloroferrate(III) _____

hexanitrocobaltate(III) _____

pentaamminechlorocobalt(III) _____

tetraiodomercurate(II) _____

6. Write names for the following coordination compounds.

$[Cu(NH_3)_4]SO_4$ _____

$K_2[PbCl_4]$ _____

$K_4[Fe(CN)_6]$ _____

$[Cr(NH_3)_4CO_3]NO_3$ _____

IONIC AND COVALENT COMPOUNDS.
IONIC REACTIONS

PRE-STUDY

Two Types of Chemical Bonds

When an active metal reacts with an active non-metallic element, the latter becomes negatively charged by the transfer of one or more electrons from the metal, which in turn becomes positively charged. In general, each element tends to assume its most stable electronic configuration, with its valence electron shell filled, as in the noble gases. Using the conventional electron-dot formulas, we write, for example,

$$\text{Na} \cdot + \cdot \ddot{\underset{\displaystyle \cdot \cdot}{\text{Cl}}}: \;\rightarrow\; \text{Na}^+ \quad :\ddot{\underset{\displaystyle \cdot \cdot}{\text{Cl}}}:^-$$

Such electrically charged atoms, or radicals, are called *ions.* The bonding force between ions, due primarily to the attraction of unlike electrical charges, results in an *ionic bond.*

When two elements of similar electronegative character react, however, they do so by the formation of a stable electron pair orbital, mutually shared

by both of the atomic nuclei. For example,

$$\text{H} \cdot + \cdot \text{H} \;\rightarrow\; \text{H}:\text{H} \qquad \text{or} \qquad \text{H}_2$$

$$2\text{H} \cdot + \cdot \ddot{\text{O}}: \;\rightarrow\; \underset{\displaystyle \overset{\textstyle \cdot \cdot}{\text{H}}}{\text{H}:\ddot{\text{O}}:} \qquad \text{or} \qquad \text{H}_2\text{O}$$

Such a bond is called a *covalent bond.* In the second example, since oxygen is more electronegative than hydrogen, the bond is quite *polar,* or *partially ionic,* in character. Note that the two types of bonds are really two extremes of a continuum—the purely ionic bond at one extreme and the purely covalent bond at the other. In between are the polar or partially ionic bonds.

Structure and Bond Type

Substances with covalently bonded atoms generally have definite molecules as the units of crystal structure.[1] Such substances, if soluble in water or other

---

[1] Omitted from consideration here are the giant molecules such as quartz $(SiO_2)_x$, and metal bonds and intermetallic compounds.

suitable solvent, give simply a mixture of electrically neutral molecules, and the solution is a nonconductor. Examples of such *nonelectrolytes* are sugar ($C_{12}H_{22}O_{11}$) and acetone (CH_3COCH_3).

Acids, bases, and salts possess either ionic bonds or bonds which are quite polar. As solids, the ions of such substances constitute the structural units in the crystal. When melted by heat, the ionic lattice structure is broken down and the mixture of independent ions is an electrical conductor. When these substances dissolve in water, the ions likewise separate as independently moving particles, and the solutions are electrical conductors. Accordingly, in equations throughout this manual[2] we write the formulas of such substances in solution, or in the molten state, as separate ions. Examples of such *electrolytes* are sodium hydroxide (NaOH), which supplies Na^+ and OH^-, and potassium sulfate (K_2SO_4), which supplies K^+ and SO_4^{2-}.[3]

The strong or active acids, bases, and most salts ionize completely in dilute aqueous solution. In solutions of weak or slightly active acids and bases, a large part of the dissolved substance is present in molecular form, thus, although the total concentration may be high, the concentration of ions is low. This accounts for their slight conductivity. Relatively insoluble salts, although regarded as strong electrolytes in that, normally, all of the substance which is dissolved is present in ionic form, will likewise supply only a low concentration of ions to a solution.

The Hydration of Ions

When the quite polar substance hydrogen chloride dissolves in the very polar substance water, there is a strong attraction of the positive proton, H^+, of the HCl to the negative unshared electron pair of the oxygen atom of water,

$$H\!:\!\overset{\cdot\cdot}{\underset{H}{O}}\!: \; + \; H\!:\!\overset{\cdot\cdot}{\underset{}{\underset{\cdot\cdot}{Cl}}}\!: \; \rightarrow \; H\!:\!\overset{\cdot\cdot}{\underset{H}{O}}\!:\!H^+ \; + \; :\!\overset{\cdot\cdot}{\underset{\cdot\cdot}{Cl}}\!:^-$$

or

$$H_2O + HCl(g) \rightarrow H_3O^+ + Cl^-$$

The resulting hydronium ion, H_3O^+, may be re-

garded simply as a hydrated proton, $H(H_2O)^+$.[4] In earlier years, Arrhenius explained such ionization as a simple dissociation of the acid:

$$HCl(aq) \rightarrow H^+(aq) + Cl^-(aq)$$

All acids in water solution react in a way similar to this, forming the hydronium ion and a negative ion.

When hydrogen chloride dissolves in a nonpolar solvent, such as one of the many organic liquids, the proton, H^+, is not attracted away from the chloride ion by the solvent. Consequently, hydrogen chloride is not dissociated in such solvents.

Most ions are hydrated in water solution. Thus, cupric ion attracts four molecules of water, $Cu(H_2O)_4^{2+}$, and aluminum ion attracts six, $Al(H_2O)_6^{3+}$. These water molecules are held by covalent bonds. We shall use these hydrated formulas in situations in which they contribute to a better understanding of the reaction, but in general, for the sake of simplicity, the unhydrated formula will be used. You should realize that in water solution the hydrated formula is always implied, particularly in the case of our frequent use of H^+ for the more precise H_3O^+, or other hydrated proton formulas.

The Role of the Solvent in Dissociation Reactions

Most of the solution reactions we have studied have been in aqueous solution. However it is important to remember that nonaqueous solvents are often used and that the dissociation of acids, bases, and salts depends very much on the properties of the solvent. For example, the dissociation of an acid involves the transfer of a proton from the acid to the solvent. Therefore the extent of dissociation of an acid will depend on the intrinsic basicity of the solvent. Although acetic acid is only partially dissociated in water, it is completely dissociated in a more basic solvent such as liquid ammonia.

The *dielectric constant* of the solvent and the *solvation energy* of ions in a solvent are also important. If two ions of opposite charge are placed in a vacuum, the force between them is given by Coulomb's law

$$\text{Force} = \frac{q_1 q_2}{r^2}$$

---

[2]See also the brief introductory discussion in Experiment 4.

[3]In designating the ions of K_2SO_4, one would use the coefficient "2"—for example, $2K^+$—only to designate a specific amount of the substance K^+, as when balancing an equation.

---

[4]Other hydrated proton structures are also postulated, such as $H(H_2O)_2^+$, $H(H_2O)_3^+$, and $H(H_2O)_4^+$. The experimental data do not provide a definite answer to the precise nature of $H^+(aq)$.

where k is a constant, q_1 and q_2 are the charges on the ions, and r is the distance between the centers of the ions. When the two ions are immersed in a material medium, the force between them is reduced in inverse proportion to the dielectric constant, ϵ, of the medium as expressed in the equation

$$\text{Force} = \frac{k}{\epsilon}\frac{q_1 q_2}{r^2}$$

Therefore the total energy expended in separating the ions will decrease as the dielectric constant increases, so that a solvent with a large dielectric constant will tend to promote the dissociation of an ionic solute more than a solvent with a small dielectric constant. The hydrocarbons (with $\epsilon < 5$) do not promote the dissociation of ionic solutes. Water (with $\epsilon \cong 80$) is a good solvent for ionic solutes.

The solvation energy is also an important factor. When an ionic solute dissociates in a solvent, the total energy of the reaction may be thought of as composed primarily of two terms: (1) the energy expended in the separation of the positive and negative ions of the crystal lattice (the *lattice energy*); and (2) the *solvation energy* liberated by the association of the ions with solvent molecules, resulting from interaction between the electric charges of the ions and the dipoles of the solvent molecules. When the solvation energy is enough to offset the energy expended in separating the positive and negative ions, the ionic solute will dissociate in the solvent.[5] A *polar* solvent, therefore, will tend to promote the dissociation of an ionic solute.

Ionic Equations

Equations for chemical reactions involving acids, bases, and salts are sometimes written with molecular formulas, as if actual molecules were the reacting particles:

$$\text{NaOH}(aq) + \text{HNO}_3(aq) \rightarrow \text{NaNO}_3(aq) + \text{H}_2\text{O}$$

This is satisfactory when we are considering simply the stoichiometric quantities of chemicals entering into a reaction. As we now explore the experimental evidence for the reactions of such ionic type

substances, it will be more in accord with the facts to write equations using formulas for the principal ions, or molecules, as they exist in the solution. We shall use this technique in all ionic type reactions hereafter. Further explanation will be helpful at this time.

Since data on electrical conductivity, as well as the chemical behavior of strong acids and bases and soluble salts, indicate that the aqueous solutions consist of individual ions, we shall write such equations in terms of the principal substances (ions or molecules) actually present before, and after, the reaction. Thus we write[6]

$$\text{Na}^+ + \text{OH}^- + \text{H}^+ + \text{NO}_3^- \rightarrow$$
$$\text{Na}^+ + \text{NO}_3^- + \text{H}_2\text{O}$$

which is called a *total ionic equation*.

Note that in this equation neither the Na^+ nor the NO_3^- has reacted—both are present as separate particles after, as well as before, the reaction. They may therefore be omitted, and the equation, now called a *net ionic equation*, becomes

$$\text{OH}^- + \text{H}^+ \rightarrow \text{H}_2\text{O}$$

Such an equation focuses attention on the essential changes: the disappearance of acid properties due to H^+ and of basic properties due to OH^-. The water formed is so slightly ionized that the concentration of these ions is too low to show their characteristic properties to an appreciable extent.

Ionic reactions occur whenever ions unite to form weakly ionized or insoluble substances. In the above neutralization of nitric acid by sodium hydroxide, the only factor causing the reaction is due to the great tendency of H^+ and OH^- to unite to form the slightly ionized water. A similar reaction occurs when the ions of a weak acid or of a weak base unite. For example, if ammonium chloride (NH_4Cl) and sodium hydroxide (NaOH) solutions are mixed, the only net change is expressed by

$$\text{NH}_4^+ + \text{OH}^- \rightarrow \text{NH}_3(aq) + \text{H}_2\text{O}$$

In the formation of the weak base ammonia the reaction is not nearly as complete as it is in the formation of water, for ammonia has a far greater tendency to ionize than has water.

---

[5]A more careful examination of the factors that determine the solubility of a solute requires the inclusion of the effect of changes in the solvent structure. The entropy change of the system is a measure of this effect, which can be strong enough that salts such as NaCl will dissolve even though the lattice energy is greater than the solvation energy.

[6]Hereafter we may omit the designation "(aq)" after the formula of ions, as these will be understood to be hydrated and in aqueous solution, unless otherwise indicated.

A typical example of ions uniting to form an insoluble substance is expressed by the equation

$$Ca^{2+} + CO_3^{2-} \rightarrow CaCO_3(s)$$

It makes no difference whether the calcium ion (Ca^{2+}) is obtained from calcium chloride, calcium nitrate, or any other soluble calcium salt. Carbonate ion (CO_3^{2-}) could be obtained equally well from sodium carbonate, ammonium carbonate, or from any other soluble well-ionized carbonate. The chemical change, as recorded by the net ionic equation, is the same in each reaction.

It is also possible to change weakly ionized substances into substances that ionize to an even lesser extent. When ammonia is neutralized by hydrochloric acid, the total ionic equation is

$$NH_3(aq) + H^+ + Cl^- \rightarrow NH_4^+ + Cl^-$$

The net ionic equation is

$$NH_3(aq) + H^+ \rightarrow NH_4^+$$

Since water is much less ionized than is *ammonia,* the reaction proceeds. However, it will not be as complete as is a neutralization in which both the acid and base are strong.

As another example, the poorly ionized carbonic acid (H_2CO_3) contains fewer carbonate ions (CO_3^{2-}) in solution than does a saturated solution of the relatively insoluble calcium carbonate ($CaCO_3$), so the reaction

$$CaCO_3(s) + 2H^+ \rightarrow Ca^{2+} + H_2CO_3(aq)$$

will take place. Furthermore, the carbonic acid is not very soluble, but breaks down into water and carbon dioxide gas.

Unless there is a tendency for the ions in a solution to unite, no reaction occurs. If dilute solutions of ammonium nitrate and sodium chloride are mixed, no reaction will occur. To write an equation

$$NH_4NO_3(aq) + NaCl(aq) \rightarrow$$
$$NaNO_3(aq) + NH_4Cl(aq)$$

is meaningless, because each of the above four salts is readily soluble and highly ionized:

$$NH_4^+ + NO_3^- + Na^+ + Cl^- \rightarrow$$
$$Na^+ + NO_3^- + NH_4^+ + Cl^-$$

The net result is *no action at all*—simply a mixing of the ions. Therefore *no equation should be written.*

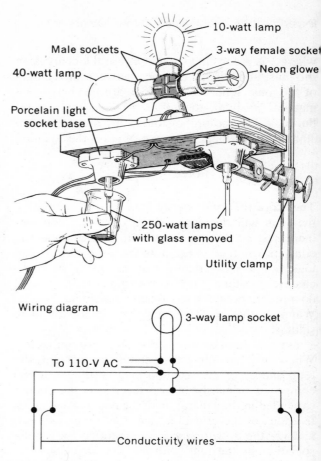

FIGURE 14-1
An apparatus to compare the electrical conductivities of various solutions. Only one bulb at a time should be screwed in.

To summarize the writing of net ionic equations, note that this technique does not mean that *all* substances should *always* be written as ions. Poorly ionized or slightly soluble substances should be written as the un-ionized molecule.

The Experimental Method. Electrical Conductivity

Solutions of ionic type substances conduct the electric current by movement of their ions as influenced by the applied potential, and by reaction at the electrodes.[7] For a given applied voltage, the amount of current depends upon the concentration of ions, and to a lesser extent upon differences in individual ion mobilities. (Hydrogen ion has an unusually high mobility—five to eight times that of many other ions. Why?) Figure 14-1 shows a

[7]With alternating current the electrode reaction is diminished, and at high frequency is practically eliminated.

simple apparatus for the comparison of electrical conductivities.

We shall compare the electrical conductivities of a number of solutions and of the products formed after certain reactions, and use the data thus obtained to interpret the character of the solutions and the course of any reactions. We shall interpret these results in terms of the net ionic equation for the reaction.

EXPERIMENTAL PROCEDURE

Special Supplies: Conductivity apparatus, source of H_2S gas.

Chemicals: 17 F $HC_2H_3O_2$, 6 F $HC_2H_3O_2$, 0.1 F $HC_2H_3O_2$, C_2H_5OH, 0.1 F $NH_4C_2H_3O_2$, 0.1 F NH_3, 0.1 F $Ba(OH)_2$, C_6H_6, $CaCO_3$ (marble chips), 0.1 F $Cu(C_2H_3O_2)_2$, 0.1 F $CuSO_4$, HCl in benzene, 0.1 F HCl, 0.1 F $HgCl_2$, $KClO_3$, 0.1 F KNO_3, NaCl, 0.1 F NaCl, 0.1 F NaOH, $C_{12}H_{22}O_{11}$, 0.1 F H_2SO_4, Zn (mossy), phenolphthalein.

NOTE: Observe the following precautions when using your conductivity apparatus.
1. Caution: Disconnect the apparatus (if 110 v) between measurements. Avoid touching the electrodes. Keep your hands dry.
2. Rinse the electrodes with distilled water between each measurement (for solids also dry them with filter paper).
3. Use *small amounts* of solution—5 to 10 ml in a 10-cm test tube, a crucible, or a 50-ml beaker held at an angle.
4. Dip the electrodes into the solution to a uniform depth of 1 cm each time, and keep the electrodes a uniform distance apart. **Use only one lamp at a time.** The neon bulb is used to test solutions of low conductivity. The area of the orange glow increases as the conductivity of the solution increases. If the glow is bright and large, unscrew the neon bulb and try the 10-watt bulb. If the 10-watt bulb glows very brightly, unscrew it and try the 40-watt bulb. The 40-watt lamp will glow with full brilliance when the electrodes are immersed in solutions of high conductivity.

---

1. Electrolytes and Nonelectrolytes. (a) By conductivity tests in accord with the preceding suggestions, explore the nature of a number of *substances and their solutions* and determine if they are largely, moderately, or poorly ionized, or if they are essentially nonelectrolytes. The pure substances can be tested first; then, if a solution is desired, add a little water, mix, and repeat the test. In each case, indicate the principal constituent particles—ions, molecules, or both. Suggested substances: distilled water, tap water, 95% ethanol (C_2H_5OH), glacial (17 F) acetic acid ($HC_2H_3O_2$), sugar ($C_{12}H_{22}O_{11}$), sodium chloride (NaCl), and dilute solutions only (not the pure substances) of 0.1 F $HgCl_2$, 0.1 F HCl, and 0.1 F NaOH.

(b) Observe the *effect of fusion* of a salt by testing the conductivity of a little solid potassium chlorate ($KClO_3$) in a crucible; then, after heating the crucible until the salt is just melted, repeat the conductivity test.

(c) Observe the *effect of the solvent* on the ionization of HCl in benzene and in water. First test a little pure benzene (C_6H_6) in a thoroughly dried test tube, and then a solution of HCl in benzene (*already prepared* on your reagent shelf—do *not* add aqueous concentrated HCl to benzene); then add 5 ml of distilled water to this same solution, mix it well, and repeat the test with the electrodes immersed further into the lower aqueous layer.

(d) Compare the *physical and chemical behavior* of 6 F HCl and 6 F $HC_2H_3O_2$ with your conductivity data. The following comparisons are suggested.
Taste: Add 6 F HCl and 6 F $HC_2H_3O_2$, respectively, *drop by drop* to 10 ml of water in separate beakers. Touch a drop of each of these diluted solutions, held on the tip of a stirring rod, to the tongue. Rinse your mouth.
Reaction on marble chips ($CaCO_3$), and on mossy zinc: Judge by the relative rates of evolution of CO_2 gas and H_2 gas, respectively.

2. Typical Ionic Reactions. Experiment with a number of ionic type reactions to determine their essential nature. By conductivity tests of the separate reactants and of the mixture after reaction, discover whether the acids, bases, and salts concerned are largely ionized (strong) or only moderately to poorly ionized (weak). For each reaction, write the total ionic equation and the net ionic equation, to interpret the observed behavior properly. This will require careful thinking on your part. The following procedures are suggested.

(a) *0.01 F HCl with 0.01 F NaOH.*[8] These may be prepared by 10-fold dilution of 5 ml of the 0.1 F solutions of each. It is an advantage to use both pairs of electrodes connected in parallel (see Figure 14-1) and to place equal volumes of each of the two solutions in contact, respectively, with the two electrode pairs, to get their total conductivity before mixing. Then mix the solutions, divide equally, and again place the two mixed solutions in contact with the two electrode pairs. (This compensates for the dilution effect of mixing two solutions.)

---

[8] Quite dilute 0.01 F solutions are suggested so you can interpret moderate conductivity changes more easily.

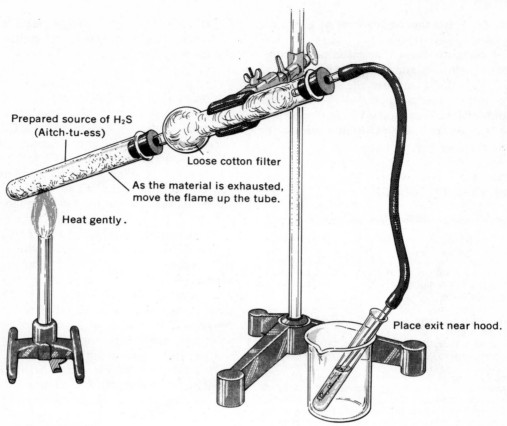

Prepared source of H₂S
(Aitch-tu-ess)

Loose cotton filter

As the material is exhausted,
move the flame up the tube.

Heat gently.

Place exit near hood.

FIGURE 14-2
A convenient method for the generation of $H_2S(g)$. (The cotton filter tube may be omitted
for brief tests.)

(b) *0.1 F HC₂H₃O₂ with 0.1 F NH₃.*[9] Test equal
volumes of the separate solutions, mix them, and
then retest the conductivity.

(c) *0.1 F H₂SO₄ with 0.1 F Ba(OH)₂.* Test the
separate solutions. Then, while the electrodes are
in the 0.1 F H_2SO_4, add a drop of phenolphthalein,
and then add the 0.1 F $Ba(OH)_2$ gradually as you
stir; when you approach the end point, add drop by
drop, until the last drop causes a permanent, very
faintly pink color. If you have added too much
base, dilute some of the acid and again approach the
end point by adding this acid drop by drop, as you
stir, until the *exact* equivalence point has been
reached. Test the conductivity of this mixture.

(d) *0.1 F CuSO₄ with 0.1 F H₂S.* First determine
the conductivity of 0.1 F $CuSO_4$ and of a solution

made by bubbling H_2S gas[10] through 10 ml of distilled water to saturate it (approximately 0.1 F).
Then pass H_2S gas directly into the 0.1 F $CuSO_4$
for 2 or 3 minutes, until the cupric sulfide (CuS)
is precipitated as completely as possible. Test the
conductivity of this solution. Consider also whether
the solid CuS contributes materially to the conductivity.

(e) *0.1 F Cu(C₂H₃O₂)₂ with H₂S gas.* Determine the conductivity of 0.1 F $Cu(C_2H_3O_2)_2$, then
saturate this solution with H_2S gas to precipitate
the CuS as completely as possible, and again test
the conductivity.

---

[9]Solutions of NH_3 in water are sometimes formulated as
NH_4OH, called ammonium hydroxide. Because there is no
convincing evidence for the existence of such a molecule, it
is preferable to speak of an aqueous ammonia solution.

[10]Consult your instructor about your best source of H_2S gas.
A cylinder of the compressed gas (in hood) or a Kipp generator
charged with FeS and 6 F HCl (in hood) is often used. A little
"Aitch-tu-ess" (a commercial mixture of sulfur, paraffin, and
asbestos) heated in a test tube fitted with a gas delivery tube is
a convenient method (see Figure 14-2).

REPORT

14

Ionic and Covalent Compounds.
Ionic Reactions

NAME

SECTION   LOCKER

INSTRUCTOR   DATE

OBSERVATIONS AND DATA

1. Electrolytes and Nonelectrolytes

(a) List the relative conductivity of each *substance and solution* tested, then write the formula of the individual species present (molecules, ions, or both) which account for this behavior. Star (*) those present at only low concentration.

Conductivity	Species	Conductivity	Species
Distilled H_2O_____	_____	$C_{12}H_{22}O_{11}(aq)$_____	_____
Tap H_2O_____	_____	NaCl_____	_____
C_2H_5OH_____	_____	NaCl(aq)_____	_____
$C_2H_5OH(aq)$_____	_____	0.1 F $HgCl_2$_____	_____
$HC_2H_3O_2$_____	_____	0.1 F HCl_____	_____
$HC_2H_3O_2(aq)$_____	_____	0.1 F NaOH_____	_____
$C_{12}H_{22}O_{11}$_____	_____		

(b) Explain your observation of the *effect of fusion* on the conductivity of $KClO_3$.

(c) Explain your observations of the *effect of the solvent* on the conductivity of HCl dissolved in benzene and HCl dissolved in water.

(d) Compare your observations of the *physical and chemical behavior* of 6 F HCl and of 6 F $HC_2H_3O_2$ with your conductivity data. Explain.

2. Typical Ionic Reactions

(a) 0.01 F HCl with 0.01 F NaOH. The relative conductivities of the solutions tested are as follows:

0.01 F HCl_____ 0.01 F NaOH_____ Mixture_____

The total ionic equation for the reaction is_____

The net ionic equation for the reaction is_____
Interpret any changes in the conductivity of the solutions, before and after mixing, in accordance with the above equations:

178

(b) 0.1 F $HC_2H_3O_2$ with 0.1 F NH_3. The relative conductivities of the solutions tested are as follows:

0.1 F $HC_2H_3O_2$——————————— 0.1 F NH_3——————————— Mixture———————————

The total ionic equation for the reaction is————————————————————————————————

The net ionic equation for the reaction is————————————————————————————————
Interpret any changes in conductivity of the solutions, before and after mixing, in accordance with the above equations:

(c) 0.1 F H_2SO_4 with 0.1 F $Ba(OH)_2$. The relative conductivities of the solutions tested are as follows:

0.1 F H_2SO_4——————————— 0.1 F $Ba(OH)_2$——————————— Mixture———————————

The total ionic equation for the reaction is————————————————————————————————

The net ionic equation for the reaction is————————————————————————————————
Interpret any changes in conductivity of the solutions, before and after mixing, in accordance with the above equations:

(d) 0.1 F $CuSO_4$ with 0.1 F H_2S. The relative conductivities of the solutions tested are as follows:

0.1 F $CuSO_4$——————————— 0.1 F H_2S——————————— Mixture———————————

The total ionic equation for the reaction is————————————————————————————————

The net ionic equation for the reaction is————————————————————————————————
Interpret any changes in conductivity of the solutions, before and after mixing, in accordance with the above equations:

(e) 0.1 F $Cu(C_2H_3O_2)_2$ with H_2S gas. The relative conductivities of the solutions tested are as follows:

0.1 F $Cu(C_2H_3O_2)_2$——————————— Mixture with H_2S gas———————————

The total ionic equation for the reaction is————————————————————————————————

The net ionic equation for the reaction is————————————————————————————————
Interpret any changes in conductivity of the solutions, before and after mixing, in accordance with the above equations:

THE CHEMISTRY OF OXYGEN.
BASIC AND ACIDIC OXIDES AND
THE PERIODIC TABLE

PRE-STUDY

In this experiment we shall study reactions of oxygen gas, and use oxygen to prepare a number of representative oxides of various elements. We shall dissolve these oxides in water, and test the chemical character of the solution formed. The following points are to be noted in this experiment: (1) What is a *catalyst*, and how does it affect the rate and completeness of a reaction? (2) Is there any difference in the vigor of reaction (rate, heat evolved, etc.) of the different elements with oxygen, which may be related to the position of the element in the periodic table? (3) Is there any relationship between the character of the water solution of these oxides and the position of the element in the periodic table?

Acids and Bases from Oxides

The oxide of an element (generalized symbol, El), reacts with water to form an addition compound in which there are hydroxide groups:

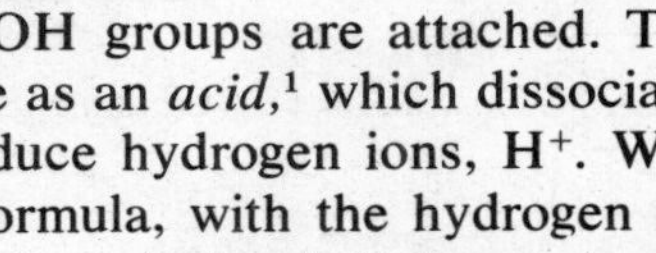

$$\text{ElO} + \text{H}_2\text{O} \rightarrow \text{El} \begin{array}{c} \text{O—H} \\ \diagdown \\ \text{O—H} \end{array}$$

(The number of OH groups, and the corresponding subscripts in the formulas, will, of course, vary with different elements.) The character of this hydroxide depends on the nature of the element to which the OH groups are attached. The product may behave as an *acid*,[1] which dissociates in solution to produce hydrogen ions, H^+. We therefore write the formula, with the hydrogen atoms first, as H_2ElO_2. In solution it ionizes according to the equation

$$H_2ElO_2 \rightleftharpoons H^+ + HElO_2^-$$

and by further ionization

$$HElO_2^- \rightleftharpoons H^+ + ElO_2^{2-}$$

---

[1] A number of important acids are not derived from oxides, e.g., the binary hydrides of the nonmetals—HCl, HBr, H_2S.

On the other hand, the compound may behave as a *base,* which dissociates in solution to produce hydroxide ions, OH^-. We therefore write the formula with the hydroxide group together. In solution, it ionizes according to the equations

$$El(OH)_2 \rightleftharpoons El(OH)^+ + OH^-$$

$$El(OH)^+ \rightleftharpoons El^{2+} + OH^-$$

An oxide that, with water, forms an acid is called an *acidic oxide* or *acid anhydride.* An oxide that, with water, forms a base is called a *basic oxide* or *basic anhydride.* A third class of oxides are the *amphoteric* oxides, such as SnO_2, ZnO, Al_2O_3, and PbO. These compounds are rather insoluble in water but their solubility is increased by the addition of either acid or base. We may determine the formula of an acidic or basic anhydride that is related to a given acid or base by subtracting water to eliminate all hydrogen atoms. For example,

$$Mg(OH)_2(s) - H_2O \rightarrow MgO(s)$$

$$2H_3BO_3(g) - 3H_2O \rightarrow B_2O_3(s)$$

EXPERIMENTAL PROCEDURE

Special Supplies: Deflagrating spoon, six glass squares, plastic basin, six 250-ml wide mouth bottles.

Chemicals: Ca (metal shavings), Cu (turnings), Fe (steel wool), Mg (ribbon), P (red), S, small pieces of charcoal, litmus paper or wide range pH paper, $KClO_3(s)$, MnO_2 (powdered), $H_3BO_3(s)$, $Na_2O_2(s)$, 60% perchloric acid, $HClO_4$ (or ClO_3OH).

1. Preparation of Oxygen.[2]

Assemble the apparatus as illustrated in Figure 15-1. Mix about 10 g of potassium chlorate with 2 g of manganese dioxide. First place a very small sample of this mixture in a large test tube and heat it just enough to melt the potassium chlorate. If it decomposes smoothly, without obvious sparks and combustion, it is safe to use.[3] Let the test tube cool, and then add the remainder of the mixture and connect the delivery tube. Fill six 250-ml wide mouth bottles with water and invert these in the plastic basin as

---

[2]Alternatively, the bottles of oxygen can be filled from an oxygen cylinder, if one is available.

[3]If organic materials are present, or if possibly the wrong chemicals are mixed, a dangerous explosion might result. It pays to treat potassium chlorate with caution and respect.

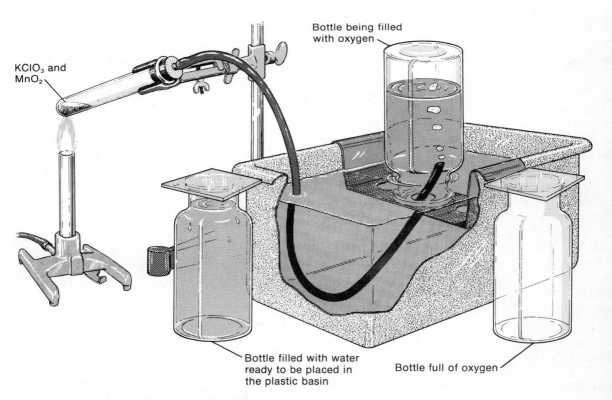

KClO₃ and MnO₂

Bottle being filled with oxygen

Bottle filled with water ready to be placed in the plastic basin

Bottle full of oxygen

FIGURE 15-1
The laboratory preparation of oxygen.

needed. Now gently heat the potassium chlorate and manganese dioxide mixture just enough to maintain a moderate evolution of oxygen. If the reaction mixture should move up the test tube, examine the gas delivery tubes and make sure they have not become plugged. Let a little of the gas escape into the air, to permit the generator to fill with pure oxygen, and then fill the six bottles with oxygen by displacement of water. Leave about 5 ml of water in each bottle. As soon as each is filled, cover it with a glass square and place it right side up on the table. Take the delivery tube out of the water before the generation of oxygen ceases or before you remove the flame. (Why?)

2. Preparation of Oxides. Prepare oxides of the following elements by burning them in oxygen gas as described, keeping the bottles covered as much as possible (see Figure 15-2). Number or label each bottle to avoid confusion. Immediately after each combustion, add 30–50 ml of water, replace the glass square, shake the bottle to dissolve the oxide formed, and set it aside for later use.

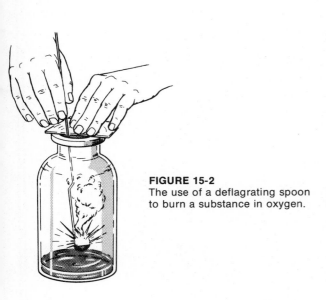

FIGURE 15-2
The use of a deflagrating spoon to burn a substance in oxygen.

(a) *Magnesium.* Ignite a 10-cm length of magnesium ribbon, holding it with the crucible tongs and at once thrust it into a bottle of oxygen. (*Do not look directly at the brilliant light*, because it can injure your eyes.)

(b) *Calcium.* Calcium metal is difficult to ignite, but burns brilliantly. *Exercise caution.* Place a shaving of calcium metal in a crucible, and ignite it in the air at the maximum temperature over a

Bunsen burner for 15 minutes, and then wash out the product with water into a beaker.

(c) *Iron.* Put a little water into a bottle of oxygen, to form a protective layer on the bottom. Heat some steel wool, holding it with the tongs, in the Bunsen flame until it ignites, and at once thrust it into this bottle.

(d) *Carbon.* Ignite a small piece of charcoal, holding it with the tongs or in a clean deflagrating spoon, and thrust the glowing charcoal into a bottle of oxygen.

(e) *Phosphorus and sulfur.* For each of these substances, prepare a deflagrating spoon by lining it with a little asbestos paper or shreds, and ignite this lining to burn out any carbon. Add a bit of the sulfur or phosphorus (use *red* phosphorus, no larger in volume than half a pea). Ignite it over the burner and then thrust it into a bottle of oxygen (the sulfur and phosphorus should be put into separate bottles). After the combustion dies down, reheat the deflagrating spoon to burn out all remaining phosphorus or sulfur. Carry out the ignition in a hood, if one is available.

3. Acids and Bases from Oxides. From each bottle in which the oxide of an element has been formed by the foregoing procedures, and to which water has been added to form a solution, remove a drop of solution on a clean glass stirring rod, and touch it to litmus or *p*H paper.[4] Correlate the chemical character of the reaction product with the position of the element in the periodic table.

The preceding tests include representative elements from Groups II, IV, V, VI, and the transition Group VIII, in the periodic table. To complete the series, let us examine an oxide or hydroxide from each of the other principal groups: 1, 3, and 7. (Why not include an element from group 0?)

The normal *alkali oxides*, such as Na_2O, are difficult to obtain. Sodium peroxide forms when sodium burns in oxygen. The final reaction product with water, however, is the same as when Na_2O reacts with water. The extra "peroxide oxygen" is liberated as free oxygen gas. Compare the two following equations:

$$Na_2O(s) + H_2O \rightarrow 2NaOH(aq)$$

$$2Na_2O_2(s) + 2H_2O \rightarrow 4NaOH(aq) + O_2(g)$$

---

[4]Ferric oxide or hydroxide is so insoluble that the solution has no effect on the litmus. The effect of one or two other hydroxides may be somewhat feeble, but it should be sufficient to indicate the acidic or basic character of the hydroxide.

In a 15-cm test tube boil a very small amount of sodium peroxide in 5 ml of water for a moment to complete the above reaction. Cool the solution, and test it with litmus. Also note the slippery, soapy feeling that is typical of such a caustic solution: *Wash your fingers thoroughly after touching the solution.*

The *oxide of boron,* B_2O_3, is not readily available, and in some forms is very insoluble. Dissolve a small amount of boron hydroxide in 5 ml of hot water. Cool the solution, and test it with litmus. Decide which formula — $B(OH)_3$ or H_3BO_3 — is preferable. How is the substance usually named?

The *oxides of chlorine,* $Cl_2O(g)$, $ClO_2(g)$ and $Cl_2O_7(g)$, are quite unstable. They react with water to form hydroxides. A solution of one of these, labeled $ClO_3(OH)$, will be available to you. Test a milliliter of this with litmus, and in your report write its formula as an acid or base according to your observations of its properties.

REPORT

15

The Chemistry of Oxygen.
Basic and Acidic Oxides and
the Periodic Table

NAME

SECTION   LOCKER

INSTRUCTOR   DATE

OBSERVATIONS AND DATA

1. Preparation of Oxygen

(a) Write the equation for the reaction by which oxygen may be prepared.

_____

(b) Compare the heating of potassium chlorate alone with the heating of the same amount of potassium chlorate mixed with some manganese dioxide for the following points.

(1) The relative temperatures at which decomposition readily takes place.

(2) The relative amounts of oxygen that may be obtained when each sample is heated until reaction is complete.

(3) Any changes taking place in the manganese dioxide.

(4) The name applied to a substance used as manganese dioxide is used in this reaction.

2. Preparation of Oxides

Describe any changes occurring during the reaction of each element with oxygen, and any distinctive characteristics of the products formed. Write the equation for each reaction.

	Observations	Equations
Magnesium:		
Calcium:		
Iron:		
Carbon:		
Phosphorus:		
Sulfur:		

184

3. Acids and Bases from Oxides

On the line opposite its periodic group, write the formula of each oxide (or hydroxide, if the oxide was not available) studied in this experiment. Indicate the reaction to litmus of its water solution, and write the equation for the formation of the acid or base.

Group	Formula of oxide	Reaction to litmus	Equation for reaction, if any
I	_____	_____	_____
II (two elements)	_____	_____	_____
	_____	_____	_____
III	_____	_____	_____
IV	_____	_____	_____
V	_____	_____	_____
VI	_____	_____	_____
VII	_____	_____	_____
VIII (transition)	_____	_____	_____

APPLICATION OF PRINCIPLES

1. Comment on the acidic or basic character of the oxide of an element, as compared to its position in the periodic table.

2. Considering the positions of the elements in the periodic table, write the formulas of three other acidic oxides and three other basic oxides. On the longer line, write the equation for reaction of the oxide with water.

Acidic oxides: _____ _____

_____ _____

_____ _____

Basic oxides: _____ _____

_____ _____

_____ _____

3. What does the term "anhydride" mean?

4. Deduce and write the formulas of the anhydrides of the following.

H_2SO_3_____ RbOH_____ HNO_2_____

H_2SO_4_____ $Ba(OH)_2$_____ HNO_3_____

$La(OH)_3$_____ HClO_____ H_2CO_3_____

$Sn(OH)_4$_____ $HClO_3$_____ H_4SiO_4_____

THE CHEMISTRY OF SOME NONMETALS OF GROUPS V, VI, AND VII

PRE-STUDY

The Oxidation States of Some Elements and Compounds of the Nonmetals

The oxidation states of some representative Group V, VI, and VII elements are shown in Figure 16-1. This chart illustrates three important points. First, each of the elements in these groups can have many formal oxidation states. This leads to a rich variety of compounds whose chemistry is complex in comparison with the chemistry of most metals in Groups I and II. Second, the electronic structure of the elements in the three groups places definite limits on the formal oxidation states accessible to the element. Nitrogen, which has five valence electrons, can share these 5 electrons with electronegative elements like oxygen to produce a formal oxidation state as high as $+5$. It can accept a share of as many as 3 electrons from less electronegative elements like hydrogen to produce a formal oxidation state as low as -3. In both the $+5$ and -3 oxidation states the nitrogen atoms have achieved an *inert gas configuration* with each nitrogen atom having a share of 8 electrons.

Oxidation state	Group V	Group VI	Group VII
+7			Cl_2O_7, $HClO_4$, ClO_4^-
+6		SO_3, H_2SO_4, HSO_4^-, SO_4^{2-}	Cl_2O_6
+5	$N_2O_5(g)$, HNO_3, NO_3^-		$HClO_3$, ClO_3^-
+4	$NO_2(g)$, $N_2O_4(g)$	SO_2, H_2SO_3, HSO_3^-, SO_3^{2-}	ClO_2
+3	$N_2O_3(g)$, HNO_2, NO_2^-		$HClO_2$, ClO_2^-
+2	$NO(g)$	$S_2O_3^{2-}$	
+1	$N_2O(g)$		Cl_2O, $HClO$, ClO^-
0	$N_2(g)$	S_8	Cl_2
−1	NH_2OH, NH_3OH^+		Cl^-
−2	N_2H_4, $N_2H_5^+$, $N_2H_6^{2+}$	H_2S, HS^-, S^{2-}	
−3	$NH_3(g)$, NH_4^+		

Oxidizing strength →

Reduction ↓ / *Oxidation* ↑

FIGURE 16-1
Compounds representing a wide variety of oxidation states of three nonmetals.

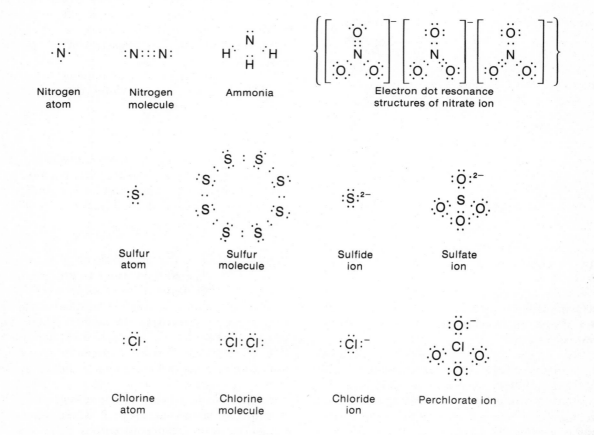

FIGURE 16-2
Electron dot structures of three nonmetallic elements and some of their compounds.

Because a neutral sulfur atom has 6 valence electrons, the formal oxidation state of sulfur ranges from +6 in sulfate ion to −2 in sulfide ion. The oxidation states of chlorine range from +7 to −1 because a neutral chlorine atom has 7 valence electrons. The addition of one electron to a neutral chlorine atom produces the very stable chloride ion, which has the same electron configuration as the inert gas argon. In an ion like perchlorate, ClO_4^-, chlorine shares its electrons with the electronegative oxygen atoms, again producing an electron configuration like that of argon.

The electron dot structures for some of the compounds mentioned above are shown in Figure 16-2.

The third important feature to note about the compounds in these groups is that the *oxidizing power* of a compound tends to increase as the formal *oxidation state* of the element increases. Nitric, sulfuric, and perchloric acids are all strong oxidizing agents, particularly if they are hot and concentrated. There is also a tendency for the compounds with the most *negative* oxidation states to be good *reducing agents*. This is particularly true of NH_3 and hydrazine, N_2H_4, as well as sulfide ion, S^{2-}. However chloride ion is not a good reducing agent and in fact all of the chlorine compounds except chloride ion are strong oxidizing agents. You can understand this concept qualitatively by keeping in mind that the nuclear charge increases in any given period as we go from Group V to Group VII elements. The increased nuclear charge makes it more difficult for electrons to be removed or for the atom to share its electrons with any but the most electronegative elements. Therefore Group VII compounds tend to be stronger oxidizing agents.

The Chemistry of Representative Elements of Groups V, VI, and VII

It would be impossible to explore the chemistry of all the elements in these three groups in the course of one afternoon: therefore we will choose one member of each group as representative of the chemistry of that particular group. First we will study the chemistry of some of the oxides of nitrogen, a Group V element, and some oxides of sulfur (Group VI) and of chlorine (Group VII).

Oxygen is so electronegative that it exists primarily in stable compounds in the formal oxidation state −2, with a few compounds (peroxides) having oxidation state −1. For these reasons we have chosen sulfur as the representative element of the Group VI elements (chalcogens).

Similarly, we have chosen chlorine as the representative element of the Group VII elements (halogens) rather than the first member of the series, fluorine, because fluorine is so electronegative that all stable fluorine compounds contain fluorine with a formal oxidation state of −1.

Nitrogen Oxides and Sulfur Dioxide in the Atmosphere

Nitric oxide is produced in large amounts in steelmaking furnaces and by reaction of nitrogen with oxygen at high temperature in automobile engines. The nitric oxide quickly reacts with oxygen to yield nitrogen dioxide, a brownish gas. Under the action of sunlight, nitrogen dioxide can be photolyzed to give NO + O. The free oxygen atom can then react with oxygen to give ozone, O_3. Ozone is a lung irritant and is also very toxic to many species of pine trees. It can also react with unburned hydrocarbons in the atmosphere to yield peroxyacetyl nitrate (PAN), which is very irritating to the eyes and also damages leafy crops such as spinach.

The combustion of coal containing sulfur produces large quantities of sulfur dioxide by the reaction

$$S + O_2 \rightarrow SO_2$$

In regions in which an appreciable amount of ozone has been formed by a photochemical reaction with nitrogen dioxide, the ozone can further oxidize the SO_2 to sulfur trioxide by the reaction

$$SO_2 + O_3 \rightarrow SO_3 + O_2$$

Sulfur trioxide is the anhydride of sulfuric acid. Thus it can combine with water to produce a fine aerosol of sulfuric acid mist, or, if ammonia is present (produced in large quantities in agricultural areas by livestock), ammonium sulfate particles can be produced. The haze that is associated with photochemical smog apparently contains a considerable amount of ammonium sulfate as well as organic matter.

Sulfur dioxide is used in agriculture to kill the wild yeasts on grapes used in winemaking. The grapes are then inoculated with a pure strain of yeast that produces a higher quality of wine than the wild strains would produce. It is also used in bleaching dried fruit, such as apricots and raisins. One can demonstrate this bleaching action easily by taking a few petals of a red rose, boiling them in 30 ml of ethanol to extract the red pigment, acidifying the extract with a drop of 6 F HCl, and bubbling SO_2 through the solution. The solution will turn pale as the sulfur dioxide reduces the red pigment to a colorless chemical form of the pigment.

SOME CHEMISTRY OF NITROGEN COMPOUNDS

Nitrogen, the first element in Group V, is an important nonmetal that forms compounds illustrating all the oxidation states from −3 to +5. In the zero oxidation state, nitrogen is particularly stable and has the electron dot structure $:N:::N:$. The commercially important compounds are those in the −3 and the +5 oxidation states. Ammonia and its salts are important as soluble fertilizers. Nitric acid is an important oxidizing agent and is used in making explosives.

The electron-dot structure of ammonia is $H:\overset{..}{\underset{H}{N}}:H$,

which adds a hydrogen ion to form the ammonium ion $\left[H:\overset{H}{\underset{H}{N}}:H\right]^+$. Ammonia reacts with water to form a weak base.

When nitric acid acts as an oxidizing agent, it may be reduced to any of the lower oxidation states. Concentrated nitric acid is usually reduced to NO_2, dilute nitric acid to NO or to lower oxidation states by very strong reducing agents.

EXPERIMENTAL PROCEDURE: NITROGEN COMPOUNDS

Chemicals: Cu (turnings), 6 F HNO_3, Na_2O_2, wood splints.

NOTE: In your written report of this experiment, include all observations of the properties of the products formed, and write the equations for all reactions.

The Preparation of Nitric Oxide by Reduction of HNO_3 with Copper: Oxidation States +2, +4, and +5. Assemble the generator shown in Fig. 16-3 and place about 3 g of copper turnings into it. Prepare to collect three 15-cm test tubes of nitric oxide by displacement of water. Add 15 ml of 6 F HNO_3 to the generator, replace the delivery tube connection,

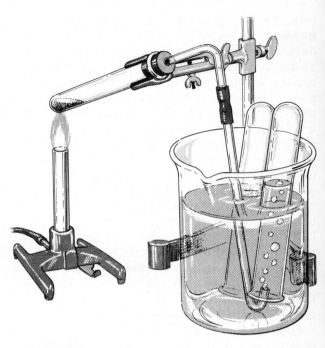

FIGURE 16-3
Apparatus for the preparation of nitrogen gas or of the oxides of nitrogen.

and warm the test tube gently to initiate the reaction. After the air has been displaced from the apparatus and the gas bubbling through is colorless, collect two full test tubes of the gas and a third test tube about half full. Do not allow the delivery tube to remain under water while the heated test tube cools; otherwise, water will be drawn into the test tube. What is the reaction for the reduction of dilute nitric acid by copper?

Test the nitric oxide in one test tube with a glowing splint to see if it supports combustion. Note the colored gas produced when the tube was exposed to the air. Write the equation for the reaction which accounts for this change.

Test the second sample of nitric oxide for solubility in water by swirling the test tube with its mouth under the water to allow contact of fresh water with the gas. Note if the water level in the test tube rises. Now take the test tube out of the water for a few seconds, and allow the oxygen of the air to react with the gas, as will be evidenced by the formation of a brown gas. Invert the test tube under the water again, and swirl it to note the solubility of the brown gas. Write the equation for the reaction occurring when this gas dissolves in water.

Mark the water level in the third test tube, which is about half-full of nitric oxide, with a wax pencil or gummed label. Set up the small oxygen generator shown in Figure 16-4, placing about 2 g of sodium peroxide (Na_2O_2) into the generator (a dry, 15-cm Pyrex test tube). Draw up a few milliliters of water into the medicine dropper. Sodium peroxide reacts vigorously with water to produce oxygen and sodium hydroxide. Add only a drop of water at a time to the peroxide whenever the flow of oxygen becomes too slow. After the air has been displaced from the generator, place the delivery tube under the marked test tube and allow 8–10 bubbles of oxygen to enter. Note whether the level of the water is lowered by the addition of oxygen. Recall that in the balanced equation for the reaction that is taking place, two volumes of nitric oxide react with one volume of oxygen to produce two volumes of nitrogen dioxide. Now swirl the test tube with its mouth under the water and note what happens to the water level as the NO_2 reacts with the water. Allow more oxygen to bubble into the tube until the gases turn brown, note the water level, and again allow the gases to dissolve in water. Repeat the process until the water level approaches the top of the tube. Remember that excess oxygen is not soluble in water. What substances are present in the water solution in the test tube? Apply a simple test to verify your answer. What part of one of the commercial processes for the production of nitric acid does this experiment illustrate?

SOME CHEMISTRY OF SULFUR COMPOUNDS

In this experiment we shall study some chemical properties of some common compounds of sulfur in oxidation states −2, +2, +4, and +6. Sulfur is a Group VI element with the electron configuration, $1s^2, 2s^22p^6, 3s^23p^4$. Its electron dot structure is :$\overset{\cdot}{\underset{\cdot}{S}}\cdot$. By sharing two electrons with two hydrogen atoms it forms the covalent compound H_2S. A saturated aqueous solution of H_2S is weakly acidic and contains small concentrations of H^+ and HS^- ions. In strongly basic solutions the concentrations of HS^- and S^{2-} ions are greater.

By sharing its electrons with more electronegative elements, such as oxygen, sulfur attains positive oxidation states. In sulfur dioxide, four of the sulfur electrons are involved in bonding with oxygen:

$$\ddot{O}\colon\!\underset{\underset{\displaystyle \overset{|}{H}}{}}{\overset{\times\!\times}{S}}\!\times + H\!\times\!\ddot{O}\colon \;\rightleftharpoons\; H\!\times\!\ddot{O}\!\times\!\overset{\times\!\times}{\underset{\displaystyle \ddot{O}\colon}{S}}\!\times\!\ddot{O}\!\times\!H$$

SO_2 is the anhydride of sulfurous acid, H_2SO_3, which is a weak acid, and forms some $H^+ + HSO_3^-$ ions. In basic solutions the equilibria

$$H_2SO_3 \rightleftharpoons H^+ + HSO_3^-$$

$$HSO_3^- \rightleftharpoons H^+ + SO_3^{2-}$$

are shifted to the right, to form more bisulfite and sulfite ions and water.

In sulfur trioxide all six of the sulfur electrons are involved in bonding:

$$\vcenter{\hbox{$:\!\ddot{O}\colon\;\;\ddot{O}.$}}\;\;\overset{\displaystyle S}{\underset{\displaystyle :\!\ddot{O}\colon}{}}\;\; + H\!\times\!\ddot{O}\colon \;\rightleftharpoons\; H\!\times\!\ddot{O}\!\times\!\overset{\displaystyle H}{\overset{\displaystyle :\!\ddot{O}\colon}{S}}\!\times\!\ddot{O}\colon$$

Sulfur trioxide is the anhydride of sulfuric acid, H_2SO_4, one of the most important industrial inorganic chemicals. It is a strong acid whose aqueous solutions contain large concentrations of $H^+ + HSO_4^-$ ions. In basic solutions the SO_4^{2-} ion is the predominant species.

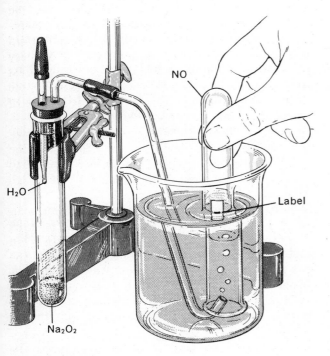

FIGURE 16-4
Generator for the preparation of a small amount of oxygen by the reaction of water with $Na_2O_2(s)$; the reaction of this oxygen with $NO(g)$.

The structures of sulfite, sulfate, and thiosulfate ions are thought to be as follows:

$$:\overset{..}{\underset{..}{O}}:\overset{xx}{\underset{xx}{S}}:\overset{..}{\underset{..}{O}}:^{2-} \qquad :\overset{:\overset{..}{O}:}{\underset{:\overset{..}{O}:}{O}}:^{2-} \qquad :\overset{:\overset{..}{S}:}{\underset{:\overset{..}{O}:}{O}}:^{2-}$$

| Sulfite | Sulfate | Thiosulfate |

Note that in the thiosulfate ion the sulfur atom which replaces the oxygen in the sulfate structure may be assigned a -2 oxidation number, and that the central sulfur atom has an oxidation number of $+6$, just as it has in sulfate. The $+2$ oxidation number assigned to sulfur in thiosulfate is obtained by finding the average of $+6$ and -2: $(+6-2)/2 = +2$.

The oxidation potentials for sulfur in the -2, 0, $+4$, and $+6$ oxidation states in acid and in basic solution are summarized in the following potential diagrams. The numerical values on the lines connecting the couples are the E^0 values in volts.

IN ACID SOLUTION

$$SO_4^{2-} \xrightarrow{+0.17} H_2SO_3 \xrightarrow{+0.45} S \xrightarrow{+0.14} H_2S$$

IN BASIC SOLUTION

$$SO_4^{2-} \xrightarrow{-0.91} SO_3^{2-} \xrightarrow{-0.66} S \xrightarrow{+0.48} S^{2-}$$

These data show that H_2SO_4 and SO_4^{2-} are mild oxidizing agents, whereas H_2S, S^{2-}, H_2SO_3, and SO_3^{2-} are strong reducing agents. Since S, SO_3^{2-}, and H_2SO_3 represent intermediate oxidation states, they can act as oxidizing agents with a species of higher potential and as reducing agents with a species of lower potential. Note how these potential diagrams are applied in the oxidation-reduction reactions occurring in this experiment.

EXPERIMENTAL PROCEDURE: SULFUR COMPOUNDS

Chemicals: Crushed roll sulfur, iron filings, 0.1 F Pb(NO$_3$)$_2$, 0.1 F SnCl$_4$, 0.1 F Zn(NO$_3$)$_2$, 0.05 F I$_2$, 0.1 F Ca(NO$_3$)$_2$, 0.1 F BaCl$_2$, 0.1 F Ba(OH)$_2$, 0.1 F Na$_2$SO$_4$, Na$_2$SO$_3$ crystals, saturated Br$_2$ water, sugar crystals, pieces of Zn, Cu, NaCl(s), KBr(s), KI(s), source of H$_2$S.

NOTE: In your written report, include all observations and equations for all reactions, and relate the changes in oxidation state to the oxidation potential diagrams.

1. Sulfides and Hydrogen Sulfide. Oxidation State -2.
(a) *Preparation of a Sulfide.* Mix approximately 3.5 g of iron filings with 2 g of crushed sulfur

in a crucible supported on a triangle. Place a lid on the crucible, and heat with a Bunsen burner until the reaction is initiated, removing the burner and lid occasionally to note whether the reaction continues with the evolution of heat. Burn off any excess sulfur, and allow the crucible to cool. Place a small piece of the compound into a small test tube. Add a few milliliters of 6 F HCl and note cautiously the products of the reaction.

(b) *Hydrogen Sulfide as a Precipitating Agent.* Metallic sulfides, other than those of the alkali and alkaline earth metals, are sparingly soluble in water. In qualitative analysis many metal ions are identified by precipitating them as sulfides. Using the source of H$_2$S gas available in your laboratory,[1] saturate 3 ml of each of the following solutions with H$_2$S gas: 0.1 F Pb(NO$_3$)$_2$, 0.1 F Zn(NO$_3$)$_2$, and 0.1 F SnCl$_4$. Record the color and formula of each precipitate.

(c) *Hydrogen Sulfide as a Reducing Agent.* Saturate each of the following solutions with H$_2$S: (1) 5 ml of warm 3 F HNO$_3$, (2) 5 ml of 0.05 F I$_2$, and (3) a freshly prepared solution of H$_2$SO$_3$ made by adding a few crystals of Na$_2$SO$_3$ and a drop of 6 F H$_2$SO$_4$ to 5 ml of water. Note the products and write balanced equations for each reaction.

2. Sulfur Dioxide, Sulfurous Acid, Sulfite Ion. Oxidation State $+4$.
(a) *Preparation of Sulfur Dioxide.* Since sulfur is in an intermediate oxidation state in SO$_2$, this compound can be prepared by oxidation of sulfides (as is done in metallurgical roasting), by oxidation of elemental sulfur, or by the reduction of hot concentrated H$_2$SO$_4$ by copper. It can also be conveniently prepared (without a change in oxidation state) from metal sulfites by the addition of a slightly volatile acid.

Dissolve about 2 g of Na$_2$SO$_3$ in 10–15 ml of water in a large test tube. Add a few drops of 6 F HCl, stir, and cautiously note the odor of the solution. Save for part (b).

(b) *Chemical Properties of Sulfurous Acid.* Divide the solution equally among three smaller test tubes. To one portion add a few milliliters of 0.1 F Ba(OH)$_2$ until the solution is basic to litmus. What is the precipitate formed? Is it soluble in 6 F HCl added a drop at a time? (Ignore a slight turbidity, which is due to air oxidation of sulfite to

---

[1] Consult your instructor about your best source of H$_2$S gas. A cylinder of the compressed gas (in hood) or a Kipp generator charged with FeS and 6 F HCl (in hood) is often used. A little "Aitch-tu-ess" (A commercial mixture of sulfur, paraffin, and asbestos) heated in a test tube fitted with a gas delivery tube is a convenient method (see Figure 14-2, Experiment 14).

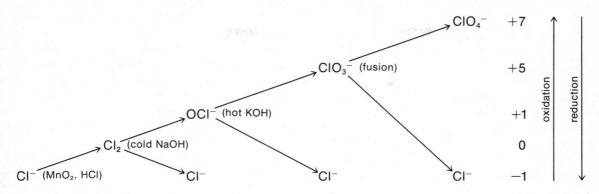

FIGURE 16-5
The interconversions of some ions containing chlorine in several different oxidation states.

sulfate.) To another portion, add 5–6 ml of saturated bromine water, drop by drop. How do you account for the decolorization that takes place? Now add a few milliliters of 0.1 F Ba(OH)$_2$ to this test tube. What is the precipitate formed? Is it soluble in 6 F HCl?

Add a little of the third portion, a few drops at a time, to a test tube containing 5 ml of 3 F H$_2$SO$_4$ and some mossy zinc. Observe the odor of the gas, and note the precipitate formed.

3. Sulfuric Acid, Sulfates. Oxidation State +6. (a) *Chemical Properties of Sulfuric Acid.* While stirring, cautiously add a few milliliters of 18 F H$_2$SO$_4$, drop by drop, to 50 ml of tap water in a small beaker. Note the temperature change. To what do you attribute this result?

Place a few drops of 18 F H$_2$SO$_4$ on a few crystals of sugar in a small evaporating dish. Repeat the test on a small piece of paper or wood (such as a match stick). How do you explain the results?

Investigate the oxidizing strength of concentrated sulfuric acid by adding 1 ml of 18 F H$_2$SO$_4$ to 1 g of each of the following salts in 10-cm test tubes: NaCl, KBr, and KI. Heat each test tube gently and cautiously note the odor, color, and acidity (with moist blue litmus) of the gases evolved. In which solutions are the halide ions oxidized to the elementary halogens?

(b) *Solubility of Sulfates.* Add 3 ml of 0.1 F Na$_2$SO$_4$ to 3 ml each of the following solutions in separate 10-cm test tubes: 0.1 F Ca(NO$_3$)$_2$, 0.1 F BaCl$_2$, 0.1 F Pb(NO$_3$)$_2$. Test the solubility of any precipitates in dilute nitric acid by adding 1 ml of 6 F HNO$_3$ to each precipitate.

SOME CHEMISTRY OF CHLORINE COMPOUNDS

Higher Oxidation States of Chlorine

Chlorine and its compounds show a marked tendency to undergo self- or auto-oxidation-reduction, in which some molecules or ions of a species are oxidized to a higher state while others are reduced to the stable −1 state. This process is called *disproportionation.*

It is possible to oxidize Cl$^-$ (oxidation state −1) to free chlorine, Cl$_2(g)$ (oxidation state 0), and then carry out a series of disproportionation reactions in which the chlorine is successively oxidized to the +1, +5, and finally the +7 oxidation states, as indicated on the flow-chart in Figure 16-5.

(a) Cl$_2(g)$ is passed into a cold basic solution, in which it is auto-oxidized to ClO$^-$, and auto-reduced to Cl$^-$. (This is the reaction that occurs in the commercial preparation of 5% NaClO bleaching solution, the bleach sold in grocery markets.)

(b) The ClO$^-$ in basic solution, when heated, is further oxidized to ClO$_3^-$, and a portion reduced back to Cl$^-$ again. (By suitable crystallization of the salts from this solution, the commercially important oxidant KClO$_3$ or the weed killers NaClO$_3$ and Ca(ClO$_3$)$_2$ may be obtained.)

(c) Maintaining KClO$_3$ crystals at a temperature just above their melting point results in further auto-oxidation of the ClO$_3^-$ to ClO$_4^-$, and reduces a portion of it to Cl$^-$. (Perchlorates are important oxidants in solid rocket fuels.)

EXPERIMENTAL PROCEDURE: CHLORINE COMPOUNDS

Chemicals: NaOCl (a commercial bleach solution of 5% NaOCl), red and blue litmus paper, 0.1 F AgNO$_3$, 6 F NaOH, 6 F HNO$_3$, 0.1 F KI, CCl$_4$, 0.1 F KBr.

Chemical Properties of the Hypochlorite Ion.   Since solid NaOCl cannot be isolated easily without decomposition, we shall test portions of the solution of a commercial bleach obtained at the grocery store. It was prepared by passing chlorine into a solution of NaOH.

(a) *Litmus Reaction.* Put several drops of NaOCl solution on red and blue litmus to note its acidity or basicity. Note any bleaching effect.

(b) *Reaction to AgNO$_3$.* To a 3-ml portion of the NaOCl solution, add 1 ml of 0.1 F AgNO$_3$. What is the precipitate? (Compare with the behavior of a drop of 6 F NaOH on 0.1 F AgNO$_3$.) Is it soluble in 6 F HNO$_3$, and does any other precipitate remain? Explain your observations.

(c) *Oxidizing Strength.* Place 2 ml of 0.1 F KI and 1 ml of CCl$_4$ in a test tube. Add some NaOCl solution by drops—shaking the test tube after each addition—and note any color change in the CCl$_4$ layer. An excess of NaOCl will remove the color, owing to further oxidation of the initial product to the colorless IO$_3^-$ ion.

Repeat the above test, using 2 ml of 0.1 F KBr instead of KI. Where would you place the ClO$^-$ with respect to Br$_2$ and I$_2$ in a scale of oxidizing strength? Acidify the test solution and shake, and note any color in the CCl$_4$ layer. Does the oxidizing strength of ClO$^-$ change when the solution is acidified?

**The Chemistry of
Some Nonmetals of
Groups V, VI, and VII**

NAME

SECTION   LOCKER

INSTRUCTOR   DATE

DATA ON NITROGEN COMPOUNDS (GROUP V)

**The Preparation of Nitric Oxide by Reduction of HNO_3 with Copper: Oxidation
States +2, +4, and +5**

Give the electron-dot structures proposed for NO.

Why is nitric oxide called an *odd* molecule?

Write the equation for the reaction of dilute nitric acid with copper._____

List the properties of NO: color_____ odor_____ ability to support combustion_____

What is the colored gas formed when NO is exposed to air?_____

Write the equation for this reaction. _____

To what extent is NO soluble in water?_____ Is NO_2 soluble in water?_____

How do you account for this?_____

Write the equation for the reaction of NO_2 with H_2O. _____

Write the equation for the production of oxygen from Na_2O_2 and H_2O.

_____

Describe the changes taking place in the test tube half-full of NO gas as oxygen is added: color of gases_____

_____ change in water level_____ Account for these observations.

When the test tube is swirled under water, the water level becomes_____and the color of the gases

becomes_____. Account for these observations.

.

What results were obtained when the above process was repeated?

What substances are present in the aqueous solution in the test tube at the end of the process?

_____

194

DATA ON SULFUR COMPOUNDS (GROUP VI)

1. Sulfides and Hydrogen Sulfide: Oxidation State −2

(a) Write the equation for the preparation of a sulfide. _____

Is the reaction endothermic or exothermic? _____

Show the reaction of ferrous sulfide with hydrochloric acid. _____

(b) Hydrogen sulfide as a precipitating agent. Note the formula and color of the insoluble sulfides prepared.

Pb^{2+} _____ _____ Zn^{2+} _____ _____ Sn^{4+} _____ _____

(c) Write the equations for oxidation of H_2S by:

(1) 3 F HNO_3 _____

(2) 0.05 F I_2 _____

(3) H_2SO_3 _____

2. Sulfur Dioxide, Sulfurous Acid, Sulfite Ion: Oxidation State +4

(a) Write the equation for the preparation of SO_2. _____

(b) Write the equation for reactions of H_2SO_3 with:

(1) $Ba(OH)_2$ _____

Solubility of precipitate in dilute HCl _____

(2) Saturated bromine water (Br_2) _____

Reaction of product with $Ba(OH)_2$ _____

Solubility of precipitate in dilute HCl _____

(3) Zn and H_2SO_4 _____

3. Sulfuric Acid, Sulfates: Oxidation State +6

(a) Physical properties of sulfuric acid

Temperature effect on dilution _____

Explanation:

Effect: on sugar _____ on paper _____

Explanation:

(b) Note the behavior of concentrated sulfuric acid on the following.

	Observations	Equations
Zn		
Cu		
NaCl		
KBr		
KI		

(c) Formulate a solubility rule for sulfates of the common metals.

DATA ON CHLORINE COMPOUNDS (GROUP VII)

Chemical Properties of the Hypochlorite Ion

(a) Is the solution in test tube D acidic or basic?_____

Does it have bleaching properties?_____

(b) Explain and write equations for the reactions with $AgNO_3$.

(c) Write the equation for the reaction
of OCl^- in basic solution with I^-_____

Write the equation for the reaction
of OCl^- in acidic solution with Br^-_____

Place the ClO^- (base), ClO^- (acid), I_2 and Br_2 in order of increasing oxidizing strength.

THE DYNAMICS OF CHEMICAL REACTIONS

HESS' LAW AND THE FIRST LAW OF THERMODYNAMICS

PRE-STUDY

Chemical thermodynamics deals with the energy changes that accompany chemical reactions. Such energy changes are guiding factors in determining (1) how fast a chemical reaction takes place — that is, the problem of chemical kinetics (see Experiment 18), and (2) how complete the reaction will be — that is, the problem of chemical equilibrium (see Experiments 19–23). *Thermochemistry* concerns energy changes manifested as the *heat of reaction, ΔH*. ΔH is the change in enthalpy of a chemical system, $H_2 - H_1$, when it changes from state 1 to state 2. ΔH equals the change in thermal energy when the process is carried out at constant pressure.

A reaction in which heat is lost by the reactants to the surroundings is said to be *exothermic*, where ΔH is negative; one in which heat is absorbed is *endothermic* where ΔH is positive. Energy changes also may be manifested as *electrical energy* measured in terms of the voltage required or produced and the amount of chemical change. Mechanical work can also be done by a system, for example when a gas expands against atmospheric pressure. The maximum amount of available energy a chemical system can produce is measured by the change in the Gibbs *free energy, ΔG*.

Enthalpy changes may be classified into more specific categories: (1) the heat of formation is the amount of heat involved in the formation of 1 mole of the substance directly from its constituent elements in their standard states; (2) the heat of combustion is the amount of heat evolved per mole of a combustible substance, such as carbon or methane, undergoing a reaction with excess oxygen; (3) the heats of solution, vaporization, fusion, and sublimation are related to the hydration of molecules or ions or changes in state; (4) the heat of neutralization is the heat evolved when 1 mole of water is produced by the reaction of an acid and a base. In this experiment, we will measure the heats of neutralization of HCl and of CH_3COOH solutions with a NaOH solution, the heat of solution of NaOH(s), and the heat of reaction of NaOH(s) with a HCl solution. The data should permit you to discover the generalization known as *Hess' law of constant heat summation*.

Enthalpy changes are usually measured in a calorimeter, a simple version of which is shown in Figure 17-1. The purpose of the calorimeter is to isolate thermally the solution under study. All of the heat liberated or absorbed by the chemical reaction is to be given to the contents of the calorimeter or absorbed from them.

The amount of heat required to raise the temperature of 1 g of a substance by 1°C is called the heat capacity of that material. (The term *specific heat* is often used for this quantity.) The units of heat capacity are cal/g °C. The unit cal is an abbreviation for calorie, the quantity of heat required to heat 1 g of water 1°C. It follows that the heat capacity of water is 1.0 cal/g °C. Heat capacities are generally obtained experimentally. The values you will need are given in the appropriate place in the text or report forms. They are to be used to compute the heats of several reactions as follows. You will measure the increase (or decrease) in temperature of a given volume of a particular solution. The amount of heat required to produce this change in temperature is calculated by multiplying the volume of the solution by its density, by its heat capacity and by the temperature change. Dimensional analysis shows that this procedure yields the number of calories released or absorbed:

Volume of solution ml × density of solution g/ml

$$\times \text{ heat capacity } \frac{\text{cal}}{\text{g°C}} \times \text{ temperature change °C}$$

$$= \text{ number of calories.}$$

Correction must be made for the heat absorbed by or evolved from the calorimeter (see Experimental procedure, part 1). Also, if the temperature difference between the calorimeter and the surroundings is appreciable, and if the insulation is insufficient, it may be advisable to take a series of temperature-time readings and to extrapolate a graph of these data back to the time of mixing, to obtain a correct temperature change, Δt°C, for the reaction (see Figure 17-2).

EXPERIMENTAL PROCEDURE

Special Supplies: 2 thermometers (1° divisions, or 0.1° divisions if available), foam plastic hot-cold cups (available from a supermarket) for use as calorimeters, corrugated cardboard for calorimeter covers.

Chemicals: 1.0 F NaOH, 1.0 F HCl, 1.0 F CH_3COOH, NaOH (pellets), phenolphthalein indicator.

1. The Water Equivalent of the Calorimeter. Prepare two calorimeters, each similar to the form illustrated in Figure 17-1, as directed. Compare your two thermometers by immersing them together in water at room temperature for 1 minute, and reading the temperature of each as nearly as possible to the nearest 0.1°C. Be careful to avoid parallax in your readings. Always use the same thermometer in the calorimeter in which the temperature change

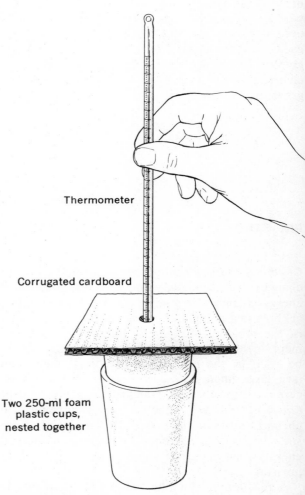

Thermometer

Corrugated cardboard

Two 250-ml foam plastic cups, nested together

FIGURE 17-1
A simple calorimeter for the measurement of the heat of a reaction.

occurs, and in all subsequent readings apply any necessary correction to the other, so that the readings of both thermometers will always correspond.

To correct for the heat lost to the calorimeter, carry out measurements in a manner similar to those to be used later. Place 50.0 ml of tap water at room temperature in one calorimeter, and 50.0 ml of tap water which has been heated to 15–20°C above room temperature in the other. With the lids and thermometers in place, make careful temperature readings (±0.1°C) of each at 1-minute intervals for 3 minutes. At the next minute interval, pour the warmer water quickly and as completely as possible into the other calorimeter, and continue the readings for the next 3 minutes. Use the thermometer in the cooler water to record subsequent temperatures. You can extrapolate the temperatures of the separate samples and of the mixture back to the time of mixing by making a graph on which you plot the temperature of each along the ordinate (vertical axis) and the mixing time for each along the abscissa (horizontal axis). A sample graph is illustrated in Figure 17-2. (If temperature changes are only slight, you need not make the graph but can instead, simply extrapolate the data.)

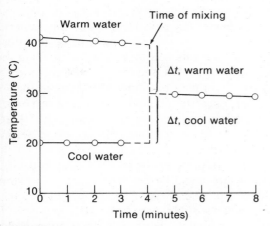

FIGURE 17-2
Typical data on temperature and time that are obtained in the determination of the water equivalent of the calorimeter.

Calculate the heat lost by the warmer water, and the heat gained by the cooler water. (Weight water × Δt°C × heat capacity. Assume the density of water = 1.0 g/ml, and the heat capacity = 1.0 cal/g °C.) The difference, representing the calories gained by the calorimeter, divided by the temperature increase of the cooler water, gives the water equivalent of the calorimeter in cal/°C. Note that the

units of the water equivalent (or heat capacity) of the calorimeter are cal/°C. It is unnecessary to express this result on a gram basis. It simply measures the amount of heat required to raise the styrofoam cup (and the thermometer) 1°C. Repeat this determination as a check, and average your results.

The following examples will illustrate how the water equivalent may be determined, and later used as a correction factor.

Example 1. Water equivalent of the calorimeter data:

Temperature of 50 ml of warmer water	37.9°C
Temperature of 50 ml of cooler water	20.9°C
Temperature after mixing	29.1°C
Heat lost by warm water: (50 g × 8.8°C × 1.0 cal/g °C) =	440 cal
Heat gained by cool water: (50 g × 8.2°C × 1.0 cal/g °C) =	410 cal
Heat lost to calorimeter	30 cal
Water equivalent of this calorimeter: (30 cal/8.2°C)	3.7 cal/°C

Example 2. Use of this water equivalent in the calculation of the heat of a reaction: During a reaction in this calorimeter, 100 ml of a solution increases in temperature by 6.5°C.

Heat gained by the water: (100 g × 6.5°C × 1.0 cal/g °C) = 650 cal	
Heat gained by the calorimeter: (3.7 cal/°C × 6.5°C) =	24 cal
Heat of the reaction =	674 cal

---

2. The Heat of Neutralization of HCl(*aq*) and NaOH(*aq*). Place 50.0 ml of 1.0 *F* HCl in one calorimeter, and 50.0 ml of 1.0 *F* NaOH in the other calorimeter. With the lids and thermometers in place, read the temperatures (±0.1°C) for 3 minutes at 1-minute intervals, quickly mix the NaOH thoroughly into the HCl solution, and continue the readings for 3 minutes at 1-minute intervals. Extrapolate the temperatures back to the time of mixing for each solution as in step 1, and calculate the heat of neutralization per mole of water produced. Refer to Figure 17-3. (The density of the 0.5 *F* NaCl produced is 1.02 g/ml, and its heat capacity is 0.96 cal/g °C.)

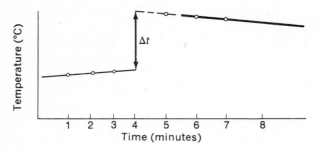

FIGURE 17-3
A time-temperature graph for a solution before and after a chemical reaction. The temperature of the solution is measured three times at one-minute intervals. The solution is gradually warming, indicating that it is below room temperature. After four minutes have elapsed, another solution, which yields an exothermic reaction, is added, causing the temperature to rise. Temperatures are measured again at one-minute intervals, beginning at the fifth minute. The temperature slowly falls between the fifth and seventh minutes, indicating that the solution is now above room temperature.

Rinse the calorimeter with water between each experiment.

3. The Heat of Neutralization of CH₃COOH(aq) and NaOH(aq). Repeat the procedure of step 2, this time using 50.0 ml of 1.0 F CH₃COOH and 50.0 ml of 1.0 F NaOH. Calculate the heat of neutralization as before. (Assume the same density and heat capacity as for NaCl in step 2.)

4. The Heat of Solution of NaOH(s). Carefully weigh (to ±0.01 g) about 2.00 g (0.05 gfw) of NaOH(s). (Because of its hygroscopic nature, weigh this, by difference, in a stoppered 25- or 50-ml Erlenmeyer flask used as a weighing bottle. Your instructor will tell you the approximate number of NaOH(s) pellets required, to assist in estimating the weight needed.) Measure 50.0 ml of distilled water into your calorimeter. With the lid and thermometer in place, read the temperature (±0.1°C) for 3 minutes at 1-minute intervals; then add the NaOH(s), replace the lid and thermometer, and gently swirl the mixture and stir it with the

thermometer to complete solution as quickly as possible. At the same time continue the temperature readings at 1-minute intervals for 4 minutes.

Because of the time required for solution and complete mixing, the proper estimate of the temperature for complete solution at the time of mixing is more difficult. Make your best estimate based on the maximum temperature attained and on an extrapolation of temperature readings back to the time of mixing.

Calculate the heat of solution per mole NaOH(s) accompanying the formation of a NaOH solution whose final concentration is 1.0 F. (Note that you have about 52 g of solution. The heat capacity of 1.0 F NaOH is 0.94 cal/g °C.

5. The Heat of Reaction of HCl(aq) and NaOH(s). Again carefully weigh about 2.00 g (to ±0.01 g) of NaOH(s), using a stoppered Erlenmeyer flask as in step 4. Measure about 55 ml of 1.0 F HCl[1] in a 100-ml graduate, and dilute this to 100.0 ml, as precisely as you can. Transfer this completely to your calorimeter and, with the lid and thermometer in place, take temperature readings each minute for 3 minutes; then add the NaOH(s), replace the lid and thermometer, gently swirl the mixture and stir it with the thermometer to complete solution as quickly as possible, and at the same time continue temperature readings for 4 minutes at 1-minute intervals. As in step 4, extrapolate these temperatures back to the time of mixing remembering that there may be some time delay in the solution reaching its maximum temperature due to the time required for the dissolution of the NaOH pellets. Calculate the heat of the reaction per mole of water formed.

Note that the density and the heat capacity of the NaCl(aq) will be about the same as in step 2. Calculate on the basis of the weight of NaOH(s) used, since a slight excess of HCl was specified to ensure complete neutralization. You may check this point with phenolphthalein.

---

[1]This amount provides a small excess of HCl to react with *all* of the NaOH used.

DATA

1. The Water Equivalent of the Calorimeter

Comparison readings of thermometers: *No. 1* _____ *No. 2* _____ Correction to use on *No. 2* _____
50.0 ml water placed in each calorimeter; temperature readings for three 1-minute intervals for separate calorimeters, mix on 4th minute, continue readings 5th, 6th, 7th minutes:

Trial 1   Warmer calorimeter: _____ _____ _____
                                          1   2   3
                                                                   (mix) _____ _____ _____ _____
                                                                              4   5   6   7
            Cooler calorimeter: _____ _____ _____
                                          1   2   3
            Temperatures at time of mixing
            (from graph or by extrapolation): Warmer _____   Cooler _____   Mixture _____

Trial 2   Warmer calorimeter: _____ _____ _____
                                          1   2   3
                                                                   (mix) _____ _____ _____ _____
                                                                              4   5   6   7
            Cooler calorimeter: _____ _____ _____
                                          1   2   3
            Temperatures at time of mixing
            (from graph or by extrapolation): Warmer _____   Cooler _____   Mixture _____

	Calculation	(1)	(2)
Heat lost by warm water		cal	cal
Heat gained by cool water		cal	cal
Heat lost to calorimeter		cal	cal
Water equivalent of calorimeter		cal/°C	cal/°C

2. The Heat of Neutralization of HCl(*aq*) and NaOH(*aq*)

50.0 ml each of 1.0 *F* HCl and 1.0 *F* NaOH, at room temperature, are placed in respective calorimeters; with temperature readings as follows:

Trial 1   HCl calorimeter: _____ _____ _____
                                       1   2   3
                                                                   (mix) _____ _____ _____ _____
                                                                              4   5   6   7
            NaOH calorimeter: _____ _____ _____
                                        1   2   3
            Temperatures at time of mixing
            (from graph or by extrapolation): HCl _____   NaOH _____   Mixture _____

Trial 2 HCl calorimeter: _____ _____ _____
                       1   2   3

                                                (mix) _____ _____ _____ _____
                                                           4   5   6   7

       NaOH calorimeter: _____ _____ _____
                              1   2   3

Temperatures at time of mixing
(from graph or by extrapolation): HCl_____ NaOH_____ Mixture_____

	Calculation	(1)	(2)
Heat gained by mixed solutions[1]		cal	cal
Heat gained by calorimeter		cal	cal
Total heat of the reaction		cal	cal
Heat of neutralization per mole of water formed		cal/mole	cal/mole

[1]For temperature gain, $\Delta t°C$, use the temperature of the mixture minus the average temperature of HCl and NaOH. The density of 0.5 F NaCl produced is 1.02 g/lite and its heat capacity is 0.96 cal/g °C.

3. The Heat of Neutralization of CH₃COOH(*aq*) and NaOH(*aq*)

50.0 ml each of 1.0 F CH_3COOH and 1.0 F NaOH, at room temperature, are placed in respective calorimeters; temperature readings as follows:

Trial 1 CH₃COOH calorimeter: _____ _____ _____
                               1   2   3

                                         (mix) _____ _____ _____ _____
                                                       4   5   6   7

       NaOH calorimeter: _____ _____ _____
                          1   2   3

Temperatures at time of mixing
(from graph or by extrapolation): CH₃COOH_____ NaOH_____ Mixture_____

Trial 2 CH₃COOH calorimeter: _____ _____ _____
                               1   2   3

                                         (mix) _____ _____ _____ _____
                                                       4   5   6   7

       NaOH calorimeter: _____ _____ _____
                          1   2   3

Temperatures at time of mixing
(from graph or by extrapolation): CH₃COOH_____ NaOH_____ Mixture_____

	Calculation	(1)	(2)
Heat gained by mixed solutions[1]		cal	cal
Heat gained by calorimeter		cal	cal
Total heat of the reaction		cal	cal
Heat of neutralization per mole of water formed		cal/mole	cal/mole

[1]The NaC₂H₃O₂ solution has about the same density and heat capacity as the 0.5 F NaCl resulting from the mixture in part 2 (See the footnote above).

4. The Heat of Solution of NaOH(s)

(Record the weight data for NaOH(s) in the table below.)

Temperature readings for 50.0 ml of water placed in a calorimeter, before and after adding NaOH(s) were as follows:

Trial 1 _____ _____ _____ (mix) _____ _____ _____ _____
         1   2   3   4   5   6   7   8

  Temperatures at time of mixing
  (from graph or by extrapolation): H$_2$O_____ Solution_____

Trial 2 _____ _____ _____ (mix) _____ _____ _____ _____
         1   2   3   4   5   6   7   8

  Temperatures at time of mixing
  (from graph or by extrapolation): H$_2$O_____ Solution_____

	Calculation	(1)	(2)
Weight NaOH(s) + flask		g	g
Weight flask		g	g
Weight NaOH(s)		g	g
Heat gained by solution[1]		cal	cal
Heat gained by calorimeter		cal	cal
Total heat of reaction		cal	cal
Heat of solution per gfw NaOH(s)		cal/gfw	cal/gfw

[1]Note the weight of solution is about 52 g (H$_2$O + NaOH). The heat capacity of the NaOH solution is about 0.94 cal/g °C. The density and specific heat are about the same as in procedure 2.

5. The Heat of Reaction of HCl(aq) and NaOH(s).

(Record the weight data for NaOH(s) in the table at the top of the next page.)

Temperature readings for 100.0 ml of solution, (55 ml of 1.0 F HCl + H$_2$O), placed in a calorimeter, before and after adding NaOH(s) were as follows:

Trial 1 _____ _____ _____ (mix) _____ _____ _____ _____
         1   2   3   4   5   6   7   8

  Temperatures at time of mixing
  (from graph or by extrapolation): HCl_____ Solution_____

Trial 2 _____ _____ _____ (mix) _____ _____ _____ _____
         1   2   3   4   5   6   7   8

  Temperatures at time of mixing
  (from graph or by extrapolation): HCl_____ Solution_____

Calculation		(1)	(2)
Weight NaOH(s) + flask		g	g
Weight flask		g	g
Weight NaOH(s)		g	g
Heat gained by solution[1]		cal	cal
Heat gained by calorimeter		cal	cal
Total heat of reaction		cal	cal
Heat of solution per gfw NaOH(s)		cal/gfw	cal/gfw

[1]The density and specific heat are about the same as in procedure 2.

INTERPRETATION OF THE DATA

1. Compare the heats of reaction per mole of water formed for: $1.0\ F$ HCl + $1.0\ F$ NaOH, $1.0\ F$ HCl + NaOH(s), and NaOH(s) + H_2O. Explain fully. These data provide an excellent illustration of *Hess' law of constant heat summation,* which is one application of the law of conservation of energy, or the *first law of thermodynamics.*

2. Compare your data for the heat of neutralization of $1.0\ F$ HCl + $1.0\ F$ NaOH with that for $1.0\ F$ CH_3COOH + $1.0\ F$ NaOH, and interpret the results in terms of the net ionic equations for the reactions.

3. Apply Hess' law to the reaction of CH_3COOH and NaOH solutions, by writing the step equations for the ionization of CH_3COOH, the neutralization of H^+ and OH^-, and the overall reaction; include in each the appropriate heat quantities. Interpret these equations.

4. The heat of neutralization of HCN(aq) with NaOH(aq) is 2900 cal/mole. Write equations for the three processes occurring in this reaction (as in problem 3 above), and include the heat term appropriate to each.

THE RATE OF
CHEMICAL REACTIONS:
CHEMICAL KINETICS

PRE-STUDY

The economic feasibility of a commercial process depends both on *how fast* the reacting substances interact and upon *how far* or completely they react. The second question "How far?" is considered in the experiments of Section VI. The first question "How fast?" or the problem of chemical kinetics, considered in this experiment, is of theoretical as well as practical interest, for the relative rates of the forward and the reverse reactions in a chemical equilibrium are important factors in determining that equilibrium. Furthermore, a study of the factors affecting the rate of a reaction often gives important information about the very nature of the reaction itself.

1. *How does the **concentration** of each of the reacting substances affect the rate?* If a reaction occurs in a single step involving only one reactant particle (ion or molecule), the rate of the forward reaction must be proportional to the concentration of this one particle; that is, if we double the

concentration the rate will be doubled, if we triple it, the rate will be tripled, and so on. Such a reaction is called a homogeneous[1] *first-order* reaction. We may express it for the generalized reaction

$$A \rightarrow \text{products}$$

by the rate expression

$$\text{rate} = k(A)$$

Throughout this discussion and this experiment we will neglect the rate of the reverse reactions. This choice is valid for the experimental conditions selected, allowing us to measure initial rates only when the concentrations of reactants are high and

---

[1] A *homogeneous reaction* is one occurring uniformly throughout *one phase* (a solution, for example), while a *heterogeneous reaction* is one occurring at the interface of *two phases* (a solid and liquid, or gas and solid, for example). The reaction rate of the latter type depends upon the extent of subdivision and surface area of the phases, and upon adsorption and other phenomena.

thus almost constant and the concentrations of products are very low.

The proportionality constant k (the specific rate constant) is simply the rate of the reaction when the concentration of A is 1 mole/liter. The parentheses around the symbol A means the "concentration of A", expressed in moles per liter, or for gases in pressure units.

For *second-order reactions*, which depend upon the interaction of *two* particles, we may write

$$A + B \rightarrow \text{products}$$

for which the second-order rate expression may be written:

$$\text{rate} = k(A)(B)$$

The rate may be measured by the decrease in concentration per unit time for either reactant, and will be proportional to the *product* of the concentrations of each. For example, if we double the concentration of A and triple the concentration of B, the rate will be increased *six*fold.

Similarly, for *third-order reactions,* such as

$$A + B + C \rightarrow \text{products} \qquad (1)$$

$$2A + B \rightarrow \text{products} \qquad (2)$$

we can write corresponding rate expressions

$$\text{rate} = k(A)(B)(C) \qquad (1')$$

$$\text{rate} = k(A)^2(B) \qquad (2')$$

In equation (2) the rate of decrease in concentration of A is twice that of B, and the corresponding rate expression supports the idea that when the coefficient in the equation is 2 ($2A$ in this example), then normally the rate would be proportional to the *square*; that is, if we triple the concentration of A, the rate will increase *nine*fold. However, for higher-order reactions, *such rate expressions based on the stoichiometric equation are seldom correct.* The above equations imply that the simultaneous collision of three particles—a highly improbable occurrence—constitutes a single reaction. The stoichiometric equation is more often the sum of several reactions, called *elementary processes,* at least one of which may be distinctly slower and is therefore the rate-controlling reaction. The rate expression would then correspond to the rate for that elementary process.

The above sets of reactions are called first-, second-, and third-order reactions respectively, depending on whether the sum of the exponents

of the concentration terms equals 1, 2, or 3. It must be emphasized that the equations given for the rates of these reactions are valid only if these reactions are *elementary processes.*

If one (or more) of the reactants is present in great excess, its relative concentration (or their concentrations) will change very little compared with the concentration change for another reactant present in smaller amount. The reaction is thus *pseudo-first-order* with respect to this latter reactant. This technique of using a large excess of a given reagent, or the related procedure of varying the concentration of only one constituent at a time and measuring the corresponding rate change, is often used as a means of experimentally studying the mechanism of a reaction.

2. *How does the **temperature** at which a reaction occurs affect the rate?* From considerations of kinetic theory, we know that only a small proportion of the many molecular collisions result in reaction. In most collisions the molecules simply rebound like billiard balls, without change. Only the more energetic molecules collide with enough energy for their electron orbitals to interpenetrate, thus forming a momentary "activated complex," and then more stable products. The proportion of such energetic molecules in a mixture depends on the absolute temperature, and increases rapidly as the temperature is increased. Thus the rate of a reaction generally increases rapidly with increasing temperature. A common rule of thumb, which is far from quantitatively true in many reactions, is that the reaction rate is doubled for each 10°C increase in temperature.

3. *How does a **catalyst** affect a reaction?* A catalyst is a substance that permits the formation of the activated complex more easily, and at a lower temperature. With a catalyst present at a given temperature, a large proportion of the molecules will possess enough energy to react, and thus the rate of the reaction is increased. Although a catalyst is never consumed in a final overall reaction, it may participate in the formation of the activated complex in an intermediate step-reaction, and then be re-formed in subsequent steps to produce the final products.

The Experimental Method

Hydrogen peroxide contains oxygen in its intermediate oxidation state of -1, and can act both as an oxidizing agent and as a reducing agent. It is

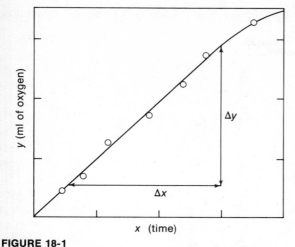

FIGURE 18-1

A graph showing a linear or direct proportional relationship between a quantity y (ml of oxygen gas), and a
quantity x (time in minutes), during the first portion of
an experiment. To obtain the slope, defined as Δy divided
by Δx, choose two points on the line and divide the *change
in y* (Δy) by the *change in x* (Δx) between these two points.
Both y and x must be measured in the same units used
in plotting y and x on the graph.

unstable and undergoes disproportionation according to the equations

$$H_2O_2(aq) \rightarrow O_2(g) + 2H^+ + 2e^- \quad \text{(oxidation)} \quad (3)$$
$$H_2O_2(aq) + 2H^+ + 2e^- \rightarrow 2H_2O \quad \text{(reduction)} \quad (4)$$
$$2H_2O_2(aq) \rightarrow O_2(g) + 2H_2O \quad \text{(total reaction)} \quad (5)$$

This decomposition normally is quite slow, but can
be catalyzed by iodide ion, I^-. Note from Table 13,
Appendix C, that H_2O_2 can oxidize I^- to I_2, or
reduce I_2 to I^-, depending on the *p*H. In an approximately neutral solution, the resulting I^- and
I_2 mixture, possibly through the intermediary of an
oxidation-reduction "activated complex," does
catalytically increase the rate of this decomposition
to the point at which we can measure, with a gas
measuring tube, the rate of formation of one of the
decomposition products, O_2 gas.

If equation (5) represents an elementary process,
it follows that the reaction will be second-order. Is
it? Is there evidence for some intermediate step-
reaction involving both H_2O_2 and I^-? To answer
this, we shall determine as quantitatively as possible, the relationship between the concentrations
of H_2O_2 and of I^- and the rate of the reaction. In
other words, your problem is to determine the
exponents, *x* and *y*, in the rate law

$$\text{rate} = k(H_2O_2)^x(I^-)^y$$

for the decomposition of H_2O_2 into $O_2(g)$ and H_2O,
when catalyzed by I^-.

Graphical Relations and Experimental Data.
Scientists frequently graph their experimental
data, as a practical means of averaging individual
measurements and of discovering fundamental relationships. For each series of measurements, you
will make a graph of the milliliters of oxygen gas
evolved (plotted on the ordinate) for each half-
minute time interval (plotted on the abscissa). The
slope of this line, or of the tangent to the curve at
a given point, will express the rate of the reaction
in milliliters per minute, ml/min. Figure 18-1 shows
how you may evaluate this slope in terms of in-
crements of milliliters and of minutes, as measured
on your graph.

EXPERIMENTAL PROCEDURE

The Rate of Decomposition of Hydrogen Peroxide

Special Supplies: 50-ml buret or gas measuring tube, leveling
bulb, thermometer, water bath, 10-ml pipet (or your 10-ml graduated cylinder), a watch with a second hand, ice.

Chemicals: 3% H_2O_2 (commercial grade), 0.10 *F* KI.

NOTE: Two students should cooperate in this experiment,
one swirling the flask, noting the half-minute time intervals,
and recording data, while the other maintains the water
levels and reads the gas volumes.

Set up the apparatus as illustrated in Figure 18-2,
on the next page. A 50-ml buret may conveniently
be used in place of a regular gas measuring tube,
which is preferable if available. A beaker with a
bent glass syphon tube, B, is a satisfactory substitute for the leveling bulb, A, if necessary. What is
the function of the gas leveling bulb? *Soft* rubber
stoppers are essential to insure gas tight connections.

1. The Effect of Concentration. You will carry
out a series of measurements with each of the following mixtures.[2]

REAGENTS	MIXTURE 1	MIXTURE 2	MIXTURE 3
0.10 *F* KI	15.0 ml	15.0 ml	30.0 ml
distilled H_2O	35.0	25.0	20.0
3% H_2O_2	10.0	20.0	10.0
total	60.0 ml	60.0 ml	60.0 ml

[2]Your instructor may designate different volumes of H_2O and
KI, or a different concentration of KI, if necessary to achieve a
convenient reaction rate. Different sources of H_2O_2 may vary in
stability, because various stabilizing agents may be used.

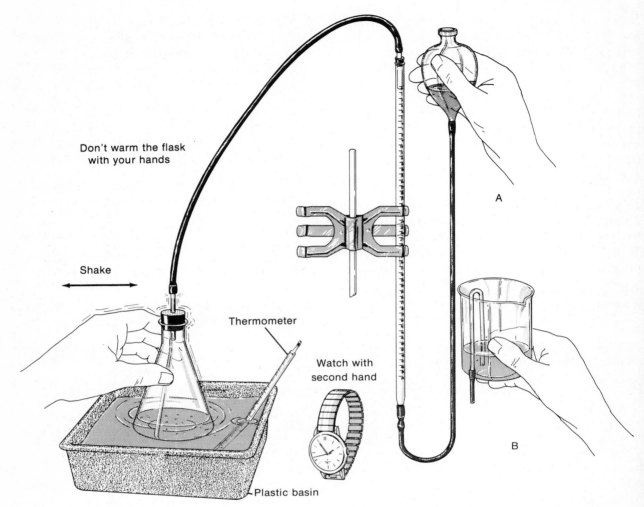

FIGURE 18-2
An apparatus for measurement of the rate of gas evolution from a reaction mixture in the flask.

Note the following precautions for careful work:
(1) Measure each volume for the mixtures as precisely as possible with a 50-ml graduated cylinder or pipet, first rinsing it with distilled water and a very little of the reagent, and filling exactly to the mark; then empty it into your reaction flask, draining the last drop as directed.

(2) Have the water in the bath 1–2 inches deep (to avoid undue splashing as you swirl the flask), and carefully adjust the temperature of this bath, and of each solution, to room temperature, so that there will be only a negligible temperature change during a run. Maintain this *same* temperature for each mixture.

(3) Hold the reaction flask near the top, to avoid warming it with your hand while swirling it.

Procedure for each mixture. (a) Carefully measure the required volume of 0.10 F KI in your graduated cylinder, and transfer it as completely as you can into the reaction flask.

(b) Rinse the graduated cylinder and fill it with the required volume of distilled water. Then measure precisely the volume of H_2O_2 required, preferably with a 10-ml pipet or in your 10-ml graduated cylinder (use two fillings for mixture 2), and transfer this carefully and completely into the water in a 50-ml graduated cylinder.[3]

(c) Hold the leveling bulb high so that the water level in the gas measuring tube is well above the

---

[3]This technique provides for a quite rapid and complete transfer and mixing of reagents at the start of the experiment.

zero mark. When all is ready, quickly add the H_2O and H_2O_2 to the reaction flask as completely as you can, insert the stopper firmly so it is gas tight, and begin swirling the mixture at a uniform, moderate rate in the bath.

(d) One student should keep the water in the leveling bulb and in the measuring tube at the same level, and read the volumes of gas to ± 0.1 ml while the other student swirls the flask, notes the time for each half-minute reading, and records the data. Begin your first reading at zero time after 1–2 ml of gas has been generated and conditions have become uniform. Continue the readings until about 20 ml of $O_2(g)$ has been generated, or for only 8 to 10 minutes with the slower reacting mixtures. When finished, insert a thermometer in the reaction mixture and check its temperature to $\pm 0.1°C$.

2. The Effect of Temperature. Adjust the temperature of the bath, and of each reagent and the distilled water, to a temperature about 10°C above room temperature, and repeat the measurements for Mixture 1 at this elevated temperature, again noting the final reaction temperature of the mixture.

If time is available, adjust the temperature of the bath and all reagents to about 10°C below room temperature by adding ice to the bath, and take another series of measurements with Mixture 1.[4]

---

[4]Such a series of measurements for a reaction at different temperatures makes possible a calculation of the activation energy for the reaction. The techniques here are not precise enough for this. For further discussion of this, and experimental directions for such a measurement with another reaction, see Experiment 25 in Frantz, Harper W., and Lloyd E. Malm, *Chemical Principles in the Laboratory*, W. H. Freeman and Company, 1966.

The Rate of Chemical Reactions: Chemical Kinetics

NAME

SECTION   LOCKER

INSTRUCTOR   DATE

DATA ON THE RATE OF DECOMPOSITION OF HYDROGEN PEROXIDE

Time interval (minutes)	Volume of oxygen gas produced (ml)				
	Mixture 1 _____°C	Mixture 2 _____°C	Mixture 3 _____°C	Mixture 1 (warmer) _____°C	Mixture 1 (cooler) _____°C
0.0					
0.5					
1.0					
1.5					
2.0					
2.5					
3.0					
3.5					
4.0					
4.5					
5.0					
5.5					
6.0					
6.5					
7.0					
7.5					
8.0					
8.5					
9.0					
9.5					
10.0					

GRAPHICAL REPRESENTATION OF DATA

Plot all of your results neatly on the following two graphs, with the volume in milliliters on the ordinate, and the time in minutes on the abscissa. (Use as large a scale as practical). Draw the best line or curve through the points for each run to average individual measurements. Label the axes and each line or curve. Determine the slope of the lines for the earlier part of each run, based on measured values of Δ ml and Δ min, neglecting the first readings and also the last ones if their points deviate from a linear relationship. (See Figure 18-1). Calculate each reaction rate in milliliters per minute. Report these calculations, including the method, on a separate, added report sheet.

---

1. The Effect of Concentration on Rate

Plot the results for Mixtures 1, 2 and 3 at room temperature on this graph. Use a code to distinguish the three sets of data.

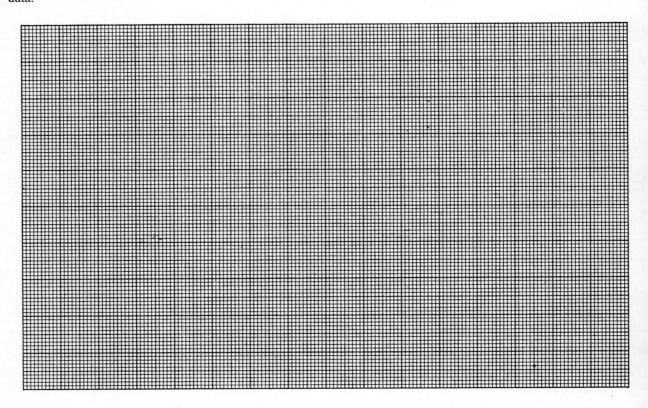

2. The Effect of Temperature on Rate

Plot the results from Mixture 1 (room temperature), Mixture 1 (warmer), and Mixture 1 (cooler) on this graph. Indicate the temperature corresponding to each line.

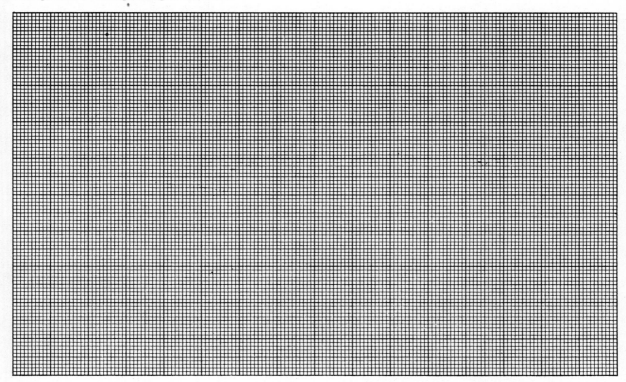

3. The Determination of Kinetic Order

Complete the following table for the data obtained at room temperature. Obtain the necessary concentrations from the experimental descriptions of each mixture. Obtain the rates from the slopes of your graphs.

Mixture	(H_2O_2) (moles/liter)	(I^-) (moles/liter)	Rate (ml/min)
1			
2			
3			

INTERPRETATION OF DATA

1. What conclusion can you draw about the dependency of the reaction rate on the concentration of H_2O_2, and also on the concentration of the catalyst, I^-, at room temperature? Is there evidence for some intermediate complex involving I^-? Write an expression for the rate law, inserting the proper exponents for the concentrations of H_2O_2 and I^-.

2. Evaluate the specific rate constant, k, in this expression. (Calculate the molarity of H_2O_2 and of I^- in each mixture, and then the rate when each of these is $1\ M$). Give the units of k.

3. Discuss in as quantitative a manner as possible the effect of changes in temperature on the rate of the reaction.

4. Why might your experimental curves show some deviation from a linear relationship (a decreasing slope) at the end of each run?

5. What percentage of the hydrogen peroxide has been decomposed in Mixture 1 (10.0 ml of 3% H_2O_2) when 20.0 ml of oxygen gas has been evolved? (Assume 20°C, 740 mm Hg pressure, and a density of 1.00 g/ml for the H_2O_2.)

CHEMICAL EQUILIBRIUM

AN INTRODUCTION
TO CHEMICAL EQUILIBRIUM

The five experiments contained in this section demonstrate a number of features of the principles of chemical equilibrium. In addition, the experiments on qualitative analysis, in Section VIII, are based upon an understanding of chemical equilibrium. The purpose of this study assignment is to provide an introduction to the principles and concepts of this topic. If you are already rather familiar with these concepts, you may find it profitable to turn to the problems at the end of this assignment and verify for yourself that you are able to work them.

QUALITATIVE DESCRIPTION
OF CHEMICAL EQUILIBRIUM

A useful way to distinguish between the concepts of chemical kinetics discussed in the preceding section and chemical equilibrium is to consider a plot of concentration versus time for a particular reaction. Consider the reaction:

$$A + B \rightarrow C \tag{1}$$

which we will take to be an elementary process.

Figure E-1 displays what may happen to the concentration of each species as time goes on, assuming that finite concentrations of the reactants A and B were present initially and had not interacted to produce any product, C.

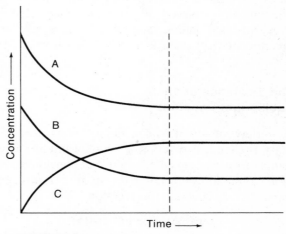

FIGURE E-1
Data on concentration and time for the chemical reaction A + B → C. At the dashed line the system has reached equilibrium.

Chemical kinetics is concerned with the behavior of this chemical system in the time span to the left of the dashed line in Figure E-1. The instantaneous rate of the reaction at any particular time is given by the slope of curve C at that point. This slope exactly equals the negative of the slope of the A and B curves at that time. These slopes are all finite.

To the right of the dashed line in the figure, the slopes are all zero and the system is said to be in chemical equilibrium. No further changes in the net concentration of any of the three substances can be observed, no matter how much longer one might care to observe the system. This does not mean that the system is static and nothing is happening. Indeed, as we will see below, the compounds are constantly being formed and dissociated on a microscopic scale. But macroscopic measurements taken in the laboratory would not reveal any net changes that had taken place as a function of the passage of time after that point represented by the dashed line in Figure E-1.

QUANTITATIVE DESCRIPTION OF CHEMICAL EQUILIBRIUM

It undoubtedly comes as no surprise to you to discover that scientists have sought methods by which they can quantitate the observations made in the foregoing discussion. Indeed, you are probably aware that quantitation is one of the chief activities of the scientist. These studies are particularly important in a course in general chemistry, both because they are concerned with a large number of phenomena observed in the laboratory and in everyday experiences, and because most of them are not studied again in upper-division chemistry courses.

The Existence of an Equilibrium Constant

A variety of studies have demonstrated that for a given chemical reaction at equilibrium at a fixed temperature, there is a fixed ratio of concentrations of all compounds present in the mixture. It has been found that for the general reaction

$$aA + bB \rightleftharpoons cC + dD \qquad (2)$$

the equilibrium constant, K, has the form

$$K = \frac{(C)^c(D)^d}{(A)^a(B)^b} \qquad (3)$$

where the parentheses denote the equilibrium concentrations of each species.

The form of this constant can be demonstrated experimentally. Let us examine an actual example — the reaction of hydrogen and iodine at an elevated temperature to form hydrogen iodide, as expressed by the equilibrium equation

$$H_2(g) + I_2(g) \rightleftharpoons 2HI(g) \qquad (4)$$

Suppose we have two glass bulbs. Bulb I contains 1 mole of hydrogen, H_2, and 1 mole of iodine, I_2. Bulb II contains 2 moles (the equivalent amount) of hydrogen iodide, HI. Both bulbs are heated and maintained at a constant temperature of 400°C, so that all substances are gases. At the start, Bulb I will be violet colored, owing to the iodine vapor present, and Bulb II will be colorless (HI has no color). After a time we will observe that both bulbs have an identical intermediate violet coloration, indicating that the forward and the reverse processes, as illustrated by the arrows in equation (4), have taken place until the same relative concentrations of $H_2(g)$, $I_2(g)$, and $HI(g)$ are present in each bulb. An equilibrium has been attained. This and many similar quantitative measurements demonstrate that equation (3) is valid for all chemical systems in equilibrium.

A second method enabling us to arrive at equation (3) is the powerful technique of considering reaction rates. Let us again consider the general reaction

$$aA + bB \underset{k_r}{\overset{k_f}{\rightleftharpoons}} cC + dD \qquad (5)$$

If these two reactions represent elementary processes, we can write rate expressions for them following the guidelines given in the pre-study of Experiment 18.

$$Rate_{forward} = k_f(A)^a(B)^b \qquad (6)$$

$$Rate_{reverse} = k_r(C)^c(D)^d \qquad (7)$$

As we saw earlier, k_f and k_r represent *specific rate constants*. They are the rates of the reaction that prevail when the concentrations of all species are 1 M.

One definition of equilibrium is that it is the condition in which the rates of the forward and reverse reactions are equal. This equivalence is a necessary condition, because any inequality in the two rates would lead to the net production or consumption of some of the compounds in the reaction. Thus this condition requires that

$$Rate_{forward} = Rate_{reverse} \qquad (8)$$

Ag⁺   Cl⁻

FIGURE E-2
Equilibria at the crystal surface in a saturated silver chloride solution. Silver ions and chloride ions are leaving the crystal surface and returning to it at equal, opposing rates.

Substitution of the rate expressions of equations (6) and (7) into equation (8) yields an interesting result.

$$k_f(A)^a(B)^b = k_r(C)^c(D)^d \qquad (9)$$

This equation may be rearranged in either of two ways to yield a constant. *By convention*, chemists always choose to place the *product* concentration terms in the numerator.

$$\frac{k_f}{k_r} = \frac{(C)^c(D)^d}{(A)^a(B)^b} \qquad (10)$$

It should be apparent that the expression on the right has both the form and the value of the equilibrium constant because these concentrations are those present at equilibrium.

$$K = \frac{(C)^c(D)^d}{(A)^a(B)^b} \qquad (11)$$

The dynamic nature of the state of equilibrium is displayed in Figure E-2. It shows the larger chloride ions and smaller silver ions constantly joining the crystal lattice (precipitation) and leaving it (dissolution). The reaction for this process is

$$AgCl(s) \rightleftharpoons Ag^+ + Cl^- \qquad (12)$$

A third route to arrive at the existence of an equilibrium constant is through thermodynamics.

We will forgo a rigorous derivation of this expression and only note that there is a relationship between the standard Gibbs free energy change for a reaction, $\Delta G°$, and the equilibrium constant, K.

$$\Delta G° = -RT \ln K \qquad (13)$$

where R is the ideal gas constant and T is the absolute temperature. The value of $\Delta G°$ is given by the change in free energy when the reaction corresponding to K is carried out holding all concentrations at 1 M. Because $\Delta G°$ is a constant at a given temperature, it follows that K is also a constant. The form of the K derived by thermodynamics is identical to equation (11).

Conventions Regarding the Equilibrium Constant

By convention, the concentrations of pure solids and pure liquids are not included in equilibrium constant expressions. We can understand the reason for this decision on the basis of thermodynamic activities or more simply if we combine the concentration of the pure substance into the original constant. For example, in the reaction

$$H_2O(l) \rightleftharpoons H^+ + OH^- \qquad (14)$$

we begin by writing the equilibrium constant in the form of equation (11).

$$K = \frac{(H^+)(OH^-)}{(H_2O)} \qquad (15)$$

However, the concentration of water is a constant and equals 55.5 M. Thus we can combine it with the original constant K and call it K_w.

$$K_w = 55.5 \, K = (H^+)(OH^-) \qquad (16)$$

In all subsequent equations, we will use K_w. It is called the ion product of water.

Similarly the dissolution of a slightly soluble salt such as barium sulfate is described by a solubility product, K_{sp}.

$$BaSO_4(s) \rightleftharpoons Ba^{2+} + SO_4^{2-} \qquad (17)$$

$$K_{sp} = K(BaSO_4(s)) = (Ba^{2+})(SO_4^{2-}) \qquad (18)$$

Two shorthand notations are used to write equilibrium constants for ionization reactions for weak acids. An accurate description of this process is

$$CH_3COOH + H_2O \rightleftharpoons H_3O^+ + CH_3COO^- \qquad (19)$$

TABLE E-1
Some examples of equilibrium constant expressions

Reaction	Equilibrium constant expressions
$H_2O(l) \rightleftharpoons H_2O(g)$	$K = P_{H_2O}$
$I_2(s) \rightleftharpoons I_2(g)$	$K = P_{I_2}$
$H_2O(l) \rightleftharpoons H^+ + OH^-$	$K_w = (H^+)(OH^-)$
$2CrO_4^{2-} + 2H^+ \rightleftharpoons Cr_2O_7^{2-} + H_2O(l)$	$K = \dfrac{(Cr_2O_7^{2-})}{(CrO_4^{2-})^2(H^+)^2}$
$Fe^{3+} + SCN^- \rightleftharpoons Fe(SCN)^{2+}$	$K = \dfrac{(Fe(SCN)^{2+})}{(Fe^{3+})(SCN^-)}$
$NH_3(aq) + H_2O(l) \rightleftharpoons NH_4^+ + OH^-$	$K_b = \dfrac{(NH_4^+)(OH^-)}{(NH_3)} = \dfrac{K_w}{K_a}$
$NH_4^+ + H_2O(l) \rightleftharpoons H_3O^+ + NH_3(aq)$	$K_a = \dfrac{(H^+)(NH_3)}{(NH_4^+)} = \dfrac{K_w}{K_b}$
$HAc(aq) \rightleftharpoons H^+ + Ac^-$	$K_a = \dfrac{(H^+)(Ac^-)}{(HAc)} = \dfrac{K_w}{K_b}$
$Ac^- + H_2O(l) \rightleftharpoons HAc(aq) + OH^-$	$K_b = \dfrac{(HAc)(OH^-)}{(Ac^-)} = \dfrac{K_w}{K_a}$
$Hg_2Cl_2(s) + 2NH_3(aq) \rightleftharpoons$ $Hg(NH_2)Cl(s) + Hg(l) + NH_4^+ + Cl^-$	$K = \dfrac{(NH_4^+)(Cl^-)}{(NH_3)^2}$
$Ag(NH_3)_2^+ \rightleftharpoons Ag^+ + 2NH_3(aq)$	$K = \dfrac{(Ag^+)(NH_3)^2}{(Ag(NH_3)_2^+)}$
$Pb(OH)_3^- + H^+ \rightleftharpoons Pb(OH)_2(s) + H_2O(l)$	$K = \dfrac{1}{(Pb(OH)_3^-)(H^+)}$

By the convention given above, we do not include the (H_2O) in writing the equilibrium constant expression, and H_3O^+ is shortened to H^+, partly because the actual number of water molecules bound to each proton is not known. The numerical value of this expression is equal to a constant called the acid ionization constant and is given the symbol K_a.

$$K_a = \frac{(H^+)(CH_3COO^-)}{(CH_3COOH)} \qquad (20)$$

By convention, concentrations of substances in the gas phase are expressed in partial pressures in atmospheres. This follows from the ideal gas equation (concentration $= n/V = P/RT$).

Examples of Equilibrium Constant Expressions

Table E-1 gives a number of examples of equilibrium constants for reactions or phase transformations carried out in this manual.

Several features displayed in the table should be noted in addition to the conventions described earlier. Often it is important to include the letters *s, l* or *g* in parentheses after the elemental or molecular formula to denote whether the substance is a solid, liquid, or gas. The concentrations of gases are given in pressure with units of atmospheres. Sometimes the symbol K_b is used to represent the dissociation of a base. As is apparent from the table, the product of K_b and K_a for a conjugate acid-base pair equals K_w.

RESPONSES OF SYSTEMS TO EXTERNAL CHANGES

In the preceding discussion, it has been emphasized that when systems reach chemical equilibrium, only certain concentrations of the reactants and products are allowed. We now raise the question, what happens if an external stress is applied to this system? The answer was provided by the French chemist Henry Louis Le Châtelier (1850–1936). His principle states that a system responds to an external stress in such a way as to relieve the stress. Such stresses are changes in temperature, pressure, or concentrations. We will only be concerned with concentration effects.

Consider the barium sulfate equilibrium

$$BaSO_4(s) \rightleftharpoons Ba^{2+} + SO_4^{2-} \qquad (21)$$

If solid $BaSO_4$ is added to water and the solution shaken for a long time, equilibrium concentrations of Ba^{2+} and SO_4^{2-} ions are obtained. If some Na_2SO_4 is added to this solution, what happens to the original ion concentrations? The immediate effect is to increase the sulfate concentration greatly. (The added Na^+ ions have no effect on the equilibrium.) This increase in concentration represents a stress on the system. The system responds by precipitating some $BaSO_4(s)$ to relieve this stress. This property of shifting the equilibrium concentrations by the addition of an ion involved in the reaction is called the *common ion effect*. In general, the addition of a common ion is an effective technique to reduce the concentration of ions liberated by a precipitate, a weak acid, or a complex ion.

Some care must be exercised, however. For instance, the Ag^+ in solution resulting from the dissolution of $AgCl(s)$ can be diminished by the addition of moderate amounts of Cl^- from, say, NaCl. However, the addition of a large amount of NaCl will not result in the formation of more precipitate but will actually decrease it, owing to the formation of $AgCl_2^-$.

ACID-BASE EQUILIBRIA

The majority of the experiments in this section are concerned with the properties of acids and bases. Therefore, these properties will be discussed more fully than those of precipitates and complex ions.

Definitions

At least three definitions of acids and bases are in common use at present. Arrhenius defined an acid as a substance that liberates H^+ upon dissolution in water, and a base as a substance that releases OH^-. Lewis described acids as electron-pair acceptors and bases as electron donors. This description is a more general one, and is particularly useful in describing organic compounds.

The Brønsted-Lowry definition is the most useful description for substances in aqueous solutions and for the chemical systems we will investigate. According to this classification, an acid is a substance that can donate a proton (H^+), and a base is any substance that can accept a proton. Thus, a Brønsted acid might be represented by HA. The ion A^- is termed the conjugate base. Likewise the conjugate acid of a Brønsted base is the base with a proton added. Some examples are given below:

Equilibria in Water

Water is the ubiquitous solvent of the general chemistry laboratory. It is of considerable importance for you to understand its acid-base properties. Equation (22) is frequently abbreviated to equation (25) by eliminating one of the water molecules as a reactant. (This is the same way in which equations (19) and (20) were formulated.)

$$H_2O \rightleftharpoons H^+ + OH^- \qquad (25)$$

The ion product for water, K_w, is given by

$$K_w = (H^+)(OH^-) \qquad (26)$$

and equals 1.00×10^{-14} at 25°C. This constant must be satisfied in every aqueous solution independent of the origin of either the H^+ or OH^-.

Example 1. Calculate the (H^+) in the following solutions: (a) 1.00 M NaOH, (b) pure water. Also compute the (H^+) in solutions in which the following conditions obtain: (c) $(OH^-) = 3.48 \times 10^{-6}$ M, (d) $(OH^-) = 1.00 \times 10^{-14}$ M.

(a) NaOH is a strong base and dissociates to give

$$(OH^-) = 1.00\ M$$

$$(H^+) = \frac{1.00 \times 10^{-14}}{(OH^-)} = \frac{1.00 \times 10^{-14}}{1.00}$$

$$= \underline{1.00 \times 10^{-14}\ M}$$

(b) In pure water, $(H^+) = (OH^-)$

$$(H^+)(H^+) = 1.00 \times 10^{-14}\ M^2$$

$$(H^+) = \underline{1.00 \times 10^{-7}\ M}$$

(c) $(H^+) = \dfrac{10.0 \times 10^{-15}}{3.48 \times 10^{-6}} = \underline{2.87 \times 10^{-9}\ M}$

(d) $(H^+) = \dfrac{1.00 \times 10^{-14}}{1.00 \times 10^{-14}} = \underline{1.00\ M}$

The range of (H^+) and (OH^-) is approximately from 10 M to 10^{-15} molar. Because of this enormous exponential range, a logarithmic system has been adopted to describe H^+ and OH^- concentrations in aqueous solutions. A small p placed before

Brønsted Acid	+	Brønsted Base	$\rightleftharpoons$	Conjugate Acid	+	Conjugate Base	
H_2O	+	H_2O	$\rightleftharpoons$	H_3O^+	+	OH^-	(22)
$HCl(aq)$	+	H_2O	$\rightleftharpoons$	H_3O^+	+	Cl^-	(23)
$CH_3COOH(aq)$	+	OH^-	$\rightleftharpoons$	H_2O	+	CH_3COO^-	(24)

a symbol indicates the negative logarithm of that quantity. Examples of the use of this notation are:

$$pH = -\log(H^+) \tag{27}$$

$$pOH = -\log(OH^-) \tag{28}$$

$$pK_w = -\log K_w \tag{29}$$

You should practice calculations involving pH, pOH and pK's until the arithmetical manipulations cause no difficulties for you.

Example 2. Calculate the pH's of the four solutions described in the preceding example.

(a) $(H^+) = 1.00 \times 10^{-14}\ M$

$$pH = -\log(H^+) = -\log(1.00 \times 10^{-14})$$
$$= -\log 1.00 - \log 10^{-14}$$
$$= 0 + 14 = \underline{\underline{14}}$$

(b) $(H^+) = 1.00 \times 10^{-7}\ M$

$$pH = -\log(H^+) = -\log(1.00 \times 10^{-7})$$
$$= \underline{\underline{7}}$$

(c) $(H^+) = 2.87 \times 10^{-9}\ M$

$$pH = -\log(H^+) = -\log(2.87 \times 10^{-9})$$
$$= -\log 2.87 - \log 10^{-9}$$
$$= -0.458 + 9.000$$
$$= \underline{\underline{8.542}}$$

(d) $(H^+) = 1.00\ M$

$$pH = -\log(H^+) = -\log 1.00$$
$$= \underline{\underline{0}}$$

Aqueous Solutions of Weak Acids

In several of the experiments that follow, you are to make quantitative calculations using equilibrium constants. A sample calculation and a derivation are given below to show you how to proceed.

Example 3. Calculate the pH of a 0.100 M HAc solution, using the acid ionization constant given in Appendix C, Table 9. The following procedures are helpful first steps in such a calculation.

Let $x =$ the concentration of HAc that dissociates.

$$HAc \rightleftharpoons H^+ + Ac^-$$

Before dissociation: 0.100 0^1 0

After dissociation: $(0.100 - x)$ x x

$$K_a = \frac{(H^+)(Ac^-)}{(HAc)} = \frac{x \cdot x}{0.100 - x} = 1.76 \times 10^{-5}$$

A general guideline is that if the concentration of weak acid is a factor of 10^3 (or more) larger than the ionization constant, x can be dropped from the (concentration of weak acid minus x) term.

$$\frac{x^2}{0.100} = 1.76 \times 10^{-5}$$

$$x^2 = 1.76 \times 10^{-6}$$

$$x = (H^+) = \sqrt{1.76} \times 10^{-3}\ M = 1.33 \times 10^{-3}\ M$$

At this point in the calculation, it is always wise to go back and check the original assumption that x can be neglected in comparison with the concentration of weak acid. In this calculation, we have neglected 0.00133 M in comparison with 0.100 M. This is an error of about 1%, which is generally acceptable.

$$pH = -\log(H^+) = -\log(1.33 \times 10^{-3})$$
$$= -\log 1.33 - \log 10^{-3} = -0.122 + 3.000$$
$$= \underline{\underline{2.878}}$$

In some calculations, it is particularly useful to use a logarithmic form of the ionization constant expression. The following derivation is for the general weak acid, HA.

$$HA \rightleftharpoons H^+ + A^- \tag{30}$$

$$K_a = \frac{(H^+)(A^-)}{(HA)} \tag{31}$$

Taking logarithms of each side, we obtain

$$\log K_a = \log \frac{(H^+)(A^-)}{(HA)}$$

$$= \log(H^+) + \log \frac{(A^-)}{(HA)} \tag{32}$$

---

[1]The (H^+) of $10^{-7}\ M$ from the dissociation of water is so small that it can be neglected here.

Transposition yields:

$$-\log(H^+) = -\log K_a + \log \frac{(A^-)}{(HA)} \quad (33)$$

$$pH = pK_a + \log \frac{(A^-)}{(HA)} \quad (34)$$

This expression is called the Henderson-Hasselbalch relation. It is particularly useful when calculating the pH of solutions containing significant concentrations of both the weak acid and its salt.

Titration Curves

Several titrations will be performed in Experiment 23. Although a knowledge of how the pH of the solution varies as titrant is added is not necessary to obtain good data, your understanding of what is occurring will be deeper if you have thought about this variation in advance.

Example 4. Compute enough points on the titration curve of 25.0 ml of 0.100 M HAc with 0.100 M NaOH to permit you to sketch the titration curve that will result from the addition of the following reagents in the quantities specified.

(a) 0 ml NaOH. At the beginning of the titration, we have just a solution of 0.100 M HAc. In example 3, the pH of this solution was computed to be 2.88.

(b) 5.00 ml NaOH. At this point, $\frac{1}{5}$ of the HAc has been neutralized according to the titration equation

$$HAc + OH^- \rightarrow H_2O + Ac^-$$

It is convenient to use the Henderson-Hasselbalch equation now. This requires a value for pK_a.

$$K_a = 1.76 \times 10^{-5}$$

$$pK_a = -\log K_a = -\log(1.76 \times 10^{-5})$$

$$= -\log 1.76 - \log 10^{-5}$$

$$= -0.246 + 5.000$$

$$= 4.754$$

The ratio, $(Ac^-)/(HAc) = \frac{1}{4}$ because of the stoichiometry described above.

$$pH = pK_a + \log(Ac^-)/(HAc)$$

$$= 4.754 + \log 0.250$$

$$= 4.754 + \log(2.5 \times 10^{-1})$$

$$= 4.754 + \log 2.50 + \log 10^{-1}$$

$$= 4.754 + 0.398 - 1.000$$

$$= 4.152$$

(c) 10.0, 15.0, 20.0 ml NaOH. Calculations similar to those immediately preceding yield pH's of 4.554, 4.906, and 5.332 respectively.

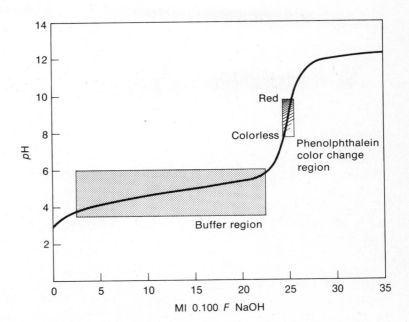

FIGURE E-3
Titration curve of 25 ml of 0.1 F acetic acid titrated with 0.1 F sodium hydroxide.

(d) 12.5 ml NaOH. This is the midpoint of the titration and is a particularly interesting point in any titration because $(A^-)/(HA) = 1$. This means that $\log(A^-)/(HA) = 0$, and

$$pH = pK_a = 4.754$$

This fact is often employed to obtain approximate values of pK_a's.

(e) 25.0 ml NaOH. At this point, hydrolysis calculations must be employed. In the pre-study for Experiment 22, this approach is discussed in considerable detail. The result of these calculations yields a $pH = 8.72$.

These five points are plotted in Figure E-3 and a smooth curve drawn through them. Notice how flat the curve is in the middle. This is the region in which HAc and Ac$^-$ act as an effective buffer.[2] At the equivalence point (at which 25.0 ml of NaOH have been added), the curve rises rather abruptly. As discussed in the pre-study for Experiment 23, this behavior requires an indicator with a pK_a near 8.7 that yields a perceptible color change over a pH interval of 2.

---

[2]A buffer solution is one which resists pH changes when either acid or base is added.

**An Introduction to
Chemical Equilibrium**

NAME

SECTION   LOCKER

INSTRUCTOR   DATE

PROBLEMS

1. Write the equilibrium constant expressions for the following reactions. Use the appropriate subscript on each K.

Reactions	Equilibrium constant expressions

$HNO_3(aq) + 3HCl(aq) \rightleftharpoons Cl_2(g) + NOCl(g) + 2H_2O(l)$

$3Mg(s) + N_2(g) \rightleftharpoons Mg_3N_2(s)$

$NH_3(g) + H_2O(l) \rightleftharpoons NH_4^+ + OH^-$

$CaCO_3(s) \rightleftharpoons CaO(s) + CO_2(g)$

$2Na_2O_2(s) + 2H_2O(l) \rightleftharpoons 4Na^+ + 4OH^- + O_2(g)$

$HAc(aq) + OH^- \rightleftharpoons H_2O(l) + Ac^-$

$H^+ + OH^- \rightleftharpoons H_2O(l)$

$2H_2O_2(aq) \rightleftharpoons O_2(g) + 2H_2O(l)$

$$AgAc(s) \rightleftharpoons Ag^+ + Ac^-$$

$$Ca^{2+} + C_2O_4^{2-} \rightleftharpoons CaC_2O_4(s)$$

$$Pb(H_2O)_4^{2+} + 2OH^- \rightleftharpoons Pb(H_2O)_2(OH)_2(s) + 2H_2O(l)$$

$$2HCO_3^- \rightleftharpoons CO_3^{2-} + H_2O(l) + CO_2(g)$$

---

2. The solubility product of $BaCrO_4$ is 1.2×10^{-10} M^2. Calculate the solubility of $BaCrO_4$ in water.

Solubility_____ M

Calculate the solubility of $BaCrO_4$ in 0.100 M Na_2CrO_4.

Solubility_____ M

REPORT E	Sheet 2	NAME

3. Calculate the (OH^-) in the following solutions:
(a) 5.00 M HCl

(OH^-) ________________ M

(b) 5.00 M NaOH

(OH^-) ________________ M

(c) Solution having 4.61×10^{-6} M H^+

(OH^-) ________________ M

(d) Solution having 1.71×10^{-11} M H^+

(OH^-) ________________ M

4. Calculate the pH of each of the following solutions:
(a) 1.72×10^{-3} M HCl

pH ________________

(b) 0.137 M NaOH

pH ________________

(c) $(OH^-) = 9.91 \times 10^{-3}$ M

pH ________________

(d) $(H^+) = 3.73 \times 10^{-11}$ M

pH ________________

5. The ionization constant of nitrous acid, HNO_2, is 4.5×10^{-4} M. Calculate the pH of a solution of 0.100 M HNO_2.

pH_____

6. The ionization constant of hydrofluoric acid, HF, is 6.9×10^{-4}. Calculate the pH when 25.0 ml of 0.100 M NaOH have been added to 50.0 ml of 0.100 M HF.

pH_____

SOME EXAMPLES OF CHEMICAL EQUILIBRIUM

In the preceding study assignment, the general principles of chemical equilibrium have been outlined, and a quantitative description provided. This experiment is designed to demonstrate some of the general principles of equilibrium in reacting systems. You will observe a number of interesting chemical reactions and the ways in which these systems respond to a variety of chemical reagents. Thus it is important for you to understand the Le Châtelier principle and the effect of a common ion on chemical equilibria as discussed in Study Assignment E.

Subsequent experiments will be concerned with the quantitative aspects of chemical equilibria based on equilibrium constant calculations. This introductory experiment is designed to show qualitatively several important features of chemical equilibria.

EXPERIMENTAL PROCEDURE

Chemicals: 6 F $HC_2H_3O_2$, 0.1 F $HC_2H_3O_2$, 1 F NH_4Cl, 0.1 F NH_3, 0.25 F $(NH_4)_2C_2O_4$, 0.1 F $FeCl_3$, 0.5 F $H_2C_2O_4$, 1 F K_2CrO_4, 0.1 F $KSCN$, 0.1 F $AgNO_3$, 3 F $NaC_2H_3O_2$, 1 F $NaC_2H_3O_2$, 5.4 F $NaCl$ (sat.), methyl orange and phenolphthalein indicators.

1. The Shifting of an Equilibrium. The Common Ion Effect. (a) *The Chromate Ion-Dichromate Ion Equilibrium.* Yellow chromate ion reacts with hydrogen ion to form first hydrogen chromate ion, and then, by condensation and loss of H_2O, orange dichromate ion:

$$2CrO_4^{2-} + 2H^+ \rightleftharpoons 2HCrO_4^- \rightleftharpoons Cr_2O_7^{2-} + H_2O$$

At present, we need consider only the overall reaction,

$$2CrO_4^{2-} + 2H^+ \rightleftharpoons Cr_2O_7^{2-} + H_2O$$

To 3 ml of 1 F K_2CrO_4 add several drops of 3 F H_2SO_4. Mix this, and observe any change. Now add several drops of 6 F NaOH, with mixing, until a change occurs. Again add H_2SO_4. Interpret the observed changes in terms of "shifting the equilibrium."

(b) *Weak Acids and Weak Bases.* In Experiment 14, we compared the relative concentrations of molecules and ions in weak acids and weak bases. Let us now examine these relative concentrations from the equilibrium point of view.

First, in order to observe the effects of acid and base on indicators, add a drop of methyl orange to 3 ml of H_2O. Then add 2 drops of 6 F HCl followed by 4 drops of 6 F NaOH. Repeat the experiment, using the phenolphthalein indicator in place of the methyl orange.

To 3 ml of 0.1 F $HC_2H_3O_2$, add a drop of methyl orange, and then add 1 F $NaC_2H_3O_2$, a few drops at a time, with mixing. Explain your observations in terms of the equilibrium equation and the "common ion" effect.

To each of two 3-ml samples of 0.1 F NH_3 add a drop of phenolphthalein. To one sample, add 1 F NH_4Cl, a few drops at a time, with mixing. To the other, add 6 F HCl, a drop at a time, with mixing. In each case, note any changes in color and in odor of the solutions. Write the equation for the equilibrium in dilute NH_3 and interpret the different observed results, including the additional overall net ionic equation for the reaction with HCl, in terms of shifting the equilibrium.

(c) *The Thiocyanatoiron(III) Complex Ion.* This ion, often called the ferric thiocyanate complex ion, is formed as a blood red substance according to the following equilibrium equation[1]

$$Fe^{3+} + SCN^- \rightleftharpoons Fe(SCN)^{2+}$$

It is not uncommon for certain ions (or ions and molecules) to react in such proportions that the resulting structures, although not electrically neutral, are stable. These structures are called complex ions and their composition may vary with the proportion and concentration of reactants. The structure and other properties of the complex ions will be studied in greater detail in Experiment 21.

Add 3 ml of 0.1 F $Fe(NO_3)_3$ to 3 ml of 0.1 F

KSCN, and dilute this by mixing with water to a volume of 35–40 ml, until the deep red color is reduced in intensity so that further changes (either to increase or decrease the color) are easily observed. To a 5-ml portion of this solution, add a small amount of 0.1 F $Fe(NO_3)_3$; to a second 5-ml portion add a little 0.1 F KSCN; to a third 5-ml portion add a few drops of 6 F NaOH ($Fe(OH)_3$ is quite insoluble); and to a fourth 5-ml portion add a small amount of additional water. Correlate your observations with Le Châtelier's principle and the equilibrium equation in the preceding paragraph.

2. The Equilibria of Saturated Solutions. (a) *Saturated Silver Acetate.* Prepare some freshly precipitated silver acetate by mixing 10 ml of 3 F $NaC_2H_3O_2$ with 25 ml of 0.1 F $AgNO_3$. Let the mixture stand a few minutes to complete the precipitation, and then filter it. Drain the precipitate, rinse it with not more than 2 ml of distilled water, and let this drain. Place a clean 15-cm test tube under the funnel, puncture the filter (Figure 19-1) with a stirring rod, and wash the precipitate into the test tube with a stream of water from the wash bottle (but do not use more than 10 ml, one-third of a test tube, of water). Warm this mixture slightly (not over about 35°C), and shake it gently for about 10 minutes to establish the equilibrium of a saturated solution according to the equation

$$AgC_2H_3O_2(s) \rightleftharpoons Ag^+ + C_2H_3O_2^-$$

Filter this solution through a dry filter, retaining most of the residue in the original test tube (save this), and divide the filtrate into two 4–6-ml portions in small test tubes. To one of these, add 6–10 drops of 3 F $NaC_2H_3O_2$, to the other add 6–10 drops of 6 F $HC_2H_3O_2$. Observe the results, which may require several minutes to be noticeable. Explain, in detail, the results in terms of the above equilibrium equation.

To the test tube containing the residual solid $AgC_2H_3O_2$ add 1–2 ml of water and several drops of 6 F HNO_3. Mix, and explain the results.

(b) *Saturated Sodium Chloride.* To 10 ml of saturated sodium chloride solution (5.4 F) add several milliliters of concentrated HCl (12 F). Note the relative Cl^- concentrations in these solutions, and explain the results in terms of the saturated solution equilibrium.

(c) *Saturated Barium Chromate.* To 3 ml of 0.1 F $BaCl_2$, add a few drops of 1 F K_2CrO_4, and then a little 6 F HCl. Explain your observations in terms of the equilibrium equations involved.

---

[1]For simplicity, we are using here the unhydrated formulas. This makes no difference in our equilibrium consideration, however, since water is always present at high constant concentration. With the addition of SCN^- to hydrated iron(III) ion, $Fe(H_2O)_6^{3+}$, a substitution of SCN^- for H_2O occurs, with possible formulas such as $Fe(H_2O)_5(SCN)^{2+}$, $Fe(H_2O)_4(SCN)_2^+$, $Fe(H_2O)_3(SCN)_3$, $\cdots$, $Fe(SCN)_6^{3-}$.

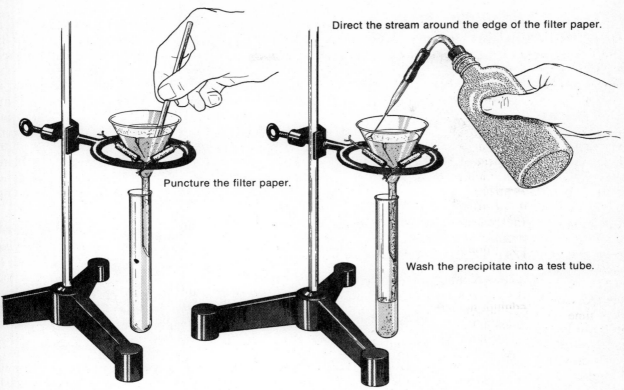

Direct the stream around the edge of the filter paper.

Puncture the filter paper.

Wash the precipitate into a test tube.

FIGURE 19-1
The technique of puncturing a filter paper and washing the precipitate into a test tube.

3. Application of the Law of Chemical Equilibrium to Analytical Procedures. In qualitative analysis Ca^{2+} is usually identified as calcium oxalate:

$$Ca^{2+} + C_2O_4^{2-} \rightleftharpoons CaC_2O_4(s)$$

What conditions will make the precipitation as complete as possible? Since Ca^{2+} is the unknown ion, the completeness of the above reaction must depend on the maximum $C_2O_4^{2-}$ concentration possible. Should we use a soluble salt, such as $(NH_4)C_2O_4$, or the moderately weak acid $H_2C_2O_4$? The ionization equations are

$$(NH_4)C_2O_4(aq) \rightarrow 2NH_4^+ + C_2O_4^{2-}$$

and

$$H_2C_2O_4(aq) \rightleftharpoons H^+ + HC_2O_4^-$$
$$\updownarrow$$
$$H^+ + C_2O_4^{2-}$$

Should the solution be made acidic or basic to achieve the maximum $C_2O_4^{2-}$ concentration? Test your reasoning by the following experiments.

Mix 10 ml of 0.1 F $CaCl_2$ with 10 ml of distilled water, and divide into two 15-cm test tubes. To one add 1 ml of 0.5 F $H_2C_2O_4$, and to the other add 2 ml of 0.25 F $(NH_4)_2C_2O_4$. Compare the results. To the $H_2C_2O_4$ mixture add 1–2 ml of 6 F HCl, and mix. Explain the results. Now to the same solution add a slight excess of 6 F NH_3, and mix. Determine whether the precipitate may be $Ca(OH)_2$ by adding a little 6 F NH_3 to some diluted $CaCl_2$ solution. What is the precipitate? Explain all these results in terms of the equilibrium equations above.

Would the acidity of the solution be of more importance in the precipitation of the salt of a strong acid or in the precipitation of the salt of a weak acid?

REPORT

19

Some Examples of
Chemical Equilibrium

NAME

SECTION   LOCKER

INSTRUCTOR   DATE

OBSERVATIONS AND DATA

1. The Shifting of an Equilibrium. The Common Ion Effect

(a) The Chromate Ion-Dichromate Ion Equilibrium.

Rewrite the equation for this equilibrium_____

Describe the color changes obtained on the addition of H_2SO_4 and NaOH, respectively, and interpret these effects.

(b) Weak Acids and Weak Bases.

The equation for the equilibrium in dilute acetic acid solution is the following:

_____.

Explain any observed changes when sodium acetate (+ methyl orange) is added.

The equation for the equilibrium in dilute ammonia solution is the following:

_____.

Note any observed odor or color changes when 0.1 F NH_3 (+ phenolphthalein) is treated with the following:

NH_4Cl_____; HCl_____

Which direction, right or left, does each reagent shift the above equilibrium? Explain fully.

(c) The Thiocyanatoiron(III) Complex Ion.

Rewrite the equation for this equilibrium, and correlate your observations with the Le Châtelier principle on the addition of the following:

(1) $Fe(NO_3)_3$ _____

_____

(2) KSCN _____

_____

(3) NaOH _____

_____

(4) Additional H_2O _____

_____

236

2. The Equilibria of Saturated Solutions

(a) Saturated Silver Acetate.

Rewrite the equilibrium equation for a saturated silver acetate solution by the addition of the following:

3 F $NaC_2H_3O_2$——————————————————; 6 F $HC_2H_3O_2$——————————————

Note the predicted result if 3 F $AgNO_3$ were added to the $AgC_2H_3O_2$ equilibrium:————————

Note the observed result when HNO_3 is added to the solid $AgC_2H_3O_2$ residue: ————————————

Interpret the observed and predicted results (or lack of results) for each of the above reagents in terms of the equilibrium equation.

(b) Saturated Sodium Chloride. What happened when 12 F HCl was added to saturated (5.4 F) NaCl? Explain this behavior in terms of the relative Cl^- concentrations in the two solutions, and in terms of the equilibrium equation $NaCl(s) \rightleftharpoons Na^+ + Cl^-$.

(c) Saturated Barium Chromate. Write the equation for the equilibrium established when K_2CrO_4 and $BaCl_2$ solutions are mixed, and explain any changes observed when this is treated with 6 F HCl.

3. Application of the Law of Chemical Equilibrium to Analytical Procedures

Solutions mixed	Relative amounts of precipitate, if any
(1) $CaCl_2(aq)$ + $H_2C_2O_4(aq)$	————————————
(2) $CaCl_2(aq)$ + $(NH_4)_2C_2O_4(aq)$	————————————
(3) Results of (1) + 6 F HCl	————————————
(4) Results of (3) + 6 F NH_3	————————————
(5) $CaCl_2(aq)$ + 6 F NH_3	————————————

$$H_2C_2O_4(aq) \rightleftharpoons H^+ + HC_2O_4^-$$
$$\updownarrow$$
$$H^+ + C_2O_4^{2-}$$

Relate the above data on the relative amounts of CaC_2O_4 formed, or dissolved, under various conditions to the equilibrium equations at the right, and answer the questions below.

$$Ca^{2+} + C_2O_4^{2-} \rightleftharpoons CaC_2O_4(s)$$

(a) Account for the difference in behavior of $H_2C_2O_4$ and $(NH_4)_2C_2O_4$ in items (1) and (2).

(b) In item (3), explain the effect of HCl in these equilibria.

(c) Considering items (4) and (5), explain the effect of adding excess NH_3 on these equilibria.

(d) Is it more important to control the H^+ concentration in the precipitation of the salt of a weak acid or the salt of a strong acid? Why?

APPLICATION OF PRINCIPLES

Predict whether the following insoluble salts would be dissolved by the addition of a strong acid, such as HNO_3. If solution occurs, write the net ionic equation for the reaction. (See Table 9, Appendix C).

Salts	Do they dissolve?	Equation for solution, or reason for no action
1. $MgCO_3$	_____	_____
2. $AgCl$	_____	_____
3. $CaSO_3$	_____	_____
4. $BaSO_4$	_____	_____
5. ZnS	_____	_____
6. $Ca_3(PO_4)_2$	_____	_____

THE MEASUREMENT OF
AN IONIZATION CONSTANT

PRE-STUDY

Acids and bases are important classes of chemical compounds. They control the pH of living systems and of many reactions carried out in the chemical laboratory. If the pH in our blood shifts by as little as 0.3 units above or below the normal range of 7.3 to 7.5, severe illness results. Therefore, if we understand the principles of how acids and bases function, we will be better informed about the functioning of biological systems, as well as of the purely chemical systems.

In this experiment, you will study the properties of one quite common weak acid, acetic acid. Acetic acid has the molecular formula

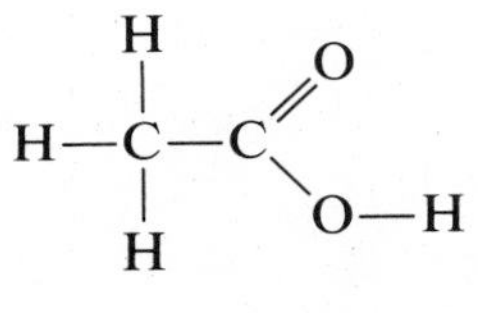

Its chemical formula is generally shortened to read CH_3—COOH or $HC_2H_3O_2$. This is often shortened even further to HAc where the Ac^- part of the formula stands for the acetate ion. The proton that dissociates is the one attached to the oxygen atom. For reasons that you will discover later in your study of organic chemistry, the hydrogen atoms bonded to the carbon atoms do not dissociate from the molecule at all in aqueous solutions.

Acetic acid dissociates according to the equation[1]

$$HAc \rightleftharpoons H^+ + Ac^-$$

---

[1]Dissociation of a weak acid in aqueous solution is a proton transfer reaction in which the proton donor (in this case HAc) transfers a proton to a proton acceptor (in this case H_2O). Thus the above reaction is more accurately expressed by

$$HAc + H_2O \rightleftharpoons H_3O^+ + Ac^-$$

For simplicity, we represent the H_3O^+ by the symbol H^+ and delete the H_2O on the left side of the equation. Because the water concentration remains virtually constant, it is included in the dissociation constant K_a.

The equilibrium constant of this reaction is 1.76 × 10^{-5} at 25°C. You are to measure the value of this constant in this experiment and compare your value with the accepted value.

The expression for this equilibrium constant[1] is

$$K_a = \frac{(H^+)(Ac^-)}{(HAc)}$$

Obviously, one way to obtain the value of K_a is to measure each of the quantities on the righthand side of this relation and carry out the indicated arithmetic. The concentrations of acetic acid, HAc, and the acetate ion, Ac^-, will be readily obtained from the quantities of NaAc and HAc used to prepare the solutions. The only difficult part is to determine the (H^+) or its derived function, the pH.

THE MEASUREMENT OF pH

The pH Meter

All modern pH meters have two electrodes, a glass electrode and a reference electrode, usually calomel (Figure 20-1). Sometimes these two electrodes are combined in a single unit called a combination electrode. However, the principle of their operation remains the same. The reference elec-

trode supplies a constant potential, $E° = +0.2$ volt from the half-reaction

$$Hg_2Cl_2 + 2e^- \rightleftharpoons 2Hg + 2Cl^- \text{ (sat.)}$$

$$E° = +0.2 \text{ volt}$$

This electrode is called the calomel electrode because calomel is the trivial name of the compound Hg_2Cl_2. The (H^+) in the solution determines the potential of the glass electrode. The potential developed across the glass membrane is proportional to the logarithm of the H^+ concentration ratio inside and outside the glass electrode. The pH meter measures the total potential across the two electrodes and displays this measurement on a scale calibrated in pH units. The use of the pH meter is the most accurate means of determining the pH of a solution. In Figure 20-1, note the construction details of the glass electrode (B) and of the calomel electrode (C). The thin glass bulb of the glass electrode is very fragile. Be extremely careful not to touch the bottom of the beaker with it.

Indicators

Indicators are a specialized class of rather large organic compounds that are weak acids or bases.

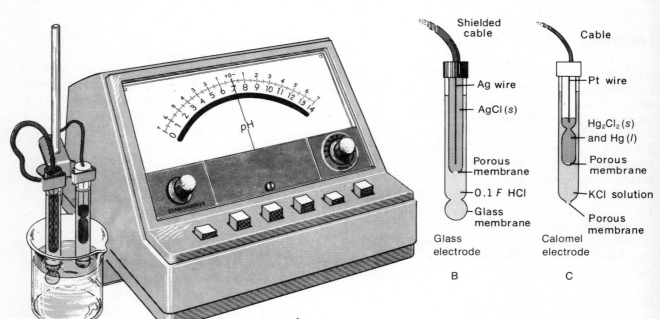

FIGURE 20-1
The modern pH meter and its electrodes: (A) the complete apparatus; (B) detail of the glass electrode, which is responsive to (H^+); (C) detail of the calomel reference electrode.

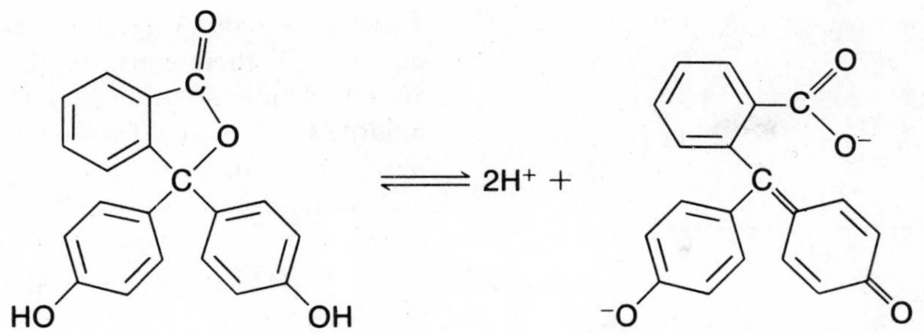

FIGURE 20-2
The acid and base forms of the indicator phenolphthalein and their associated colors.

They display one color when the acidic group is protonated and another color (or no color at all) when they are ionized. Such compounds that absorb light in the visible part of the spectrum invariably contain a large number of alternating single and double carbon-carbon bonds. An example of an indicator is phenolphthalein, whose structure and colors are given in Figure 20-2.

If most of the phenolphthalein molecules are present as HIn,[2] the solution will be colorless. If most of them are present as In⁻, the solution will be red. Because the indicator is a weak acid, the pH of the solution determines which form predominates. We write the equilibrium constant expression for HIn and then a derived form of the same equation, which you should prove is correct.

$$K_a = \frac{(H^+)(In^-)}{(HIn)}$$

$$pH = pK_a + \log \frac{(In^-)}{(HIn)}$$

The pK_a of phenolphthalein is about 9. Thus at $pH = 10$, where the ratio of (In^-) to $(HIn) = 10$, a majority of all molecules have lost a proton and the solution will be red. At $pH = 8$, where the ratio (In^-) to $(HIn) = 0.1$, most molecules are present in the colorless HIn form and the solution will be colorless. It is important to realize that, at all pH's above 10, the solution will be red and, at all pH's below 8, the solution will be colorless. Thus this indicator may be used to detect a change in pH as the pH shifts through the 8–10 range. They will be

used in this way in Experiment 23 for volumetric analysis.

In this experiment, indicators will be used for the alternate purpose of determining that the pH is greater than or less than a given value. Remember that in general you will determine only a range of possible pH values with an indicator. In some instances, you may be able to specify a more narrow range of pH's if the solution pH is quite close to the pK_a of the indicator. Thus, if the pH is 9, the ratio of (In^-) to (HIn) is 1.0, equal amounts of (HIn) and (In^-) are present and the solution will be light pink. Such determinations require careful observation and the use of prepared standard solutions for the purpose of comparison.

Table 8 in Appendix C provides a list of a number of important indicators and the pH ranges in which color changes occur. You will use a number of these indicators in the first part of this experiment. This experience will give you practice in determining an approximate pH of a solution rapidly if no pH meter is available. It will demonstrate for you the way in which indicators can be used to locate the end point of the titration of a weak acid for which the pH will be some value other than 7.0. Figure 20-3 demonstrates the correct technique for careful comparison of the color of two solutions.

Preparation of Standards and Dilution of Solutions

As we have already mentioned, it is necessary to prepare solutions of known pH to compare carefully with the colors of the unknown solution. The solutions for pH range 4–11 will be supplied to you.

---

[2]As is evident from Figure 20-2, phenolphthalein is a diprotic acid. For simplicity in this discussion, we refer to this indicator as HIn however.

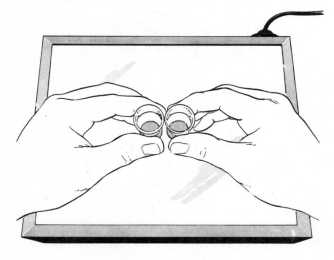

Label the H⁺ concentration of each tube.

FIGURE 20-3
The correct technique for the observation and comparison of indicator colors.

You will be able to prepare your own standard solutions for pH ranges 1–3 and 12–13. The simplest way to prepare these is to dilute solutions of 0.10 M HCl for the 1–3 range, and of 0.10 M NaOH for the 12–13 range.

Dilutions are based on the principle that the concentration after dilution is proportional to the ratio of the original volume to the final volume of the solution. The example, based upon the first dilution you will perform, demonstrates this idea.

Example. Calculate the pH of a solution prepared by diluting 5.0 ml of HCl of pH 1 to 50.0 ml.

The solution originally contained

$$5.0 \ \text{ml} \times \frac{0.1 \ \text{mmole H}^+}{\text{ml}} = 0.50 \ \text{mmole H}^+$$

The symbol mmole stands for millimole and is useful when dealing with milliliter quantities of solution. The concentration of H⁺ in a pH 1.0 solution is of course 0.10 moles/liter, or 0.10 mmole/ml.

Because only water has been added to the 5.0 ml of HCl, there must be 0.50 mmole H⁺ in the 50 ml of new solution and the new (H⁺) = 0.50 mmole H⁺/50 ml solution = 0.010 M. Thus the new pH = 2.0.

A simpler but less obvious calculation is

$$(\text{H}^+)_{\text{final}} = \frac{V_{\text{orig}}}{V_{\text{final}}} \cdot (\text{H}^+)_{\text{orig}}$$

$$= \frac{5.0}{50} (0.10) = 0.010 \ M.$$

The reason that you cannot make the solutions for pH 4–11 by dilution is that both the (H⁺) and (OH⁻) are so small that the dissolution of a trace amount of HCl, NH_3, CO_2, or other gaseous acids and bases would drastically alter the pH of the solution. The solutions provided for these pH's are buffers. A buffer contains a high concentration (0.01–1.0 M) of both a weak acid and its salt, or of a weak base and its salt. Thus a good buffer for pH 4.8 would be a solution 1 M in HAc and 1 M in NaAc. The addition of an acid to this solution converts some of the Ac⁻ → HAc. However, because the pH varies only as the log of the ratio of (Ac⁻) to (HAc), the pH does not change rapidly (see equation (34), Study Assignment E). Similarly, the addition of base causes some of the HAc to be converted to Ac⁻. But again, within limits, the pH shift is rather small.

EXPERIMENTAL PROCEDURE

Chemicals: 1.0 F $HC_2H_3O_2$, 1.0 F $NaC_2H_3O_2$, 0.10 F $HC_2H_3O_2$, 0.1 F NH_4Cl, 0.1 F NH_3, 0.1 F H_3BO_3, 0.02 F (sat.) $Ca(OH)_2$, 0.1 F HCl, saturated $Mg(OH)_2$, 0.1 F H_3PO_4, 0.1 F $NaHSO_4$. Indicator solutions: Orange IV (Tropeolin 00), methyl orange, methyl red, bromthymol blue, phenolphthalein, alizarin yellow R, indigo carmine.[3] Buffer solutions pH 4 to pH 11.

1. The Preparation of Comparison pH Standards.

Use 0.1 F HCl for a pH 1 solution. Dilute precisely 5.0 ml of 0.1 F HCl with distilled water, in a graduated cylinder to 50.0 ml, and mix this well to give a pH 2 solution. Then by a second 10-fold dilution of 5.0 ml of this, prepare a pH 3 solution. On the basic side, use 0.1 F NaOH for a pH 13 standard, and make a 10-fold dilution of this for a pH 12 standard. For the intermediate ranges, pH 4 to pH 11, standard buffer mixtures are available.

---

[3]If the indigo carmine indicator is not a deep blue color, it must be discarded and a fresh solution prepared.

2. The Colors of Indicators in Acid and in Basic Solutions. Gain experimental familiarity with two indicators in the acid range as follows. Prepare a series of clean, labeled 10-cm test tubes containing 5 ml each, respectively, of your acid standards or of the buffer solutions, ranging from pH 1 to pH 6. To each of these, add 1 drop of orange IV indicator. Mix each. Observe these lengthwise through the solution toward white paper (Figure 20-2), and record the colors which are characteristic at each pH. Keep the three or four tubes that show the color change range for comparisons throughout the experiment.

Repeat the procedure with a series of pH 1 to pH 6 5-ml samples to which you add a drop of methyl orange indicator. Again carefully observe and record the colors, and keep four labeled tubes that show the color range. Enter your data as directed, in the spaces provided in the report form.

3. The Use of Indicators in the Measurement of pH. (a) *Unknowns.* Label two clean test tubes, for two unknown solutions. Test 5-ml portions of these with indicators.[4] If necessary, match the colors carefully with your comparison pH standards and buffer series. Report these pH values to your instructor at once.

(b) *Typical Acids, Acid Salts, and Bases.* In similar manner, determine the pH of the following: 0.1 F H_3BO_3, 0.1 F H_3PO_4, 0.1 F NH_3, 0.1 F $NaHSO_4$ (i.e., HSO_4^-), 0.02 F $Ca(OH)_2$, saturated $Mg(OH)_2$, 0.1 F NaH_2PO_4 (i.e., $H_2PO_4^-$), and 0.1 F $KHC_4H_4O_6$ (i.e., $HC_4H_4O_6^-$, the hydrogen tartrate ion). For each, record the indicators used, the observed color, and the pH. Determine the relative strength of each, and list the solutions according to decreasing H^+ concentration, with the smallest pH first. If any two of these have about the same pH, bracket them together.

4. The Ionization Constant of Acetic Acid. Your instructor will demonstrate the use of the pH meter

if this instrument is available in your laboratory. If it is not, you will use indicators. Instructions for each method of determining approximate pH values follow.

With a pH Meter. (a) Take a 10 ml sample of 1.0 F HAc to the pH meter and measure its pH. Convert this value to the corresponding (H^+). Think about the source of H^+ and Ac^- in a HAc solution considering the equilibrium $HAc \rightleftharpoons H^+ + Ac^-$. Compute the (Ac^-) and (HAc) in this solution, and enter the values on the report sheet. Calculate the ionization constant of acetic acid.

(b) Repeat step (a), using 0.1 F HAc instead of 1.0 F HAc.

(c) Add 24.0 ml of 1.0 F NaAc to 10.0 ml of 1.0 F HAc, and mix well. Measure the pH of this solution as you did in steps (a) and (b). Calculate the (H^+) from the measured pH. Calculate the (Ac^-) from the concentration of NaAc after dilution, noting that NaAc completely dissociates in solution and ignoring any tendency of the Ac^- to combine with H_2O to give HAc. Similarly compute the (HAc) from the amount added, taking dilution into account and ignoring any ionization of the HAc. Compute K_a as before.

(d) Dilute 10.0 ml of the solution employed in (c) with 40.0 ml of distilled water, mix thoroughly, and measure the pH with the pH meter. Calculate each concentration as before and again calculate K_a.

(e) Thoroughly mix 5.0 ml of 1.0 F NaAc with 28.0 ml of 1 F HAc. Again measure the pH of this solution and compute the ionization constant of acetic acid.

With Indicators. If a pH meter is not available for this part of the experiment, the (H^+) of each solution may be determined by means of the indicator techniques described in steps 1 through 3. Use Orange IV for part (a), Orange IV and methyl orange for part (b), methyl red for parts (c) and (d), and methyl orange for part (e). With care, you should be able to locate the pH value of each solution to within 0.3–0.5 pH units, although the preparation of buffers of pH 3.5 and 4.5 may be necessary to achieve this accuracy.

---

[4]The approximate pH range may be checked first, if desired, with universal indicator paper.

REPORT

20

The Measurement of
an Ionization Constant

NAME

SECTION   LOCKER

INSTRUCTOR   DATE

OBSERVATIONS AND DATA

1. Preparation of Comparison pH Standards

Show your calculations for one of the dilutions that you carried out.

2. The Colors of Indicators in Acid and Base Solutions

Summary of observed relative colors with orange IV and methyl orange indicators. Enter your data from part 2 of the experimental procedure in the spaces provided below, giving the two limiting colors for the lower and upper pH values. Also enter any intermediate colors that can be observed for the middle pH values.

pH	Molarity		Orange IV	Methyl orange
1	10^{-1} M H^+	10^{-13} M OH^-		
2	10^{-2} M H^+	10^{-12} M OH^-		
3	10^{-3} M H^+	10^{-11} M OH^-		
4	10^{-4} M H^+	10^{-10} M OH^-		
5	10^{-5} M H^+	10^{-9} M OH^-		
6	10^{-6} M H^+	10^{-8} M OH^-		

Summary of the color-change range and limiting pH values for other indicators. Use the data from Table 8 of Appendix C to complete the table below.

Indicator	Acid color and pH	Intermediate color and pH	Basic color and pH
Methyl red			
Bromthymol blue			
Phenolphthalein			
Alizarin yellow R			
Indigo carmine			

3. The Use of Indicators in the Measurement of pH

(a) Unknowns:

Sample	Indicators used and colors	pH	H^+ conc.	OH^- conc.	Instructor's approval
Unknown 1					
Unknown 2					

(b) Typical acids, acid salts, and bases:

Arrange by decreasing H^+ Conc.:

H_3BO_3					
H_3PO_4					
NH_3					
$NaHSO_4$					
$Ca(OH)_2$					
$Mg(OH)_2$					
NaH_2PO_4					
$KHC_4H_4O_6$					

4. The Ionization Constant of Acetic Acid

(a) 1.0 F HAc

pH = _____

Calculate the corresponding (H^+), compute the concentrations of Ac^- and HAc, and calculate the ionization constant.

_____ M H^+ _____ M Ac^- _____ M HAc $\quad K_a = \dfrac{(H^+)(Ac^-)}{(HAc)} =$ _____

(b) 0.10 F HAc

pH = _____

Calculate the corresponding (H^+), compute the concentrations of Ac^- and HAc, and calculate the ionization constant.

_____ M H^+ _____ M Ac^- _____ M HAc $\quad K_a = \dfrac{(H^+)(Ac^-)}{(HAc)} =$ _____

(c) 24.0 ml 1 F NaAc + 10.0 ml 1 F HAc

$pH =$ _____

Calculate the (H^+), the concentrations of Ac^- and HAc, and the ionization constant.

_____ M H^+ _____ M Ac^- _____ M HAc $K_a =$ _____

(d) 10.0 ml of mixture (c) + 40.0 ml H_2O

$pH =$ _____

Calculate the (H^+), the concentrations of Ac^- and HAc, and the ionization constant.

_____ M H^+ _____ M Ac^- _____ M HAc $K_a =$ _____

(e) 5.0 ml 1 F NaAc + 28.0 ml 1 F HAc

$pH =$ _____

Calculate the (H^+), the concentrations of Ac^- and HAc, and the ionization constant.

_____ M H^+ _____ M Ac^- _____ M HAc $K_a =$ _____

(f) Examine your five values of K_a as determined in (a) through (e). If one of them deviates significantly from the others, you may discard it. Then average your values.

$$K_a, \text{average} =$$

The known K_a of acetic acid is 1.76×10^{-5} at 25.0°C. Calculate your percentage error:

$$\% \text{ error} = \frac{K_a, \text{known} - K_a, \text{average}}{K_a, \text{known}} \times 100\% =$$

List two possible reasons for the % error you have obtained.

1.

2.

PROBLEMS

1. The OH^- concentration of 0.50 F ammonia, NH_3, is 9.5×10^{-4} M. Write (a) the equilibrium equation for its ionization, (b) the equilibrium constant expression, and (c) solve for the value of K.

1. _____

2. The accepted value for the ionization constant of acetic acid is 1.76×10^{-5}. Using this figure, calculate the H^+ concentration in 0.50 F $HC_2H_3O_2$. (Let $x = (H^+)$, and also $(C_2H_3O_2^+)$, then $0.50 - x = (HC_2H_3O_2)$. However, since x is small, $(HC_2H_3O_2)$ approximately $= 0.50$ M.)

2. _____

3. Calculate the acidity (H^+ concentration) of a carbonated beverage which is 0.10 F in H_2CO_3. (Practically all of the H^+ comes from the first stage of ionization, $H_2CO_3(aq) \rightleftharpoons H^+ + HCO_3^-$, for which $K_1 = 4.4 \times 10^{-7}$.

3. _____

4. Explain the relative H^+ concentrations you observed in parts 4(c) and 4(d) in the experimental procedure.

4. _____

EQUILIBRIA OF COORDINATION COMPOUNDS

PRE-STUDY

The Hydration of Ions

We have already studied the most common class of coordination compounds. In Experiment 14, we noted that all ions in solution attract polar water molecules to them to form ion-dipole bonds. Thus, the cupric ion, Cu^{2+}, does not exist in that form in water but rather as the hexaaquo ion, $Cu(H_2O)_6^{2+}$. The origin of this attraction lies in the polar nature of the water molecule. Figure 21-1 shows the bent nature of the water molecule, which gives rise to the high dipole moment (a measure of the separation of charge in a molecule) of water.

The bond angle formed by the one oxygen and the two hydrogen atoms is known to be 105°. It arises from the sp^3 hybrid orbitals the oxygen employs to bond the two hydrogen atoms. The symbol $\leftrightarrow$ indicates the negative direction of the dipole moment and the small delta symbols (δ) indicate where partial charges exist in the molecule.

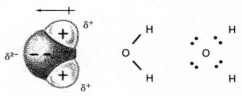

FIGURE 21-1
Three representations of the water molecule.

The negative side of the water molecule is strongly attracted to cations. In general, the greater the charge on the cation and the larger the cation, the greater the number of coordinated water molecules. Examples are $H(H_2O)_x^+$,[1] $Be(H_2O)_4^{2+}$, $Cu(H_2O)_6^{2+}$, $Al(H_2O)_6^{3+}$, $Fe(H_2O)_6^{3+}$.

Anions attract the positive side of the water molecule and are also hydrated. However, the attraction is not as large, the hydrates are less stable, and fewer water molecules are coordinated.

---

[1] The precise composition of the hydrated proton is not well known. The subscript "x" may vary from 1 to 4.

Coordination Compounds

These hydrates are one example of the type of substances called coordination compounds. Other neutral, but polar, molecules such as ammonia, NH_3, and also a number of ions, such as OH^-, Cl^-, CN^-, S^{2-}, $S_2O_3^{2-}$, and $C_2O_4^{2-}$ likewise can form similar, very stable coordination groupings about a central ion. Such coordination compounds result from the replacement of the water molecule from the hydrated ion by these other molecules or ions when they are present in the solution at high concentration, to form a still more stable bond. The resulting coordination compound may be a positively or negatively charged ion (a *complex ion*), or it may be a neutral molecule, depending on the number and kind of coordinating groups attached to the central ion.

What bonding forces hold the atoms of such a complex coordination compound together? Where large differences of electronegativity are present, as in AlF_6^{3-}, these forces are largely ionic. In the majority of cases the bonding is mainly covalent, often with partial ionic character. The type of bond orbitals determines the spatial geometry, and affects the stability of a given complex. Study the examples in Figure 21-2.

Ammonia Complex Ions. Some of the important ammonia complexes are the following:

$Ag(NH_3)_2^+$ $Cu(NH_3)_4^{2+}$ $Ni(NH_3)_4^{2+}$

$Au(NH_3)_2^+$ $Cd(NH_3)_4^{2+}$ $Ni(NH_3)_6^{2+}$

$Cu(NH_3)_2^+$ $Zn(NH_3)_4^{2+}$ $Co(NH_3)_6^{2+}$

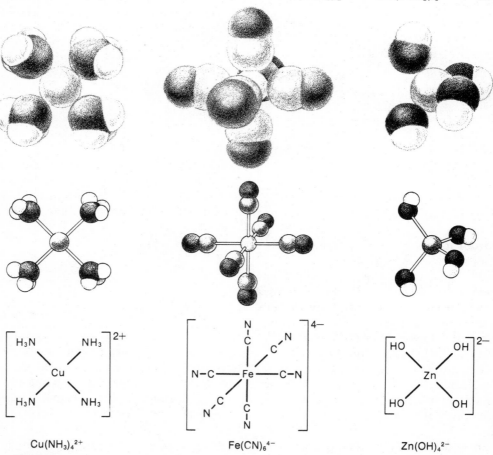

$Cu(NH_3)_4^{2+}$ $Fe(CN)_6^{4-}$ $Zn(OH)_4^{2-}$

FIGURE 21-2
Diagrams showing the spatial arrangement of the coordinating groups about a central ion in the formation of a complex ion. Above each conventional formula, the "structural formula" represents each covalent electron-pair bond by a single line. The spatial geometry is determined by the type of orbitals represented: sp^3 (tetrahedral) by $Zn(OH)_4^{2-}$, dsp^2 (square planar) by $Cu(NH_3)_4^{2+}$, and d^2sp^3 (octahedral) by $Fe(CN)_6^{4-}$. The "ball and stick" models indicate these geometric patterns more clearly, and the "space filling" models in the top row portray the atoms according to their accepted ionic diameters and bond lengths.

These complexes are formed by the addition of ammonia to a solution containing the hydrated cation. Ammonia molecules are bound by the cation one at a time as the concentration of ammonia increases. At low concentrations of ligand, smaller numbers of ammonia molecules may be bound. For instance, the two NH_3 molecules bound to the Ag^+ bind in successive steps. The equilibrium constant is known for each step of the following reaction sequence.

$$Ag^+ + NH_3 \rightleftharpoons Ag(NH_3)^+$$

$$K_1 = \frac{(Ag(NH_3)^+)}{(Ag^+)(NH_3)} = 1.6 \times 10^3$$

$$Ag(NH_3)^+ + NH_3 \rightleftharpoons Ag(NH_3)_2^+$$

$$K_2 = \frac{(Ag(NH_3)_2^+)}{(Ag(NH_3)^+)(NH_3)} = 6.8 \times 10^3$$

The equilibrium constant for the overall reaction can be obtained by the addition of the two reactions and the corresponding multiplication of their equilibrium constants.

$$Ag^+ + 2NH_3 \rightleftharpoons Ag(NH_3)_2^+$$

$$K = K_1 K_2 = \frac{(Ag(NH_3)^+)}{(Ag^+)(NH_3)} \cdot \frac{(Ag(NH_3)_2^+)}{(Ag(NH_3)^+)(NH_3)}$$

$$= \frac{(Ag(NH_3)_2^+)}{(Ag^+)(NH_3)^2} = 1.1 \times 10^7$$

In part 5 of the Experimental Procedure, you will determine the $K_{dissociation}$ for the $Ag(NH_3)_2^+$ complex ion.[2] Because the (NH_3) is so high under the experimental conditions that are employed, the $Ag(NH_3)^+$ species may be neglected.

Hydroxide Complex Ions, or Amphoteric Hydroxides. The hydroxides of most metals are relatively insoluble in water. Thus when sodium hydroxide is added to a metal ion in solution, such as lead ion, a precipitate is formed:

$$Pb(H_2O)_4{}^{2+} + 2OH^- \rightleftharpoons$$
$$Pb(H_2O)_2(OH)_2(s) + 2H_2O$$

or, using the simple unhydrated metal ion formula,

$$Pb^{2+} + 2OH^- \rightleftharpoons Pb(OH)_2(s)$$

By Le Châtelier's principle, excess hydroxide ion

would give more complete precipitation. Instead, the precipitate dissolves. This is explained by the tendency of lead ion to form a more stable coordination compound[3] with excess hydroxide ion:

$$Pb(H_2O)_2(OH)_2(s) + OH^- \rightleftharpoons$$
$$Pb(H_2O)(OH)_3{}^- + H_2O$$

or

$$Pb(OH)_2(s) + OH^- \rightleftharpoons Pb(OH)_3{}^-$$

Other ions react similarly; for example, hydrated aluminum ion, $Al(H_2O)_6{}^{3+}$, reacts to form the hydroxide precipitate, $Al(H_2O)_3(OH)_3$, or, with excess OH^-, the hydroxide complex ion $Al(H_2O)_2(OH)_4{}^-$. Traditionally, chemists use the unhydrated formulas in ordinary chemical equation just because they are simpler, except where it is important to emphasize the hydrated structure.

The reactions to form these hydroxide complex ions are entirely reversible. The addition of acid to the above strongly basic $Pb(OH)_3{}^-$ solution reacts first to reprecipitate the hydroxide,

$$Pb(OH)_3{}^- + H^+ \rightleftharpoons Pb(OH)_2(s) + H_2O$$

and then, with excess acid,

$$Pb(OH)_2(s) + 2H^+ \rightleftharpoons Pb^{2+} + 2H_2O$$

Such metal hydroxides, which may be dissolved by an excess of either a strong acid or a strong base, are called *amphoteric hydroxides*.

The more important metal ions whose hydroxides are amphoteric are given in Table 21-1.

TABLE 21-1
Some important amphoteric hydroxides

Simple ion[1] (acid solution)	Precipitate	Hydroxide complex ion[2] (strongly basic solution)
Pb^{2+}	$Pb(OH)_2$	$Pb(OH)_3{}^-$, plumbite ion
Zn^{2+}	$Zn(OH)_2$	$Zn(OH)_4{}^{2-}$, zincate ion
Al^{3+}	$Al(OH)_3$	$Al(OH)_4{}^-$, aluminate ion
Cr^{3+}	$Cr(OH)_3$	$Cr(OH)_4{}^-$, chromite ion
Sn^{2+}	$Sn(OH)_2$	$Sn(OH)_3{}^-$, stannite ion
Sn^{4+}	$Sn(OH)_4$	$Sn(OH)_6{}^{2-}$, stannate ion

[1]Such a highly charged ion as Sn^{4+} probably does not exist as such. In strong HCl solution, stannic salts dissolve as the chloride complex, $SnCl_6{}^{2-}$.
[2]Formerly these ions were written in the anhydrous form: $PbO_2{}^{2-}$, $ZnO_2{}^{2-}$, $AlO_2{}^-$, $CrO_2{}^-$, $SnO_2{}^{2-}$, and $SnO_3{}^{2-}$. These formulas may be derived from the hydroxide complex ion formulas simply by subtracting the appropriate number of H_2O molecules or H_3O^+ ions.

[2]The overall reaction may be written as a formation or as a dissociation, with $K_{formation} = \dfrac{1}{K_{dissociation}}$

[3]There is some uncertainty as to whether the lead hydroxide complex (and also the stannous hydroxide complex) will coordinate further to form $Pb(OH)_4{}^{2-}$ (and $Sn(OH)_4{}^{2-}$).

EXPERIMENTAL PROCEDURE

Chemicals: 1 F NH_4Cl, 1 F NH_3, 15 F (conc.) NH_3, $CuSO_4$ · $5H_2O(s)$, 0.1 F $CuSO_4$, 0.1 F $NaCl$, 0.1 F $AgNO_3$, 0.1 F $Zn(NO_3)_2$, phenolphthalein, alizarin yellow R, and indigo carmine indicators.

1. The Formation of Complex Ions with Ammonia. To 3 ml of 0.1 F $CuSO_4$, add a drop of 6 F NH_3. Mix this. (Write the equation.) Continue to add NH_3 a little at a time, with mixing, until a distinct change occurs. Is this result contrary to the law of Le Châtelier? Obviously the OH^- concentration was increasing while the $Cu(OH)_2$ dissolved. How must the Cu^{2+} concentration have changed? Did it increase or decrease? (Save this solution.)

To learn which of the substances present in an ammonia solution (NH_4^+, OH^-, NH_3, H_2O) is responsible for the above change, try adding the following: (a) 1 ml of 1 F NH_4Cl to 1 ml of 0.1 F $CuSO_4$, (b) 2 drops (an excess) of 6 F $NaOH^4$ to 2 ml of 0.1 F $CuSO_4$, (c) ammonia gas by placing several crystals of $CuSO_4$ · $5H_2O(s)$ in a small dry beaker, and also placing at one side in the beaker a piece of filter paper moistened with concentrated (15 F) NH_3. Cover with a watch glass, and observe any changes. From this evidence, write an equation to show the formation of this new substance when excess NH_3 is added to Cu^{2+}.

To 1 ml of this cupric ammonia complex ion solution, add 6 F HNO_3 in excess. Explain the result and write the equation.

2. The Formation of Amphoteric Hydroxides. To 5 ml of 0.1 F $Zn(NO_3)_2$, add 6 F $NaOH$ by drops, with mixing, until the precipitate which first forms just redissolves. Avoid undue excess of $NaOH$. Divide this into two portions; test one portion with alizarin yellow R and the other with indigo carmine indicator. Estimate the approximate OH^- concentration (for later comparison in part 3). Now to one portion, add 6 F HCl, by drops, until a precipitate forms (What is it?) and then redissolves as more HCl is added. Interpret all these changes as related to Le Châtelier's law, and as to the relative concentration of the various constituents (the zinc in its various forms, H^+, and OH^-), both by words and by net ionic equations.

3. The Reaction of Zinc Ion with Ammonia. When ammonia is added gradually to Zn^{2+}, does the precipitate of zinc hydroxide that first forms redissolve

as zincate ion, $Zn(OH)_4^{2-}$, owing to the excess base added, or does it redissolve as $Zn(NH_3)_4^{2+}$, owing to the NH_3 molecules added? To test this point, to 3 ml of 0.1 F $Zn(NO_3)_2$, add 6 F NH_3 by drops, with mixing, until the precipitate that first forms just redissolves. Divide this mixture, test one portion with phenolphthalein, and the other portion with alizarin yellow R. Estimate the approximate OH^- concentration, and compare this with the corresponding situation in part 2 above, where $NaOH$ was used. What can you conclude as to the possibility of forming $Zn(OH)_4^{2-}$ by adding NH_3 to a zinc salt solution? Explain. Write the equation for the equilibrium which you have verified.

4. Some Chloride Complex Ions. (a) To 2 ml of 0.1 F $CuSO_4$ add 2 ml of 12 F (conc) HCl and then dilute this with about 5 ml of water. Write equations, and interpret the color changes you observed.

(b) To 1 ml of 0.1 F $AgNO_3$ add 3 ml of 12 F (conc) HCl, and then agitate this well for several minutes to redissolve the precipitated $AgCl$. Now dilute this with about 5 ml of distilled water. Write equations, and interpret the changes you observed.

5. The Equilibrium Constant of an Ammonia Complex Ion. The dissociation of silver diammine complex ion is represented by the equilibrium

$$Ag(NH_3)_2^+ \rightleftharpoons Ag^+ + 2NH_3 \qquad (1)$$

and the corresponding equilibrium constant expression

$$\frac{(Ag^+)(NH_3)^2}{(Ag(NH_3)_2^+)} = K_{dissociation} \qquad (2)$$

If you add sufficient Cl^- gradually to an equilibrium mixture of Ag^+ and NH_3, represented by equation (1), so that you can just barely begin precipitation of $AgCl(s)$, a second equilibrium is established simultaneously without appreciably disturbing the first equilibrium. This may be represented by the combined equations

$$Ag(NH_3)_2^+ \rightleftharpoons Ag^+ + 2NH_3 \qquad (3)$$
$$+$$
$$Cl^-$$
$$\updownarrow$$
$$AgCl(s)$$

By using a large excess of NH_3, you can shift equation (1) far to the left, with reasonable assurance that the Ag^+ is converted almost completely to

---

[4]This provides an excess of much stronger OH^- than that of the NH_3 solution. Strong $NaOH$ shows some amphoteric effect (see part 2) with cupric salts, but is far from complete.

$Ag(NH_3)_2{}^+$ rather than to the first step only, $Ag(NH_3)^+$. From the measured volumes of NH_3, Ag^+, and Cl^- solutions used, the concentrations of the species in equation (1) may be determined and the value of $K_{dissociation}$ calculated.

To prepare the solution,[5] place 3.0 ml of 0.1 F

---

[5]If desired, some improvement in precision may result by using larger volumes—20.0 ml each of 0.1 F $AgNO_3$ and 1 F NH_3. Then dilute 10.0 ml of 0.1 F NaCl to about 50.0 ml in your 50-ml graduate, mix this well, and note the exact volume. Add first about 15 ml, then very small portions, to the Ag^+-NH_3 mixture, stirring as you do so, until a very faint permanent milky precipitate of AgCl remains. Note the total volume of 0.02 F NaCl used.

$AgNO_3$ (measure it accurately in a 10-ml graduate) in a 15-cm test tube. Add 3.0 ml (also carefully measured) of 1 F NH_3. Now prepare some 0.02 F NaCl by diluting 2.0 ml of 0.1 F NaCl to 10.0 ml in your 10-ml graduate. Mix this thoroughly and note the exact volume. Then, from a medicine dropper, add it to the mixture of $AgNO_3$ and NH_3, about 1–1.5 ml at first, and then drop by drop until a very faint, permanent milky precipitate of AgCl remains. Return any excess NaCl from the medicine dropper to the graduate, and note the exact volume used. From these data, $K_{dissociation}$ can be calculated.

REPORT

21

**Equilibria of
Coordination Compounds**

NAME

SECTION   LOCKER

INSTRUCTOR   DATE

OBSERVATIONS AND DATA

1. The Formation of Complex Ions with Ammonia

The net ionic equation for the reaction
of excess $CuSO_4$ with NH_3 is_____

The predicted effect on the above reaction of adding
excess NH_3 (based on Le Châtelier's principle) is_____

List the observed results when:

Excess NH_3 is added to $CuSO_4$ solution.   _____

NH_4Cl and $CuSO_4$ solutions are mixed.   _____

Excess $NaOH$ and $CuSO_4$ solutions are mixed. _____

$CuSO_4 \cdot 5H_2O(s)$ is exposed to NH_3 gas.   _____

Considering the equilibrium equation $NH_3 + H_2O \rightleftharpoons NH_4^+ + OH^-$, explain which substance (NH_3, NH_4^+, or OH^-)
causes the deep blue color, and give the equation for the reaction.

The observed effect, and the net ionic equation, for the reaction of HNO_3 on this deep blue solution is:

2. The Formation of Amphoteric Hydroxides

The net ionic equation for the reaction
of excess $Zn(NO_3)_2$ with $NaOH$ is_____

The predicted effect on the above reaction
of adding excess $NaOH$ (based on Le Châtelier's theorem) is_____ _____

Explain in your own words why $Zn(OH)_2(s)$ dissolves with excess OH^-, and write the net ionic equation for the
reaction.

Color with alizarin   Indigo OH^- concen-
yellow R_____   carmine_____   tration_____

256

Explain the effect of adding a moderate amount of HCl to this strongly basic solution, and give the net ionic equation for the reaction.

What further change occurs when excess HCl is added? (Give the equation and explain.)

3. The Reaction of Zinc Ion with Ammonia

Note your observations on the addition of 6 F NH_3 to $Zn(NO_3)_2(aq)$, by drops, to redissolve the precipitate:

Color with
phenolphthalein_____

Alizarin
yellow R_____

OH^- concen-
tration_____

Which coordination compound, $Zn(OH)_4^{2-}$ or $Zn(NH_3)_4^{2+}$, forms when Zn^{2+} reacts with excess NH_3 solution? Compare with part 2; explain fully.

The equation for this equilibrium
complex ion is therefore as follows: _____

4. Some Chloride Complex Ions

(a) Explain the successive changes observed when concentrated HCl, then H_2O, is added to a $CuSO_4$ solution; give the equations for the reactions.

(b) Explain the changes observed when concentrated HCl, then H_2O, is added to a $AgNO_3$ solution, and give the equations.

5. The Equilibrium Constant of an Ammonia Complex Ion

(Indicate your calculations for each step, in the spaces provided.)

Volumes of   0.1 F
solutions: $AgNO_3$_____

1 F
NH_3_____

0.02 F
$NaCl$_____

Total
volume_____

(a) Concentration of $Ag(NH_3)_2^+$:
(Assume all the silver to be present as the complex ion, ignoring the trace of free Ag^+ remaining.)

(b) Concentration of Cl^-:
(Ignore any trace of Cl^- removed as $AgCl(s)$)

REPORT 21	Sheet 2	NAME

(c) Concentration of Ag^+:
(Use the Cl^- concentration above, and the solubility product relationship, $(Ag^+)(Cl^-) = 2.8 \times 10^{-10}$.)

____

(d) Concentration of free NH_3:
(First calculate the NH_3 concentration as if none combined with Ag^+, then subtract twice the concentration of $Ag(NH_3)_2^+$ found above.)

____

(e) Use the values found in (a), (c), and (d) to calculate the value of the equilibrium constant:

____

APPLICATION OF PRINCIPLES

1. Which reagent, NaOH or NH_3, will enable you to precipitate the *first named ion* from a solution containing each of the following pairs of ions, and leave the second ion in solution? Give also the formula of the precipitate and the exact formula of the other ion in solution.

	Reagent	Precipitate	Ion in solution
(a) Al^{3+}, Zn^{2+}			
(b) Cu^{2+}, Pb^{2+}			
(c) Pb^{2+}, Cu^{2+}			
(d) Fe^{3+}, Al^{3+}			
(e) Ni^{2+}, Sn^{2+}			
(f) Sn^{2+}, Ni^{2+}			
(g) Mg^{2+}, Ag^+			

2. When ammonia is added to $Zn(NO_3)_2$ solution, a white precipitate forms, which dissolves on the addition of excess ammonia; but when ammonia is added to a mixture of $Zn(NO_3)_2$ and NH_4NO_3, no precipitate forms at any time. Suggest an explanation for this difference in behavior.

3. Calculate the OH^- concentration in (a) 1 F NH_3, and (b) a solution which is 1 F in NH_3 and also 1 F in NH_4Cl. ($K_{NH_3} = 1.8 \times 10^{-5}$, see Appendix C, Table 9.)

(a) _____

(b) _____

4. Suppose you are given the following experimentally observed facts regarding the reactions of silver ion.
   (a) Ag^+ reacts with Cl^- to give white $AgCl(s)$.
   (b) Ag^+ reacts with ammonia to give a quite stable complex ion, $Ag(NH_3)_2^+$.
   (c) A black suspension of solid silver oxide, $Ag_2O(s)$, shaken with NaCl solution, changes to white $AgCl(s)$.
   (d) $AgCl(s)$ will dissolve when ammonia solution is added, but $AgI(s)$ does not dissolve under these conditions. Write equations for any net reactions in the above cases, and then, based on these observations, arrange each of the substances AgCl, AgI, Ag_2O, and $Ag(NH_3)_2^+$ in such an order that their solutions with water would give a successively decreasing concentration of Ag^+.

   (1) _____   _____

   (2) _____   _____

   (3) _____   _____

   (4) _____   _____

HYDROLYSIS EQUILIBRIA

PRE-STUDY

This experiment is designed to explain two peculiar observations. First, if you could measure the pH of pure, freshly distilled water, your value would be 7.0. If you dissolved enough sodium acetate in this water to give a 0.10 F solution, the pH would become 8.7. We can only conclude that a reaction must have occurred to produce OH^-.

The second observation is that if we were to titrate 50.0 ml of 0.20 M HAc with 50.0 ml 0.200 M NaOH, we would have added an amount of NaOH equivalent to the HAc present according to the titration reaction

$$HAc + OH^- \rightarrow Ac^- + H_2O \qquad (1)$$

The concentration of the NaAc produced is 0.10 M. The pH of this solution at the end-point of a titration is also 8.7.

A more careful observation of the above two situations reveals that even though the 0.1 M NaAc solution has been prepared by two entirely different methods, the end result is still the same. We could hardly expect the pH's of identical solutions to differ.

Therefore, at the equivalence point in the titration of a weak acid (the point at which the amount of base added is identical to the amount of acid present) the pH of the solution is greater than 7: this result has important consequences for the selection of an indicator to observe such an equivalence point. We will explore this question of selection in some detail when we conduct the several titrations in Experiment 23 on volumetric analysis.

In this experiment, we will determine the pH of the solutions of several salts and even use our sense of smell in order to determine the reason for the above observations and to determine the extent to which such reactions proceed.

The best way to begin is to note the way equation (1) is written. The reaction is written with one arrow pointing to the right. Because this experiment is included in a section on equilibria phenomena, we should immediately think of the fact that the reaction must proceed to the left as well. This reaction produces OH^-, leading to a pH that is greater than 7. Because the reaction is with H_2O, it is called a hydrolysis reaction.

Hydrolysis always involves the water equilibria. For example, in a $0.1\ F\ NaC_2H_3O_2$ solution, we have a *competition of the proton* from the water for either of the two bases, OH^- or $C_2H_3O_2^-$. The position of the equilibria will depend upon the relative values of the constants K_w and $K_{HC_2H_3O_2}$. This is shown by the expanded equation

$$HOH \leftrightharpoons H^+ + OH^- \qquad (2)$$
$$+$$
$$C_2H_3O_2^-$$
$$\updownarrow$$
$$HC_2H_3O_2(aq)$$

The formation of any weak acid, such as $HC_2H_3O_2$, will correspondingly decrease the H^+ concentration, shift the water equilibrium to the right, and form more OH^-, as indicated by the hydrolysis equation

$$C_2H_3O_2^- + HOH \rightleftharpoons HC_2H_3O_2(aq) + OH^- \qquad (3)$$

This is just the reverse of the neutralization equation (1), above.

How completely does $HC_2H_3O_2$ neutralize OH^- in equation (1)? Or, to what extent does $NaC_2H_3O_2$ hydrolyze according to equation (3)? The hydrolysis constant expression corresponding to equation (3) is[1]

$$K_h = \frac{(HC_2H_3O_2)(OH^-)}{(C_2H_3O_2^-)} \qquad (4)$$

This constant, K_h, may be evaluated in terms of the competitive tendencies to form $HC_2H_3O_2$ or HOH, as follows. If both numerator and denominator in equation (4) are multiplied by (H^+), and the terms rearranged, we have

$$K_h = \frac{(HC_2H_3O_2)}{(H^+)(C_2H_3O_2^-)} \times \underline{(H^+)(OH^-)} = \frac{K_w}{K_{HC_2H_3O_2}}$$

$$= \frac{10^{-14}}{1.8 \times 10^{-5}} = 5.6 \times 10^{-10} \qquad (5)$$

---

[1]The factor (H_2O) is omitted here and included with K_h, since the concentration or activity of water is practically constant in aqueous solutions.

If, in $0.1\ F\ NaC_2H_3O_2$, x mole/liter each of $HC_2H_3O_2$ and of OH^- are present at equilibrium, $(0.1 - x)$ mole/liter of $C_2H_3O_2^-$ will remain. Then, by substitution in equation (5),

$$K_h = \frac{(x)(x)}{(0.1 - x)} = 5.6 \times 10^{-10}$$

Since K_h is very small, x is also small, and $(0.1 - x) =$ almost 0.1. Then $x^2 = 5.6 \times 10^{-11}$, and $x = 7.5 \times 10^{-6}\ M\ OH^-$ (or $HC_2H_3O_2$).

The fraction hydrolyzed is $7.5 \times 10^{-6}\ M\ OH^-$ out of a possible $0.1\ M\ OH^-$ if hydrolysis was complete according to equation (3). This is

$$\frac{7.5 \times 10^{-6}\ M\ OH^-}{0.1\ M\ OH^-} = 7.5 \times 10^{-5},$$

or about 0.0075% hydrolyzed. This is not much, but the solution is definitely basic.

Analogously, the hydrolysis of a cation of a weak base produces an acidic solution. If *both* a weak acid and a weak base are produced by hydrolysis, the extent of hydrolysis will be much greater because of the effect of each reaction on the other, but the solution remains more nearly pH 7, as the H^+ and the OH^- produced tend to neutralize one another. Here, the hydrolysis constant, K_h, may be calculated from the relationship

$$K_h = \frac{K_w}{K_{acid}K_{base}} \qquad (6)$$

The hydrolysis of polyvalent anions, such as CO_3^{2-}, occurs in steps, as follows:

$$CO_3^{2-} + HOH \rightleftharpoons HCO_3^- + OH^- \qquad (7)$$

$$HCO_3^- + HOH \rightleftharpoons H_2CO_3 + OH^- \qquad (8)$$

Here the OH^- formed by the first step represses the second step so that it occurs to a very slight extent, and no CO_2 results from the small amount of H_2CO_3 formed.

The hydrolysis of polyvalent cations likewise occurs in steps. For Fe^{3+}, using the simple, unhydrated formula, the first step reaction would be

$$Fe^{3+} + HOH \rightleftharpoons Fe(OH)^{2+} + H^+ \qquad (9)$$

The second and third steps in hydrolysis would occur to lesser extents, forming additional hydrogen ion and $Fe(OH)_2^+$, and, for the third step, $Fe(OH)_3(s)$.

If Fe^{3+} and CO_3^{2-} are mixed in solution, the H^+ and OH^- respectively produced by each hydrolysis reaction neutralize one another, and shift each equilibrium to the completion of all the step by step processes; the final products are $Fe(OH)_3(s)$ (not $Fe_2(CO_3)_3(s)$), and $CO_2(g)$:

$$2Fe^{3+} + 6HOH \rightleftharpoons 2Fe(OH)_3(s) + 6H^+$$
$$3CO_3^{2-} + 6HOH \rightleftharpoons 3H_2CO_3(aq) + 6OH^-$$

$$2Fe^{3+} + 3CO_3^{2-} + 6HOH \rightleftharpoons 2Fe(OH)_3(s)$$
$$+ 3H_2CO_3(aq)$$
$$\downarrow$$
$$3H_2O + 3CO_2(g)$$

EXPERIMENTAL PROCEDURE

Special Supplies: pH meter (optional).

Chemicals: Al (powder), $Al_2(SO_4)_3(s)$ or alum, Al_2S_3, $NH_4C_2H_3O_2(s)$, 1 F $NH_4C_2H_3O_2$, $(NH_4)_2CO_3(s)$, $NH_4Cl(s)$, 1 F NH_4Cl, 1 F NH_3, 1 F $HC_2H_3O_2$, 1 F $NaC_2H_3O_2$, $NaHCO_3(s)$, 1 F Na_2CO_3, 1 F NaCl, S (powd.). Indicator solutions: orange IV, methyl orange, methyl red, bromocresol purple, bromthymol blue, phenolphthalein, alizarin yellow R, indigo carmine. (Buffer tablets of appropriate pH may be used for the preparation of demonstration indicator color-range sets.)

PRELIMINARY NOTE: Review Experiment 20 on the colors of various indicators throughout their respective color change pH ranges; see also Table 8, Appendix C. Sample tubes covering the color change pH range for certain indicators may be available in the laboratory for your color comparisons with your own test samples. Results may be checked with the pH meter, if available.

1. Hydrolysis of the Salts of a Strong Base and a Weak Acid.
Measure the pH of the following solutions: 1 F NaCl,[2] 1 F NaAc and 1 F Na_2CO_3. If a pH meter is available, use it to measure the pH of 5- to 10-ml portions of these solutions. If a pH meter is not available, indicators may be used to obtain satisfactory data on the degrees of hydrolysis of each of the three salts. Start out using indicators in the central range of pH 5–9. If the pH is outside of this range, select other indicators as your evidence dictates.

Record your pH values in the first table of the Report Form and calculate the corresponding (H^+) and (OH^-). Answer the questions and write the appropriate ionic reactions. Remember that the (H^+) is very low in water and thus the predominant reactant species is water and not the hydronium

ion. Calculate the degree of hydrolysis of the acetate and carbonate sodium salts. Look up the pK of HAc and the pK_2 of H_2CO_3 in Table 9 of Appendix C. Make a general statement relating the degree of hydrolysis to the strength of the weak acid formed by the hydrolysis reaction.

2. Hydrolysis of the Salts of a Weak Base.
(a) *Hydrolysis of the salt of a weak base and a strong acid.* Measure the pH of 1 F NH_4Cl, using the pH meter or appropriate indicators. Write the net ionic equation that corresponds to the observed behavior.

(b) *Hydrolysis of the salts of a weak base and a weak acid.* As described in the pre-study section, the pH of the solutions of such salts cannot be used as the criteria for the extent of hydrolysis. The cation of the weak base will react with water to produce the weak base plus H^+, and the anion of the weak acid will liberate OH^-. If the pK's of a weak base and weak acid are about the same, the solution will remain about neutral. An extensive amount of reaction may have occurred and yet the pH meter indicates no change. In hydrolysis of the ammonium salts, we are endowed with a natural instrument — the nose —, which, if used judiciously, can be employed safely and with sufficient accuracy to determine the relative extents of hydrolysis.

Therefore, rather than measuring the pH of several such ammonium salts, carefully note the odors of several apparently dry bottles of the salts. The air contains enough water vapor to cause hydrolysis to occur on the surface of the salt crystals. The NH_4^+ combines with this water to yield NH_3 and H_3O^+. The odor of ammonia is a familiar one because it is present in a number of window cleaners. Use the procedure demonstrated in Figure 1 of the Introduction to avoid inhaling too much of this rather strong odor.

Cautiously smell the bottles of NH_4Ac and $(NH_4)_2CO_3$. As a reference, also smell the bottle of NH_4Cl. Keep the stoppers or lids on these bottles so that the next student will have a uniform comparison. List the results in the report form and explain them.

Another salt of a weak base and a weak acid is aluminum sulfide, Al_2S_3. Look up the solubility product of $Al(OH)_3$ in Table 12, Appendix C, and the pK_2 of H_2S in Table 9, Appendix C. These values should suggest the degree of hydrolysis you might expect from this salt.

Your instructor will provide you with a small piece of Al_2S_3. Add it to 10 or 15 ml of water in a small beaker or test tube. Note any evidence of the

---

[2]Traces of impurities may cause errors in interpreting the behavior of NaCl in water. If you are uncertain of the value you obtain, have your instructor dissolve a gram of reagent grade NaCl in 20 ml of freshly distilled or deionized water.

formation of a gas or precipitate. Write the equation for this hydrolysis reaction, remembering that the salt is present this time at the beginning as a solid, not as ions in solution. What would you conclude about the relative degree of hydrolysis in any reaction in which both the acid and base formed are extremely weak, or in which both the products are removed from the sphere of action by precipitation or gas evolution?

3. An Application of Hydrolysis.
Practically all baking powders consist of a mixture of sodium hydrogen carbonate with some solid substance that furnishes H^+ when water is added. In alum baking powders this H^+ is furnished by the hydrolysis of alum, $KAl(SO_4)_2 \cdot 12H_2O$.

Place about 3 g of solid alum or of aluminum sul-

fate in a dry beaker, and place 3 g of solid sodium hydrogen carbonate in a second dry beaker. Mix together a little of each in a third dry beaker. Is there a reaction? Now add 10 ml of water to each beaker and note the results. Test samples of the two separate salt solutions with litmus paper and suitable indicators to determine their approximate pH.[3] Finally, mix the two salt solutions. Explain the results, and write net equations for all reactions involved.

---

[3] The slight ionization of HCO_3^- as a very weak acid is counterbalanced by its hydrolysis, so that a hydrogen carbonate ion solution is slightly basic. On boiling, it changes to CO_3^{2-} and $CO_2(g)$:

$$HCO_3^- \rightleftharpoons H^+ + CO_3^{2-} \quad \text{(ionization)}$$
$$HCO_3^- + HOH \rightleftharpoons OH^- + H_2CO_3(aq) \quad \text{(hydrolysis)}$$
$$2HCO_3^- \rightarrow CO_3^{2-} + H_2O + CO_2(g) \quad \text{(on boiling)}$$

REPORT 22	Hydrolysis Equilibria	NAME	
		SECTION	LOCKER
		INSTRUCTOR	DATE

OBSERVATIONS AND DATA

1. Hydrolysis of the Salts of a Strong Base and a Weak Acid

Solution	Indicators used and colors (if employed)	pH	(H^+)	(OH^-)
1 *F* NaCl				
1 *F* NaAc				
1 *F* Na_2CO_3				

With sodium acetate, the hydrolysis is due to the formation of____________________.

The ion in sodium acetate involved in the hydrolysis therefore is____________________.

Write the net ionic equations for the hydrolysis, if any, of the following salts:

NaCl____________________

NaAc____________________

Na_2CO_3 (first step only)____________________

Degree of Hydrolysis: NaCl__________ NaAc__________ Na_2CO_3__________

If the 1 *F* NaAc solution hydrolyzed completely according to the reaction you have written above, the (OH^-) would be:

__________*M*

You have found the actual (OH^-) and, if all has gone well, your actual value is smaller than the complete hydrolysis value you have just found. Therefore you can calculate the percent of the salt hydrolyzed from the ratio of these two values.

Method of calculation:

__________%

Similarly, calculate the percent of the Na_2CO_3 that hydrolyzed in the 1 F Na_2CO_3. Remember that you need consider only the first step in this hydrolysis.

Method of calculation:

_____%

State the general principle relating the degree of hydrolysis and the strength of the weak acid formed by the hydrolysis reaction.

---

2. Hydrolysis of the Salts of a Weak Base

(a) Hydrolysis of the Salt of a Weak Base and a Strong Acid. Note your observations and give the net ionic equation.

Solution	Indicators used and colors (if employed)	pH	(H⁺)	(OH⁻)
1 F NH₄Cl				

Net ionic equation: _____

(b) Hydrolysis of the Salts of a Weak Base and a Weak Acid. List the three ammonium salts tested in order of strength of ammonia odor, from the strongest to the weakest.

_____   _____   _____

Write the net ionic reactions that correspond to the observed reactions, and use these equations to explain the observed order.

Give the net equation for the hydrolysis of aluminum sulfide: _____

This substance, Al_2S_3, is so hygroscopic (it readily reacts with water) that it is generally not found in chemical stockrooms. Explain this.

List four reasons why the hydrolysis of Al_2S_3 proceeds essentially to completion.

3. An Application of Hydrolysis

The observed results of mixing solid $Al_2(SO_4)_3$ and $NaHCO_3$, adding water, and also of mixing the separate solutions were the following:

The approximate pH of $Al_2(SO_4)_3(aq)$_____and of $NaHCO_3(aq)$_____

The net ionic equation for the hydrolysis
reaction occurring in $Al_2(SO_4)_3(aq)$ is_____

Combine this equation with the equation for the neutralization of $NaHCO_3(aq)$, namely,

$$\_\_\_\_H^+ + \_\_\_\_HCO_3^- \rightarrow \_\_\_\_H_2O + \_\_\_\_CO_2(g)$$

so as to eliminate H^+ and give the overall net ionic equation for the whole process.

VOLUMETRIC ANALYSIS.
THE TITRATION OF ACIDS AND BASES

PRE-STUDY

An important aspect of science is the ability of the scientist to measure precisely. Indeed the precision of measurement is one of the principal distinguishing features between the physical sciences, the biological sciences, and the social sciences. One of the purposes of this experiment is to acquaint you with a technique frequently used by chemists to determine, with a precision of one part in a thousand, the amount of a substance present in an unknown. The analytical balance, which costs between 600 and 700 dollars, is the only other instrument usually available in the General Chemistry laboratory that is capable of this precision. The only tool need for volumetric analysis is a twenty- or thirty-dollar buret.

Volumetric analysis relies on the buret to measure accurately the amount of one chemical compound that must be added to react totally with an unknown compound present in a solution. The unknown solution may contain Cl^-. A solution, called the titrant, of a known concentration of Ag^+ can be added to the Cl^- solution until all of the chloride is precipitated. There are two requirements for a successful titration. The volume of Ag^+ solution must be carefully measured, and there must be some signal to tell the experimenter when to stop the addition. A buret is used for the accurate measurement and an indicator — such as Ag_2CrO_4 — turns color to signal when the end point is reached.

In this experiment, we will titrate a number of acids and bases. A buret will be used to measure accurately the volume of acid added. An acid-base indicator, as described in Experiment 20 will be used to detect the end point. An important part of this experiment will be for you to apply here the information you learned about indicators in Experiment 20 and the study of hydrolysis you performed in Experiment 22.

Acid-Base Titrations

We first consider the titration of 40.00 ml of 0.1000 M NaOH with 0.1000 M HCl. The reaction occurring is

$$H^+ + OH^- \rightleftharpoons H_2O \tag{1}$$

Because the pK of water is 14, we know the reaction will go almost completely to the right. At the point at which we have added 40.00 ml of the acid, the numbers of H^+ and OH^- present will be identical. Thus all of the OH^- will have been converted to H_2O. Also, there will be no extra H^+ present. The only source of H^+ or OH^- will be the dissociation of water. We know that the pH of such a solution will be 7.0.

The indicator used to detect this pH should be one that changes color near pH 7 or, alternatively, one that has a pK near 7. Table 8 of Appendix C indicates that bromthymol blue would be a good choice. Actually, the pH change is so large at the end point of this titration that phenolphthalein with a pK of 9 can be used successfully also.

If the acid and base are both known as in the titration of HCl with NAOH, the use of molarities and moles is satisfactory and simple. At the end point, the following relationship holds:

$$\text{No. moles } H^+ = \text{No. moles } OH^-$$

As discussed in Study Assignment B, the technique for converting from concentrations to amounts is to multiply concentration by volume. Thus:

$$\text{ml HCl} \times \text{Molarity HCl}\left(\frac{\text{mmole}}{\text{ml}}\right) =$$

$$\text{ml NaOH} \times \text{Molarity NaOH}\left(\frac{\text{mmole}}{\text{ml}}\right)$$

The calculation is particularly simple in the present reaction:

$$40.00 \text{ ml HCl} \times 0.1000 \frac{\text{mmole } H^+}{\text{ml}} =$$

$$40.00 \text{ ml NaOH} \times \frac{0.1000 \text{ mmole } OH^-}{\text{ml}}$$

Generally the concentration of one of the solutions is unknown. The purpose of the titration is to determine this concentration as illustrated in the following example.

Example. What is the molarity of an NaOH solution if 43.25 ml of it are required to neutralize 50.00 ml of 0.500 M HCl?

The preceding discussion shows that for this particular reaction, the molarity is

$$V_{HCl} \cdot M_{HCl} = V_{NaOH} \cdot M_{NaOH}$$

$$M_{NaOH} = M_{HCl} \times \frac{V_{HCl}}{V_{NaOH}} = \frac{0.500 \times 50.00}{43.25}$$

$$= 0.578 \ M \text{ NaOH}$$

The identical calculation can be made if the acid was HAc and the base was NaOH. The net ionic reaction is

$$HAc + OH^- \rightarrow Ac^- + H_2O \tag{2}$$

Again the K of this reaction is exceedingly large and the reaction proceeds virtually to completion. Because 1 mole of HAc reacts with 1 mole of OH^-, we can write

$$M_{HAc} \times V_{HAc} = M_{NaOH} \times V_{NaOH}$$

Thus we can determine the concentration of acetic acid in a sample of vinegar, for instance, by titration with an NaOH solution of known concentration.

The one big difference in these two titrations is the pH of the solution when an amount of base identical to the amount of acid has been added. This is called the equivalence point of the titration. We saw that the pH was 7 for the HCl reaction. The pH in the HAc titration will not be 7 because of the hydrolysis reaction. As we saw in Experiment 22, the pH of the solution at the equivalence point will be 8.7 if the $(Ac^-) = 0.1 \ M$. Because the pH does not change as rapidly with increasing amounts of NaOH in a weak acid titration as it does in the strong acid titration, it is necessary now to choose an indicator with a pK near 8.7. Phenolphthalein seems like a good choice.

The same techniques and reasoning are applied to the titration of a base with an acid. The titration of sodium bicarbonate with an acid is represented by the equation

$$H^+ + HCO_3^- \rightarrow H_2CO_3 \tag{3}$$

An indicator is selected that corresponds to the pH of a solution of carbonic acid. The stoichiometry again is known: one mole of acid for each mole of bicarbonate.

The concept of equivalents and the corresponding concentration unit normality, which is equivalents/liter, are useful in the event that one of the reactants, the acid or the base, is unknown, or the stoichiometry is unknown. An equivalent in an acid-base reaction is that amount of the substance which releases or reacts with one mole of hydrogen ions. Thus the number of equivalents of acid in a titration can be determined by the number of equivalents of base required to neutralize all of the hydronium ions from the acid. One may then compute the equivalent weight by dividing the mass of the acid (obtained from the initial weighing) by the number of equivalents (obtained from the titration).

The concept of equivalents suffers from the fact that the equivalent weight of a substance depends on the reaction being carried out. For instance, sulfuric acid has an equivalent weight of 98 in reaction (4) below and an equivalent weight of 49 in reaction (5). These are not ionic equations.

$$H_2SO_4 + NaOH \rightarrow NaHSO_4 + H_2O \quad (4)$$

$$H_2SO_4 + 2NaOH \rightarrow Na_2SO_4 + 2H_2O \quad (5)$$

Because of this ambiguity, we believe you will find it easier to use moles and molarity in all situations in which the stoichiometric equation is known.

Standard Solutions

In order for the concentration of an unknown acid or base to be determined as described in the preceding discussion, the concentration of one of the solutions must be known. Such known solutions are called standard solutions. There are two kinds of these.

Primary standards are compounds that can be directly weighed out and dissolved to a given volume to give a solution of accurately known concentration. To qualify as a primary standard, the material must be at least 99.9% pure, have a definite composition, be quite soluble in water, and its composition should not change upon drying. In this experiment, you will use oxalic acid, $H_2C_2O_4 \cdot 2H_2O$, as a primary standard. With the exception of the last criterion, it satisfies the list of qualifications rather well. Since you will not be drying the sample, we can safely assume that both waters of hydration are present in the crystals. Oxalic acid has the structural formula

This formula resembles that of two acetic acid molecules that have been squeezed together and their methyl groups removed. It should not be

surprising, then, that both hydrogens are acidic and readily react with a base. Their pK's are 1.4 and 4.3. The molecular weight of $H_2C_2O_4 \cdot 2H_2O$ is 126.07.

EXPERIMENTAL PROCEDURE

Special Supplies: Two 50-ml burets, double buret clamp, 250-ml volumetric flask, boiling chips.

Chemicals: Oxalic acid crystals ($H_2C_2O_4 \cdot 2H_2O$), phenolphthalein, methyl orange, bromthymol blue. Unknowns as needed, or samples brought from home—such as vinegar, citric fruit juices, baking soda, and unknown ammonium salts—as directed by your instructor.

1. Preparation of a Standard Oxalic Acid Solution. Carefully weigh a 150-ml beaker to a precision of ± 0.001 g. Put 7–9 g of pure oxalic acid crystals, $H_2C_2O_4 \cdot 2H_2O$, into the beaker, and again weigh accurately.

With the aid of a stirring rod, carefully transfer the weighed crystals, without loss, to a clean 250-ml volumetric flask. Rinse the last powder from the beaker into the flask with distilled water, and then add 150–175 ml of distilled water (avoid filling the flask into the neck), and gently mix the solution by swirling it in the flask until solution is complete. (Save time here by preparing your NaOH solution for part 2 while these crystals are dissolving). When all of them are dissolved, fill the flask carefully to the mark with additional distilled water, adding the last portion by drops. Stopper the flask, and mix the solution well by repeated inversion and swirling. Transfer this standard acid to a clean, dry, labeled flask, and stopper it to avoid evaporation. (If the flask is wet, avoid dilution of the acid, by rinsing the flask beforehand with two small portions of your acid, and discarding these washings.)

From your exact weight of acid, and from the molecular weight of $H_2C_2O_4 \cdot 2H_2O$, calculate the number of moles you have weighed out and then the molarity of your solution. Label the bottle with a description of the acid, its molarity, and your name and date.

2. Preparation of an NaOH Solution for Standardization. Dilute 80 ml of the approximately 6 M desk reagent of NaOH with 880 ml of water. (Desk reagents are not made up with quantitative accuracy. Hence your diluted solution will be only approximately the molarity desired, and must be standardized by titration with your standard acid.) Place your basic solution in a bottle or flask, close with a rubber stopper, and mix very thoroughly by repeatedly inverting and swirling the solution. Calculate the approximate molarity of your NaOH, based on the dilution ratio, and include this in your report. Label the solution.

3. Standardization of the NaOH Solution.
Thoroughly clean your two burets, using a detergent solution and a long buret brush. A clean buret will drain freely without forming droplets on the inside surface. Rinse burets thoroughly with tap water, then once with deionized water, and allow to drain.

Examine the functioning of the stopcock. See that it turns freely to deliver a full stream of water and that it does not leak.

Rinse one buret with a 5-ml portion of your NaOH solution, and discard the rinsing solution. Repeat this rinsing procedure. Run at least a portion of each rinse through the stopcock and the tip so that all parts of the inside of the buret will have come in contact with the NaOH solution.

Fill the buret with the NaOH solution above the 0-ml mark. Allow the solution to drain out through the stopcock, and tip to insure that they are full. The liquid level may be stopped at the 0.00-ml mark or at any point somewhat below this.

Fill the second buret with the standard oxalic acid solution after following the same rinsing procedures as for the first one. Remember that, after the burets are filled, both will look the same. You must devise a procedure for distinguishing them.

Touch off the remaining drop at the tip of each buret on the side of a beaker. Read each meniscus to the nearest tenth or fifth of the smallest division on your buret. Often, it is helpful to hold a white card with a wide black mark on it behind the buret to see the meniscus more clearly. With practice, you should be able to read the buret with a precision of ±0.02 ml.

Now you are ready to conduct your first titration. Run about 20 ml of the $H_2C_2O_4$ solution into a clean, rinsed 250-ml Erlenmeyer flask. Touch the last drop of the buret tip to the inside of the flask, and rinse it into the flask with deionized water. Add 2 drops of the phenolphthalein indicator solution.[1] To begin titrating the acid with the NaOH, open the stopcock of the buret almost all the way and stir the solution by swirling the flask. As the

end point is approached, the faint pink color existing in parts of unmixed NaOH solution will persist for several seconds. At this point, you must decrease the rate of addition until you are adding drops one at a time and thoroughly mixing the solution before the addition of the next drop. As you develop your skill, you will be able to approach the end point so slowly that you will be able to touch the tip of the buret to the inside of the flask and add fractional drops that can then be rinsed down into the solution. The end point is reached when a faint pink color persists throughout the solution after it has been swirled. If the pink color fades after the solution has been standing, it will be because carbon dioxide has been absorbed.

If you have the misfortune to go past the end point by more than 1 drop, add enough additional acid to remove the pink color. Then continue the titration with the NaOH, adding the base more slowly this time.

Read and record the final buret reading for each solution.

Repeat the titration. Calculate the molarity of the NaOH solution for the two titrations. If the molarities agree within 1% proceed to the next part. If they do not, repeat the titration once more.

In calculating the molarity of the sodium hydroxide solution, you must remember that there are two titratable protons per oxalic acid molecule, and therefore 2 moles of NaOH are required to titrate one mole of $H_2C_2O_4$. The stoichiometric equation (not the net ionic equation) is

$$H_2C_2O_4 + 2NaOH \rightarrow$$

$$2Na^+ + C_2O_4^{2-} + 2H_2O \quad (6)$$

4. Titration of Unknowns.
Perform as many of the following experiments as time permits, or as directed by your instructor. One report form is provided for the first analysis. Prepare additional report forms suitable for presenting the results of parts (b), (c), and (d) if they are needed.

(a) *Determination of Acetic Acid in Vinegar.* Obtain a light colored sample of vinegar. If your sample has a dark color, it should be treated with absorbent charcoal to remove the coloring so that the faint color of the indicator can be observed readily. Rinse a buret and fill it with the vinegar.

Rinse and fill a second buret with your standardized NaOH solution.

Carry out two, or preferably three, titrations as you did in part 3, using phenolphthalein as the indicator.

---

[1] If you have some difficulty in seeing the color pink, you may want to use thymol blue as the indicator. It has about the same pK_a as phenolphthalein, but it displays a change from yellow to blue as the pH is increased from 7.5 to 9. A sheet of white paper placed under the flask will make the color change more visible.

The principal acid present in vinegar is acetic acid, CH_3COOH. Only the hydrogen bonded to the oxygen atom will react with base. Federal law requires that vinegar contain 4% acetic acid by weight. Calculate the weight percent CH_3COOH in your sample and see whether the federal standard has been met. The grams of acetic acid per liter of solution are obtained from the molarity of the NaOH and the volumes of base and vinegar required for the titration. The grams of solution per liter of solution may be obtained from the density of vinegar, which can be taken as 1.005 g/ml. These two numbers may be combined to yield weight percent acetic acid.

(b) *Determination of Citric Acid in Citrus.* Citric acid gives the tart taste to citrus juices. Its structural formula is

$$
\begin{array}{c}
CH_2-COOH \\
| \\
HO-C-COOH \\
| \\
CH_2-COOH
\end{array}
$$

which we can abbreviate as $H_3C_6H_5O_7$ to indicate that there are three acidic protons that can react with base.

$$H_3C_6H_5O_7 + 3OH^- \rightarrow 3H_2O + C_6H_5O_7{}^{3-} \quad (7)$$

The concentration of citric acid varies widely in various citrus juices and it may be of interest to you to compare your results with those of other members in the class, so that you note the differences between orange juice and grapefruit juice, fresh juice and canned juice, orange juice and orangeade, and so on. Because of this large variation however, it will probably be necessary to carry out a trial titration to see approximately what ratio of the volumes of acid and base will be satisfactory.

Squeeze out about 100 ml of lemon, orange, or grapefruit juice, or use commercially prepared juices as directed. If significant amounts of pulp are present, remove the pulp by squeezing through cheesecloth. Determine the density of the juice by means of a hydrometer or by weighing a carefully measured volume of juice.

Carry out duplicate titrations as above, using 5 drops of phenolphthalein as the indicator. If you are titrating orange juice, keep a sample of the original juice in a flask beside the one you are using for the titration. This comparison sample will help you detect the first appearance of the pink color in the highly colored solution.

Calculate the percent citric acid in your sample as you calculated the percent acetic acid in vinegar for part (a). Remember the stoichiometry of the reaction as expressed in equation (7).

(c) *Determination of the Purity of Baking Soda.* Baking soda is a familiar ingredient in many cake recipes. It should be pure sodium bicarbonate, $NaHCO_3$, that releases bubbles of CO_2 upon reaction with acid. The released gas causes the cake to rise. In this experiment, you will determine the purity of a sample of commercial baking soda.

Weigh out, to ± 0.001 g, two samples of baking soda of about 1.5 g each into numbered Erlenmeyer flasks, using one of the three spilling methods illustrated in Experiment 2, Figure 2-3. Add about 20 ml of deionized water to dissolve the samples.

Obtain about 100 ml of a standardized 0.25 M H_2SO_4 solution from your instructor. (Your oxalic acid solution cannot be used for this experiment as a titrant, because the end point of the titration of a weak acid, $H_2C_2O_4$, and a weak base, $HCO_3{}^-$, extends over several ml of oxalic acid titrant.) Fill a buret with the sulfuric acid solution, add two drops of methyl orange indicator to each bicarbonate solution and titrate to the end point. The methyl orange end point is the point at which the yellow solution just turns orange.

Calculate the number of moles of bicarbonate in each sample, remembering that the reaction you are studying is

$$H_2SO_4 + 2NaHCO_3 \rightarrow$$

$$Na_2SO_4 + 2H_2O + 2CO_2 \uparrow \quad (8)$$

From the molecular weight of $NaHCO_3$, calculate the weight percent $NaHCO_3$ in your sample of baking soda.

(d) *The Molecular Weight of a 1:1 Ammonium Salt.* A classic reaction in the determination of the concentration of proteins in solution is the Kjeldahl method, which was developed in the Carlsberg Laboratory in Copenhagen, Denmark. The method consists of converting the nitrogen present in the protein into ammonium ions, adding an excess of NaOH solution, distilling off the released NH_3 into an excess of standard HCl and back-titrating the excess acid with standard NaOH. The reactions are

$$NH_4{}^+ + OH^- \xrightarrow{\Delta} NH_3 \uparrow + H_2O \quad (9)$$

$$NH_3 + H^+ \longrightarrow NH_4{}^+ \quad (10)$$

followed by

$$H^+ + OH^- \longrightarrow H_2O \quad (11)$$

The difference between the number of moles of standard HCl added and the number of moles of NaOH required for neutralization equals the number of moles of ammonium ion originally present. Kjeldahl then could calculate the amount of protein present if the percent nitrogen in the protein was known.

You will follow a variation of this procedure. You will be provided with the unknown already present as the ammonium salt. This salt has the formula NH_4X, where X may be a variety of anions. The ammonium salt is added to an excess of standard NaOH solution, and this solution is boiled to remove the ammonia. Then the unreacted NaOH is titrated with the standard oxalic acid solution.

Obtain from your instructor a numbered vial containing an unknown 1 : 1 ammonium salt. Weigh three samples of about 0.800 g ±0.001 g each into three numbered 250-ml Erlenmeyer flasks.

Deliver a precisely measured volume of 40–50 ml of your standardized NaOH solution from a buret into each of the three flasks. Add 2 or 3 boiling chips to each flask and boil the solutions gently (in a hood), being careful not to lose any solution by spattering. Continue this procedure for about 10 minutes until you have determined, with the use of moistened red litmus paper, that NH_3 is no longer present in the vapor above the mouth of the flask.

Using 2–3 drops of bromthymol blue as the indicator, titrate the excess base in the three cooled flasks with your standardized oxalic acid solution. At the end point, the solution changes from blue to green.

To obtain the number of moles of ammonium ion, subtract twice the number of moles of oxalic acid required for neutralization from the number of moles of NaOH originally present.

Calculate the molecular weight of your unknown ammonium salt. If your results are satisfactory, your instructor will identify the salt and you can compute your percentage error.

REPORT

23

**Volumetric Analysis.
The Titration of
Acids and Bases**

NAME		
SECTION		LOCKER
INSTRUCTOR		DATE

DATA AND CALCULATIONS

1. Preparation of a Standard Oxalic Acid Solution

Give your data for the following:

Weight of oxalic acid + beaker   _____g

Weight of beaker   _____g

Weight of oxalic acid   _____g

Calculate the number of moles of oxalic acid, showing how you made the calculation.

_____moles

Calculate the molarity of oxalic acid solution, showing how you made the calculation.

_____M $H_2C_2O_4$

2. Preparation of an NaOH Solution for Standardization

Calculate the approximate molarity of your NaOH solution from the volumes of NaOH and water you used. Show how you made the calculation.

_____M NaOH

3. Standardization of the NaOH Solution

Give the following data for each titration that you have made.

	1	2	3
$H_2C_2O_4$ buret, 2nd reading	_____ml	_____ml	_____ml
1st reading	_____ml	_____ml	_____ml
Volume of oxalic acid solution used	_____ml	_____ml	_____ml
NaOH buret, 2nd reading	_____ml	_____ml	_____ml
1st reading	_____ml	_____ml	_____ml
Volume of NaOH solution used	_____ml	_____ml	_____ml

Give the molarity of the NaOH solution for each titration. Show your calculations for one of them.

_____ M _____ M _____ M

Calculate your mean value for M. If you reject one of your values, give specific experimental reasons for not including it.

Mean value = _____ M NaOH

---

4. Determination of Acetic Acid in Vinegar

Vinegar sample unknown number: _____

Give the following data for each titration.

	1	2	3
Vinegar buret, 2nd reading	_____ ml	_____ ml	_____ ml
1st reading	_____ ml	_____ ml	_____ ml
Volume of vinegar	_____ ml	_____ ml	_____ ml
NaOH buret, 2nd reading	_____ ml	_____ ml	_____ ml
1st reading	_____ ml	_____ ml	_____ ml
Volume of NaOH solution	_____ ml	_____ ml	_____ ml
Molarity of standard NaOH from part 3			_____ M

Calculate the molarity of vinegar for each titration. Show one calculation.

_____ M _____ M _____ M

Give the mean value. If you have discarded any values, explain why.

Mean value: _____ M

Calculate the grams of CH_3COOH per liter of vinegar, using the mean value of the molarity of CH_3COOH.

_____g CH_3COOH/liter

Calculate the grams of solution per liter of vinegar.

_____g soln/liter

Calculate the weight percent CH_3COOH in your sample.

_____%

What is the difference between your value and the federal standard of 4%? Comment on the significance to you as a consumer and to the manufacturer as a producer of this variation.

If you perform additional titrations for citric acid or ammonium concentration, prepare additional report forms similar in format to this one.

EXERCISES

1. Washing down the sides of the flask with water during the titration produces which of the following effects: (a) a high result; (b) a low result; (c) no effect?

_____

2. A student delivers about 20 ml of solution from a buret in the titration. The percentage uncertainty in the measurement if he reads the buret to the nearest 0.02 ml is (a) 1%, (b) 0.1%, or (c) 0.5%?

_____

3. For the baking soda determination of Part 4c, in which all the sample is dissolved in the flask and titrated, comment on the importance of the volume of distilled water added: (a) it must be carefully measured in order that the normality can be calculated; (b) the volume makes no difference because the water does not affect the number of equivalents. Explain your answer.

_____

PROBLEMS

1. Demonstrate that the correct indicator was chosen to mark the end point of the titration for *one* of the titrations described in Part 4 of this report. You will need to understand the concept of hydrolysis discussed in Experiment 22 and to look up the appropriate ionization constants of the indicator and the acid in Tables 8 and 9 of Appendix C.

2. 24.60 ml of 0.185 M HNO_3 are titrated with 27.35 ml of a KOH solution. What is the molarity of the KOH?

3. A 0.345-g sample of pure $H_2C_2O_4 \cdot 2H_2O$ crystals is dissolved in water, and titrated with 24.50 ml of a sodium hydroxide solution. What is the molarity of the NaOH?

4. 0.280 g of KOH will just neutralize what volume of 0.200 F H_2SO_4?

5. 0.750 g of a sample of commercial lye, NaOH, is dissolved in water and titrated with 32.00 ml of 0.500 M HCl. What is the percent purity of the lye sample?

6. 30.0 ml of 0.300 F H_3PO_4 is mixed with 90.0 ml of 0.200 F KOH, and the mixture is evaporated. Will the salt that crystallizes out be K_3PO_4, K_2HPO_4, or KH_2PO_4? Show your calculations.

7. How many grams of each of the following are required to just react with 300 ml of 0.250 M HNO_3?

    (a) $Mg(OH)_2$ _____ g   (c) Na_2CO_3 _____ g

    (b) CaO _____ g   (d) $NaHCO_3$ _____ g

REPORT 23	Sheet 3	NAME

8. To what volume must each of the following be diluted to give the concentration called for?

(a) 10.0 ml of 6.00 M HNO_3, to give a 0.200 M solution ______________ml

(b) 9.00 ml of concentrated (18.0 F) H_2SO_4 to give a 0.300 F solution ______________ml

(c) 150 ml of glacial acetic acid (17.0 F) to make it 6.00 F ______________ml

9. 80.0 ml of 0.200 M NaOH is mixed with 20.0 ml of 0.600 M HCl. What is the concentration of the remaining OH^-? (Final volume 100 ml.)

______________M

10. Complete the blank spaces in the following chart for each substance, utilizing the data given and any needed atomic weights.

Substance	Formula weight	Equiv weight	Vol (ml)	F	N	No. of moles	No. of equiv	Grams
KOH	56		800				0.400	
HXO_3		63			0.150	0.600		
H_2SO_4	98	49		3.00		0.500		
H_2SO_4	98				0.500		0.250	24.5
$Al_2(SO_4)_3$		57		0.050				34.2

Use this space for calculations:

PREPARATION AND IDENTIFICATION OF INORGANIC COMPOUNDS AND IONS

SOME INORGANIC PREPARATIONS

All of the reagents that you can find on the shelves of a chemistry storeroom were prepared from various raw materials and purified before they were bottled and sold. Inorganic preparations are therefore very important; without them, the chemical industry and research would come to a halt. The following procedures are given to provide you with further experience in methods and techniques for making some inorganic compounds. Carry out the preparations as you are directed by your instructor.[1] The compounds you make might be used in subsequent procedures to synthesize more inorganic compounds, or they might be used as pure chemical reagents for various purposes.

---

[1]It is suggested that about two or three preparations can be carried out in a three-hour laboratory period.

EXPERIMENTAL PROCEDURES

A. Sodium Hexanitrocobaltate(III), $Na_3Co(NO_2)_6$

Many metal ions, particularly transition metal ions, will form *complexes* with negative ions or with small neutral molecules that have regions of high electron density on electronegative atoms such as N or O. The formation of metal complexes can be put to many practical uses. The cyanide complex of silver, $Ag(CN)_2^-$, is an example that has important application in the extraction of silver metal from ores. In this experiment the $Co(NO_2)_6^{3-}$ ion will be prepared.

Special Supplies: compressed air, or aspirator or other source of vacuum, 125-ml filter flask, 4.25-cm Buchner funnel, and 4.25-cm filter paper.

Chemicals: $Co(NO_3)_2 \cdot 6H_2O(s)$, $NaNO_2(s)$, 50% acetic acid, acetone, 95% ethanol, 0.1 F $LiNO_3$, 0.1 F KNO_3, 0.1 F NH_4NO_3

Dissolve 5 g hydrated cobalt(II) nitrate, $Co(NO_3)_2 \cdot 6H_2O$, and 15 g sodium nitrite, $NaNO_2$, in 20 ml of hot water. Cool until slightly warm and add 5 ml of 50% acetic acid, drop by drop, stirring constantly during the addition. Allow the solution to cool, and pass air through the solution for 20 minutes by bubbling compressed air through the solution or by using the apparatus for drawing air (Figure 24-1), which you have connected to an aspirator or other source of vacuum. Filter and discard any residue. Add 27 ml of 95% ethanol to the filtrate to decrease the solubility of the salt, and allow the solution to stand for about 40 minutes. Filter, wash the solid, first with 95% ethanol and then with acetone, and then dry.[2]

Weigh the product. Prepare a saturated solution of sodium hexanitrocobaltate(III) in a test tube by adding small portions of the salt to 10 ml of water and shaking until no more salt dissolves. Add 2 ml of this solution to 2 ml of a 0.1 F solution of Li^+ (i.e., LiCl or $LiNO_3$). Record the result.

Add 2 ml of the sodium hexanitrocobaltate(III) solution to a solution containing 0.1 F K^+. Record the result.

---

[2]The precipitate is very finely divided and the filtration is much faster if you use a Buchner funnel and filter flask connected to a source of vacuum. See Figure i-17 of the Introduction.

Vacuum ←

Air →

125-ml Erlenmeyer flask

FIGURE 24-1
Apparatus for drawing air through a solution.

Finally, add 2 ml of the sodium hexanitrocobaltate(III) solution to a solution containing 0.1 F NH_4^+. Record the result.

B. Preparation of Lead

Lead was one of the first metals to be used by civilized man because it is easily prepared from its ores. The preparation of lead described below illustrates chemical reactions that are commonly used in "winning," or smelting, metals from their ores. Only a few metals are found in the free metallic state. Since most metals are found in combination with oxygen (as oxides), or with sulfur (as sulfides), they must be reduced to obtain the free metal.

Special Supplies: iron or glazed porcelain crucible and cover, clay triangle.

Chemicals: $PbCO_3$ (lead carbonate) or $2PbCO_3 \cdot Pb(OH)_2$ (basic lead carbonate), powdered charcoal.

Grind a 15 g sample of $PbCO_3$ or $2PbCO_3 \cdot Pb(OH)_2$ (basic lead carbonate) to a fine powder with a mortar and pestle. Place the powder in an iron or glazed porcelain crucible that is supported in a clay triangle, and begin heating. After a few minutes, begin heating the crucible strongly, and continue to heat until all of the material has changed color.

Allow the crucible to cool, and then empty its contents into a mortar. Add about 1 g of finely powdered charcoal to the mortar, and grind thoroughly. Transfer the contents to the crucible used previously and sprinkle about 0.5 g charcoal over the mixture. Place the cover on the crucible and heat strongly for 10 minutes. Remove the cover and stir the contents to cause the small globules of metal to coalesce. Replace the cover and heat for a few minutes longer. When heating is completed, allow the crucible to cool for about 10 minutes. Then remove the cover, and cautiously add a few ml of water. Place the contents of the crucible in a 100-ml beaker, and wash the carbon from the lead with a stream of water from a wash bottle. Dry and weigh the lead product.

C. Cupric Ammonium Sulfate Hydrate and Tetraamminecopper(II) Sulfate Hydrate

These two salts are considered together to emphasize their differences in structure. The first is a *double salt* and is prepared simply by cooling a concentrated solution containing cupric ions, ammonium ions, and sulfate ions. It crystallizes in the monoclinic crystal lattice and has the formula $Cu(NH_4)_2(SO_4)_2 \cdot 6H_2O$, or $CuSO_4 \cdot (NH_4)_2SO_4 \cdot 6H_2O$. In this, as in all cupric salts in aqueous solution, four of the six H_2O molecules are a part of the hydrated cupric ion, $Cu(H_2O)_4^{2+}$, so the formula may be written

$$Cu(H_2O)_4(NH_4)_2(SO_4)_2 \cdot 2H_2O$$

In the tetraammine complex, which is crystallized from a concentrated ammonia solution, four NH_3 molecules have replaced the H_2O molecules around the cupric ion, so the crystal lattice contains $Cu(NH_3)_4^{2+}$ ions and SO_4^{2-} ions, and it has the formula

$$Cu(NH_3)_4SO_4 \cdot H_2O$$

which is analogous to that for hydrated cupric sulfate, $Cu(H_2O)_4SO_4 \cdot H_2O$, or $CuSO_4 \cdot 5H_2O$.

Special Supplies: A desiccator containing freshly ignited $CaO(s)$ is desirable. A vacuum source and filter flask and Buchner funnel will speed the filtration.

Chemicals: $CuSO_4 \cdot 5H_2O(s)$, $(NH_4)_2SO_4(s)$, 95% ethyl alcohol, ether.

Cupric Ammonium Sulfate Hydrate. Calculate the weights of 0.1 mole $CuSO_4 \cdot 5H_2O(s)$ and 0.1 mole $(NH_4)_2SO_4(s)$. Weigh out these amounts, and dissolve the salts in 40 ml of hot water in a beaker. Cover with a watch glass, and set aside to cool slowly to form crystals of the double salt

$$Cu(NH_4)_2(SO_4)_2 \cdot 6H_2O(s)$$

The solubility of most salts decreases as the temperature decreases. Therefore, cooling the solution with ice will increase the yield. Filter the crystals, and dry them in air on a filter paper.

Tetraamminecopper(II) Sulfate Hydrate. Weigh out 0.05 mole (12.5 g) of $CuSO_4 \cdot 5H_2O(s)$, crush any large crystals in a mortar, and dissolve this in a mixture of 12 ml of H_2O and 20 ml of 15 F NH_3. Now add, a little at a time, with mixing, 20 ml of 95% ethyl alcohol. Allow this to stand, and finally cool the mixture thoroughly by immersing the beaker in a cooling bath of ice and water to promote crystal growth. Filter the crystals of $Cu(NH_3)_4SO_4 \cdot H_2O$ (best done with a Buchner funnel). Wash the crystals once with a mixture of equal volumes of alcohol and 15 F NH_3 then with alcohol alone, and finally with a little ether. The crystals may be quickly air-dried, and then should be bottled and stoppered at once to prevent loss of $NH_3(g)$. (They are best dried in a desiccator over freshly heated calcium oxide, CaO.)

D. Sodium Thiosulfate

This important salt, the "hypo" fixing bath used in photographic work, is easily prepared by the reaction of sodium sulfite with free sulfur. The S atom adds directly onto the unshared electron pair of the sulfite ion and thus has the same electron structure as a sulfate, in which one of the O atoms has been replaced by an S atom:

Chemicals: $Na_2SO_3(s)$ or $Na_2SO_3 \cdot 7H_2O(s)$, $S(s)$ (crushed roll sulfur).

Weigh 15 g of $Na_2SO_3(s)$ (or 30 g of the hydrate $Na_2SO_3 \cdot 7H_2O(s)$) and 4 g of crushed roll sulfur (flowers of sulfur are less satisfactory), and mix these with 65–70 ml of water in a beaker. Cover with a watch glass to prevent excessive evaporation, and heat to boiling very gently for an hour or longer until most of the sulfur has dissolved. Filter the solution while hot, and allow it to crystallize as it cools. You can overcome this solution's tendency to become supersaturated by rubbing the inner walls of the beaker with a stirring rod or seeding the cool solution with a single small crystal of $Na_2S_2O_3 \cdot 5H_2O(s)$. Filter the crystals, and dry them on filter paper. (A second crop of crystals may be obtained by evaporating the filtrate to half its volume and again cooling.)

The properties of sodium thiosulfate may be observed as follows. Dissolve 2–3 g of your prepared salt in 10 ml of water. First, to half of this add a few drops of 6 F HCl. Note the instability of the free acid, $H_2S_2O_3(aq)$, and identify the decomposition products formed. (Compare with the electronic structure.) Second, add the other portion of your $Na_2S_2O_3$ solution to a precipitate of AgBr made by mixing 1 ml of 0.1 F $AgNO_3$ and 1 ml of 0.1 F KBr. The complex ion formed is $Ag(S_2O_3)_2^{3-}$. This is the reaction of "hypo" in a photographic fixing bath. Unexposed (unreduced) AgBr is removed by the hypo to "fix" the photographic image.

REPORT

24

Some Inorganic
Preparations

NAME

SECTION   LOCKER

INSTRUCTOR   DATE

DATA AND OBSERVATIONS

A. Sodium Hexanitrocobaltate(III)

1. Weight of dried $Na_3Co(NO_2)_6$: _____ g

   Description of product: _____

2. Calculate the number of moles of reagents used in this synthesis:
   (a) $Co(NO_3)_2 \cdot 6H_2O$

   _____ moles

   (b) $NaNO_2$

   _____ moles

   Which reagent is in excess of the stoichiometric amount?

   _____

3. Calculate the % yield you obtained, based on the quantity of the *limiting* reagent available.

$$\text{Percent yield} = \frac{\text{g product obtained}}{\text{g product (theory)}} \times 100.$$

   _____ % yield

4. Result of the addition of 2 ml of saturated sodium hexanitrocobaltate(III) ion to the following substances. (Where you observed a chemical reaction, write an equation to interpret the reaction you observed.)

(a) 0.1 F Li^+ _____.

(b) 0.1 F K^+ _____.

(c) 0.1 F NH_4^+ _____.

5. What are the oxidation states of cobalt in the following:

   in $Co(NO_3)_2 \cdot 6H_2O$ _____ ?   in $Co(NO_2)_6^{3-}$ _____ ?

6. Write a chemical reaction to explain the reason for bubbling air through the reaction mixture during the synthesis.

7. Look up the ionic radii (crystallographic radii) of NH_4^+ and the Group I metal ions (Li^+, Na^+, K^+, Rb^+, and Cs^+) in Table 3 of Appendix C. Arrange them in order of increasing size. Next to each ionic radius on the list, note whether it gave a precipitate with $Co(NO_2)_6^{3-}$. What predictions would you make about the solubility of the Rb^+ and Cs^+ salts of $Co(NO_2)_6^{3-}$ on the basis of your data?

B. Preparation of Lead

1. Weight of lead produced: _____g

   Description of product: _____

   Expected weight of lead based on the number of moles of starting material (either $PbCO_3$ or $2PbCO_3 \cdot Pb(OH)_2$):

   _____g

2. Calculate the percent yield.

$$\text{Percent yield} = \frac{\text{g product obtained}}{\text{g product (theory)}} \times 100.$$

   _____percent yield

3. Write a chemical reaction to explain what happens when the lead carbonate (or basic lead carbonate) is heated in the first step. In the smelting industry this process is called "roasting".

4. Write a chemical reaction for the reaction of carbon (charcoal) with the product of the "roasting" step. This step is called reduction. What is the reducing agent?

5. The treatment of sulfide ores of many metals follows a similar sequence. Metallic iron can be produced from iron pyrites, FeS_2, by roasting in air to produce iron(III) oxide, followed by reduction with "coke" (a form of carbon produced by heating coal). Write balanced chemical reactions for the production of iron from the pyrites by this two step reaction:

REPORT 24	Sheet 2	NAME

C. Cupric Ammonium Sulfate Hydrate and Tetraamminecopper(II) Sulfate Hydrate

1. Weight of dried $Cu(NH_4)_2(SO_4)_2 \cdot 6H_2O$: ______________ g

   Description of product: ______________________________

2. Calculate the number of moles of reagents used in this synthesis.

   $CuSO_4 \cdot 5H_2O$

   ______________ moles

   $(NH_4)_2SO_4$

   ______________ moles

   Which reagent is in excess of the stoichiometric amount?

   ____

3. Calculate the percent yield you obtained, based on the quantity of the *limiting* reagent available.

$$\text{Percent yield} = \frac{\text{g product obtained}}{\text{g product (theory)}} \times 100$$

   ______________ percent yield

4. Weight of dried $Cu(NH_3)_4SO_4 \cdot H_2O$: ______________ g

   Description of product ______________________________

D. Sodium Thiosulfate

1. Weight of dried $Na_2S_2O_3 \cdot 5H_2O$: _____g

   Description of product: _____

2. Calculate the number of moles of reagents used in this synthesis:

   $Na_2SO_3(Na_2SO_3 \cdot 7H_2O)$

   _____moles

   S

   _____moles

   Which reagent is in excess of the stoichiometric amount?

   _____

3. Calculate the % yield you obtained, based on the quantity of the *limiting* reagent available.

$$\text{Percent yield} = \frac{\text{g product obtained}}{\text{g product (theory)}} \times 100.$$

   _____percent yield

4. What do you observe when 6 F HCl is added to a solution of sodium thiosulfate?

   Write a chemical equation to interpret your observation.

5. What do you observe when sodium thiosulfate solution is added to a solution containing a precipitate of AgBr?

   Write a chemical equation to interpret your observation.

IDENTIFICATION OF SILVER, LEAD, AND MERCUROUS IONS

PRE-STUDY

In Section VI, the studies on chemical equilibrium, we found that the various metal ions behave differently regarding the solubility of their salts and bases and regarding the relative stability of the complex ions formed with NH_3 and with OH^-. In Experiments 25 through 30 of this section we shall review these ionic equilibria and use the chemical reactions we have studied to develop systematic procedures to separate the ions and to identify them. First of all, in Experiments 25–27, we will work with a selected group of metal ions, including Ag^+, Pb^{2+}, Hg_2^{2+}, Fe^{3+}, Al^{3+}, Zn^{2+}, Ca^{2+}, Ba^{2+}, K^+, and NH_4^+. The same principles can be applied

to the analysis of anions and in Experiment 28 we will explore ways to detect the presence of NO_3^-, Cl^-, Br^-, I^-, SO_4^{2-}, SO_3^{2-}, S^{2-}, PO_4^{3-}, and CO_3^{2-}. In Experiment 29 a general scheme for the analysis of cations and anions in a general unknown is developed, and Experiment 30 is a chemical "crossword" puzzle in which unknown solutions are identified using only the unknown solutions themselves! In Experiment 31, the last experiment in this section, the separation of Fe^{3+}, Co^{2+}, Ni^{2+}, and Cu^{2+} by paper chromatography is explored.

Qualitative analysis has remained an important part of the laboratory experience in general chemistry for a number of years, in spite of the fact that these methods have been replaced by sophisticated instrumental methods in practical analysis. We believe that qualitative analysis has two useful purposes: (1) it provides an ideal context for the illustration and application of the principles of ionic equilibria, such as acid-base equilibria and solubility and complex ion equilibria, and (2) it provides a logical and systematic framework for the discussion of the descriptive chemistry of the elements. Its utility aside, qualitative analysis is fun.

Ag⁺, Pb²⁺, and Hg₂²⁺ Ions

The precipitating reagent for this group is chloride ion in an acid solution. Only three of all the common metal ions form a precipitate on the addition of this reagent Ag^+, Pb^{2+}, and Hg_2^{2+}. The mercurous ion is not the simple ion Hg^+, but consists of two atoms held together by a covalent bond, with a double positive charge.

Ammonia solution gives different results with each of the three metal ions, as shown by the preliminary experiments which follow. The behavior of mercury salts is peculiar and needs some explanation. Mercury exhibits three oxidation states: zero in free mercury, Hg^0, +1 in mercurous ion, Hg_2^{2+}, and +2 in mercuric ion, Hg^{2+}. In some situations the intermediate mercurous ion is unstable, part of the mercury being reduced to the metal and part being oxidized to Hg^{2+}, thus:

$$Hg_2^{2+} \rightleftharpoons Hg^0(l) + Hg^{2+}$$

When ammonia is added to mercuric chloride solution a white "ammonolysis" product, $HgNH_2Cl$, is formed. This is analogous to the partial hydrolysis of $HgCl_2$ to form a basic salt. Compare the two equations

$$HgCl_2(aq) + HOH \rightleftharpoons$$
$$Hg(OH)Cl(s) + H^+ + Cl^-$$

$$HgCl_2(aq) + HNH_2(aq) \rightleftharpoons$$
$$Hg(NH_2)Cl(s) + H^+ + Cl^-$$

If mercurous chloride is treated with ammonia, part of it is oxidized to the white mercuric aminochloride, above, and part is reduced to black, finely divided mercury:

$$Hg_2Cl_2(s) + 2NH_3(aq) \rightarrow$$
$$Hg(NH_2)Cl(s) + Hg(l) + NH_4^+ + Cl^-$$

The mixed precipitates appear black or dark gray.

EXPERIMENTAL PROCEDURES

Chemicals: 1 F NH_3, 0.1 F $Pb(NO_3)_2$, 0.1 F $HgCl_2$, 0.05 F $Hg_2(NO_3)_2$, 0.1 F KBr, 1 F K_2CrO_4, 0.1 F KI, 0.1 F $AgNO_3$, 0.1 F $NaCl$, 1 F $Na_2S_2O_3$.

A. Typical Reactions of the Ions

1. Solubility of the Chlorides. Prepare a sample of each of the three chlorides by adding 2 ml of 0.1 F $NaCl$ to 2 ml each of 0.1 F $AgNO_3$, 0.05 F $Hg_2(NO_3)_2$, and 0.1 F $Pb(NO_3)_2$. If a precipitate fails to form in any of these (why?), add 1 ml of 6 F HCl. Let the precipitates settle, decant and discard the supernatant liquid. Add to each about 2 ml of distilled water. Heat each nearly to boiling and shake. Results? Cool under the cold water tap. Results? Save for part 2.

2. Behavior with Ammonia. To each of the three precipitates, add 1 ml of 6 F NH_3. Shake these solutions to see if the precipitates will dissolve.

Compare the behavior of the Hg_2Cl_2 precipitate with ammonia with that of $HgCl_2$. To do this, add 1 ml of 6 F NH_3 to 1 ml of 0.1 F $HgCl_2$ solution.

Divide the ammoniacal solution of AgCl into two unequal portions. To the smaller, add a slight excess of 6 F HNO_3. Result?

3. Relative Solubility of Silver Salts and Stability of Complex Ions. To the larger portion of the ammoniacal AgCl solution add 1 ml of 0.1 F KBr. Result? To this add 1 ml of 1 F $Na_2S_2O_3$ (sodium thiosulfate), and shake. The complex ion formed is $Ag(S_2O_3)_2^{3-}$. (This is the reaction of "hypo" in fixing the developed film in photography.) To this solution add a little 0.1 F KI. From the results you can evaluate the relative Ag^+ concentrations in the following equilibrium situations: 0.1 F $AgNO_3$, saturated AgCl, saturated AgBr, saturated AgI, $Ag(NH_3)_2^+$, $Ag(S_2O_3)_2^{3-}$. Write net ionic equations for the reactions observed, and list each of the above in order of decreasing Ag^+ concentration.

B. Analysis of a Known Solution for Ag⁺, Pb²⁺, and Hg₂²⁺

First: Study the following Procedure for the Analysis, and also the flow chart, which is completed for this first group of ions. Note the precipitates obtained and the exact ions remaining in the solution.

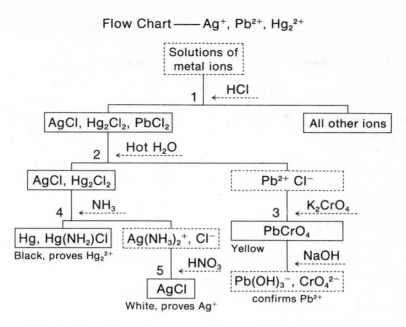

Flow Chart———Ag^+, Pb^{2+}, Hg_2^{2+}

Key: *Steps* in small numerals.   *Precipitates* or *residues* in solid-line boxes.
     Reagents on broken-line arrow.   *Solutions* of ions in broken-line boxes.

It will be a useful guide for outlining the essential separation steps and final identification tests for each ion.

Second: In your experiment report, write the net ionic equations for the reactions at each successive step of the procedure, as numbered in the flow chart. (Your instructor may require this work before issuing you an unknown.)

Third: Prepare a known solution containing Ag^+, Pb^{2+}, and Hg_2^{2+}, and analyze this according to the procedure below, following your results on the flow chart. Also study the typical analysis summary (Table 25-1), which is completed for this first group, and is similar to the one you will prepare as a record of your actual observations in the analysis of your unknown solutions in part C.

TABLE 25-1
Analysis summary of each step, observation, and conclusion for a sample known to contain all three ions of the group.
Appearance: A colorless solution.

Step	Sample	Reagents	Observations	Conclusions
1	Known sample	Dilute HCl	White precipitate	Ag, Pb^{2+}, Hg_2^{2+} (one or more) present
2	Precipitate 1	Hot H_2O	Some precipitate remains	Ag^+ and/or Hg_2^{2+} present
3	Hot H_2O filtrate	K_2CrO_4	Yellow precipitate	Pb^{2+} present, confirmed by dissolving in NaOH
4	Precipitate 2, insoluble in hot H_2O	NH_3 on the filter	Turns black	Hg_2^{2+} present
5	NH_3 filtrate	HNO_3 (excess)	White precipitate	Ag^+ present
				Summary: Ag^+, Pb^{2+}, and Hg_2^{2+} present

Procedure for the Analysis

Precipitation of the Group. Step 1. To 3 ml of the solution to be tested, add 6 F HCl drop by drop, and mix, until any precipitation that occurs is complete. Avoid a large excess. (No precipitation indicates absence of Ag^+, Hg_2^{2+}, and of much Pb^{2+}.) Filter into a 15-ml test tube, and test the filtrate for complete precipitation with a drop of HCl. If the test is positive, add additional HCl to the filtrate and refilter. After all the liquid has drained through the filter, spray the precipitate with 1–3 ml of distilled water from your wash bottle to wash it. Let this drain. (The filtrate still contains any other cations whose chlorides are soluble — and these will be analyzed for in Experiments 26, 27, and 29.)

Test for Lead Ion. Steps 2 and 3. Heat 10–15 ml of water to boiling; then pour 3–4 ml of this water over the residue on the filter, catching the hot filtrate in a small test tube. (Wash the remaining residue by pouring 10 ml of boiling water over it, discarding these washings.) Add several drops of 1 F K_2CrO_4 to the filtrate. A yellow precipitate, soluble when 6 F NaOH is added, proves Pb^{2+}.

Test for Mercurous Ion. Step 4. To the residue from the hot water treatment on the filter, add 1 ml of 6 F NH_3 and then 2 ml of water, collecting the filtrate in a small test tube. A black residue on the filter proves Hg_2^{2+}.

Test for Silver Ion. Step 5.[1] If the ammonia filtrate from the Hg_2^{2+} test is not perfectly clear, it may be due to colloidal Pb(OH)Cl coming through the filter. In this case, refilter the solution as often as necessary through the same filter, until a perfectly clear filtrate is obtained. Then acidify the filtrate with a little 6 F HNO_3, and mix. (If in doubt, test for acidity with litmus.) A white precipitate, AgCl, proves Ag^+.

C. Analysis of Unknown Solutions for Ag^+, Pb^{2+}, Hg_2^{2+}

Obtain one or more unknown solutions from your instructor, and analyze them by the above procedure. *At the same time,* keep a record of each step, including your actual observations for negative as well as for positive tests, by completing an analysis summary (with five headings: Step, Sample, Reagents, Observations, and Conclusions) like the example for the known solution in Table 25-1. This will constitute your report for each unknown.

---

[1]You will conduct Steps 6–10 in Experiment 26, and Steps 11–14 in Experiment 27.

Identification of Silver, Lead, and Mercurous Ions

NAME

SECTION   LOCKER

INSTRUCTOR   DATE

DATA

A. Typical Reactions of the Ions

1 and 2. Solubility of the Chlorides. Their Behavior with Ammonia. In the spaces write the formula(s) of the principal substance(s) present (ions or molecules), when the chlorides are treated as indicated. Also indicate colors.

	AgCl	$PbCl_2$[1]	Hg_2Cl_2
Hot water			
NH_3 (excess)			

[1]Look up the relative solubilities of $PbCl_2$ and $Pb(OH)_2$ to help you decide whether a reaction occurs with NH_3.

3. Relative Solubility of Silver Salts and Stability of Complex Ions. Write net ionic equations for the successive reactions that take place when a $AgNO_3$ solution is treated, in turn, with HCl, excess NH_3, KBr, $Na_2S_2O_3$, and KI.

From the above data, list the following equilibrium solutions in the order of *decreasing* Ag^+ concentration in the mixture: 0.1 F $AgNO_3$, saturated AgCl, saturated AgBr, saturated AgI, $Ag(NH_3)_2^+$, $Ag(S_2O_3)_2^{3-}$.

B. Analysis of a Known Solution for Ag^+, Pb^{2+}, Hg_2^{2+}

Write below, in order, the net ionic equation for *all* reactions occurring in the systematic procedure for the analysis of the ions of this group. (Some of these repeat those in part A.)

C. Analysis of Unknown Solutions for Ag^+, Pb^{2+}, Hg_2^{2+}

Unknown no._____ Appearance:_____ Ions found:_____

Unknown no._____ Appearance:_____ Ions found:_____

Analysis Summary: Report your actual observations, negative as well as positive, for each step of the analysis of each unknown, and record above the ions found.

Step	Sample	Reagents	Observations	Conclusions

APPLICATION OF PRINCIPLES

1. After considering the observed facts listed below, arrange at the right, in order of *decreasing* Cu^{2+} concentration, the following substances: saturated $CuCO_3$, 1 F $Cu(NH_3)_4^{2+}$, saturated CuS, 1 F $CuCl_2$.

   (a) Mixing $CuCl_2$ and Na_2CO_3 solutions will form a precipitate of $CuCO_3$.
   (b) Passing H_2S gas into $Cu(NH_3)_4^{2+}$ solution will precipitate CuS.
   (c) $CuCO_3$ solid will dissolve in NH_3 solution to form a deep blue solution.

   Explain *why* you arranged the items in the order you did.

   _____

   _____

   _____

   _____

IDENTIFICATION OF FERRIC, ALUMINUM, AND ZINC IONS

PRE-STUDY

Reactions of Fe^{3+}, Al^{3+}, and Zn^{2+} Ions

The removal and separation of these three metal ions from our selected group of ten cations (see Experiment 25), after removal of the insoluble chlorides, depends upon their distinctive behavior with NH_3 and with $NaOH$. *Review the chemistry of complex coordination compounds with NH_3 and OH^-, Experiment 21.* Zinc ion is conveniently removed and identified by its white insoluble sulfide in a basic or slightly acid solution. The control of pH is important in all of these separations.

Buffer Action

Buffer substances in a solution are used to control the relative H^+ and OH^- concentrations within certain limits. These consist of weak acids or weak bases, together with their salts. Thus, in the Al^{3+} test that follows, a mixture of NH_3 and NH_4Cl is used to produce a basic solution that is much less basic than NH_3 alone. Consider the equilibrium equation, and the corresponding equilibrium constant expression

$$NH_3 + H_2O \rightleftharpoons NH_4^+ + OH^-$$

$$K_{NH_3} = \frac{(NH_4^+)(OH^-)}{(NH_3)} = 1.8 \times 10^{-5}$$

Excess NH_4^+ in the mixture shifts the equilibrium to the left resulting in a lower OH^- concentration. While $1\ F\ NH_3$ alone contains about $0.004\ M$ OH^-, a simple calculation shows that if NH_4Cl is added to make the NH_4^+ and NH_3 concentrations equal, the resulting OH^- concentration then will be only $1.8 \times 10^{-5}\ M$, a 200-fold reduction.

Likewise, if an acid solution still weaker than CH_3COOH is desired, the addition of CH_3COONa

will very materially decrease the H^+ concentration, as is evident from the equilibrium

$$CH_3COOH(aq) \rightleftharpoons H^+ + CH_3COO^-$$

Sulfide Precipitation

In a solution containing the weak acid H_2S, the resulting S^{2-} concentration is very dependent on the H^+ concentration:

$$H_2S(aq) \rightleftharpoons H^+ + HS^-$$
$$HS^- \rightleftharpoons H^+ + S^{2-}$$

This fact is utilized in more complete qualitative schemes to separate the metal ions into groups by the degree of insolubility of their sulfides. In our simple scheme, precipitation of Zn^{2+} as $ZnS(s)$ in a basic NH_3 solution is satisfactory, since only alkali and alkaline earth ions remain, whose sulfides are all readily soluble.

EXPERIMENTAL PROCEDURE

Chemicals: 0.1 F $Al(NO_3)_3$, 0.1 F $Fe(NO_3)_3$, 0.1 F $Zn(NO_3)_2$, "aluminon reagent," saturated H_2S solution or 0.1 F NH_4HS (both freshly prepared).

A. Typical Reactions of the Ions

1. Ammonia and Hydroxide Ion Complexes. Place 1-ml samples of 0.1 F $Al(NO_3)_3$, 0.1 F $Fe(NO_3)_3$, and 0.1 F $Zn(NO_3)_2$ in separate test tubes. To each add 1 drop of 6 F NH_3, then continue to add more, drop by drop, to determine whether the hydroxide precipitate first formed redissolves with excess NH_3. (NOTE: The 0.1 F $Fe(NO_3)_3$ contains excess HNO_3, which must first be neutralized before *any* precipitate appears.) Repeat these tests, using 6 F NaOH instead of 6 F NH_3. Summarize all the results in a suitable chart in your report.

2. Red Lake Formation with $Al(OH)_3$. The light flocculent precipitate of aluminum hydroxide is often difficult to observe. The dye called "aluminon" (ammonium aurin tricarboxylate) adsorbs onto the precipitate to form a characteristic red "lake," which makes the identification of the aluminum hydroxide easier.

To 1 ml of water, add 2–3 drops of 0.1 M Al^{3+} solution. Add 2 drops of aluminon reagent, then

3–5 drops of 6 F NH_3. Note the appearance of the precipitate. Now make the solution acid with 0.5 ml 6 F HCl, and again just basic with 6 F NH_3. Let the mixture stand a moment, and note the characteristic color and flocculation of the precipitate (which now should be red). Also note whether the solution is colored. A good red color is not obtained if the solution is too basic. Explain how the above treatment guarantees a very slightly basic solution.

B. Analysis of a Known Solution for Fe^{3+}, Al^{3+}, and Zn^{2+}

First: Study the following procedure for analysis until you understand each step and the reason for adding each reagent. Then complete a flow chart for these operations, in a style similar to that of the flow chart in Experiment 25. A skeleton flow chart is provided in part B of the report form: simply insert the proper formulas of reagents, precipitates or residues, ions in solution, and so on.

Second: In your experiment report, write the net ionic equations for the reactions occurring at each step of the procedure, as numbered in the flow chart. (Your instructor may require that you do this before you receive an unknown.)

Third: Prepare a known solution containing Fe^{3+}, Al^{3+}, and Zn^{2+}, and analyze according to the procedure described in the next paragraph, following your results on the flow chart. *At the same time,* complete an analysis summary similar to Table 25-1 (a form for this summary is provided in part B of the report form).

Procedure for the Analysis

Precipitation of $Fe(OH)_3$ and $Al(OH)_3$. Step 6.[1] To 3 ml of the solution to be tested, add an excess (about 1 ml) of 6 F NH_3. Mix the solution, then centrifuge it. (The use of a centrifuge, if available, is more convenient for most separations of precipitate and solution. See Figures 26-1 and 26-2. *Obtain instructions from your instructor.* If filter paper is used, modify the directions accordingly.) Wash the precipitated $Al(OH)_3$ and $Fe(OH)_3$ with 1–2 ml of water, combining this with the filtrate, which contains any zinc as $Zn(NH_3)_4^{2+}$.

---

[1]Directions for Steps 1–5 are given in Experiment 25; directions for Steps 11–15, in Experiment 27.

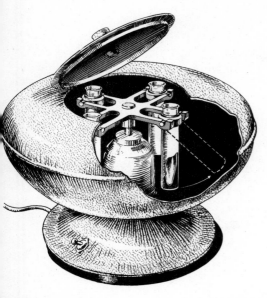

Tubes swing to dotted position when centrifuge is running.

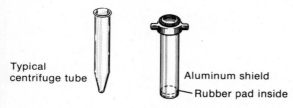

Typical
centrifuge tube

Aluminum shield
Rubber pad inside

Opposite pairs of tubes should be filled with equal
amounts of liquid to prevent excessive vibration.
FIGURE 26-1
The construction and operation of the centrifuge.

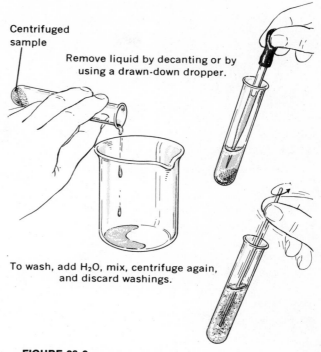

Centrifuged
sample

Remove liquid by decanting or by
using a drawn-down dropper.

To wash, add H_2O, mix, centrifuge again,
and discard washings.

FIGURE 26-2
The procedure for washing a precipitate that has been
centrifuged.

Separation and Test for Fe^{3+}. Steps 7 and 8. Treat
the mixed precipitate with about 0.5 ml of 6 F
$NaOH^2$ followed by 1 ml of water. Centrifuge and
wash with 1 ml of water, combining the washings
with the filtrate of $Al(OH)_4^-$. Dissolve the $Fe(OH)_3$
precipitate with 1–3 drops of 6 F HCl, warm, if
necessary, to complete the reaction, and add 2 ml
of H_2O and then 1–3 drops of 0.1 F KSCN. A red
color of $Fe(SCN)^{2+}$ proves Fe^{3+}.

Test for Al^{3+}. Step 9. To the $Al(OH)_4^-$ filtrate
add 6 F HCl until neutral, and then add about 0.5
ml excess. Add 1–2 drops of aluminon reagent, and
just make basic again with 6 F NH_3. Mix this and
let it stand. A flocculent precipitate, colored a
characteristic red by the dye, proves the presence

of Al^{3+}. If color and precipitate are indefinite, you
may again make the solution acid with HCl, then
basic with NH_3, to build up the NH_4^+ concentration so the solution becomes less basic.

Test for Zn^{2+}. Step 10. To the filtrate from step 6,
containing any zinc present as $Zn(NH_3)_4^{2+}$, add
1–3 drops of 0.1 F NH_4HS or 2 ml of saturated
H_2S water. A white precipitate of ZnS proves the
presence of Zn^{2+}. (When a general unknown is
being analyzed, centrifuge the ZnS precipitate
and save the solution for Experiment 27. Unless
this solution is analyzed at once, it should be
acidified with HCl and boiled to remove H_2S to
avoid oxidation to sulfate ion and precipitation of
any $BaSO_4$.)

C. Analysis of Unknown Solutions

Obtain one or more unknown solutions from your
instructor, and analyze them by the above procedures. *At the same time,* keep a record of each
step, including your actual observations for negative as well as for positive tests, by completing an
analysis summary in Part C of the report form.

---

[2]If filter paper is used, add the NaOH and H_2O directly on
the filter, collecting the filtrate in a test tube. The basic solution
may be passed over the filter a second time for more complete
solution of any aluminum as $Al(OH)_4^-$.

REPORT	Identification of Ferric,	NAME	
26	Aluminum, and Zinc Ions	SECTION	LOCKER
		INSTRUCTOR	DATE

DATA

A. Typical Reactions of the Ions

1. Ammonia and Hydroxide Ion Complexes.

In the spaces provided write the formulas of the precipitates formed, or new ions formed in solution, if any. Also indicate any characteristic colors, etc., when each ion is treated with the reagent in the left-hand column.

Reagents	Fe^{3+}	Al^{3+}	Zn^{2+}
F NH$_3$			
F NH$_3$ (excess)			
F NaOH			
F NaOH (excess)			

2. Red Lake Formation with $Al(OH)_3$

Equation for the reaction of Al^{3+} with NH_3_____

Explain how the addition, first of an excess of HCl, then of a slight excess of NH_3, insures that the solution will not become too basic to form a satisfactory adsorption compound of the dye with the precipitate.

B. Analysis of a Known Solution for Fe^{3+}, Al^{3+}, Zn^{2+}

Complete the following flow chart and analysis summary for your known mixture of Fe^{3+}, Al^{3+}, and Zn^{2+}.

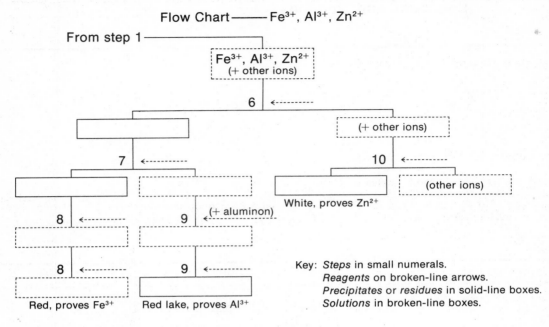

Flow Chart ——— Fe^{3+}, Al^{3+}, Zn^{2+}

From step 1

Fe^{3+}, Al^{3+}, Zn^{2+}
(+ other ions)

6

(+ other ions)

7   10

(other ions)

White, proves Zn^{2+}

8   9 (+ aluminon)

8   9

Red, proves Fe^{3+} Red lake, proves Al^{3+}

Key: *Steps* in small numerals.
Reagents on broken-line arrows.
Precipitates or *residues* in solid-line boxes.
Solutions in broken-line boxes.

Complete the analysis summary for your *known* sample, as in Table 25-1.

Step	Sample	Reagent(s)	Observations	Conclusions
6				
7				
8				
9				
10				
			Summary:	

Write below, in order, the net ionic equations for *all* reactions occurring in the systematic analysis of the ions of this group.

REPORT 26	Sheet 2	NAME

C. Analysis of Unknown Solutions for Fe^{3+}, Al^{3+}, Zn^{2+}

Unknown no.________ Appearance:________ Ions found:________

Unknown no.________ Appearance:________ Ions found:________

Analysis Summary: Report your actual observations, negative as well as positive, for each step of the analysis of each unknown, and record above the ions found.

Step	Sample	Reagents	Observations	Conclusions

APPLICATION OF PRINCIPLES

1. A solution is 0.50 F in HAc and 0.25 F in NaAc. (With this excess of Ac^- assume that the molarity of HAc equals its formality.)

   (a) Write the equilibrium equation for the ionization of acetic acid.

   (b) Write the equilibrium constant expression for acetic acid ($K_{HAc} = 1.8 \times 10^{-5}$);

   (c) Calculate the H^+ concentration in this buffer solution.

---

2. A solution is 1.8 F in NH_4Cl, and 0.010 F in NH_3. Will the solution be basic, acidic, or neutral? Show calculations. (With this excess of NH_4^+, assume that the molarity of NH_3 equals its formality. K_b for NH_3 is 1.8×10^{-5}.)

IDENTIFICATION OF ALKALINE EARTH AND ALKALI METAL IONS

PRE-STUDY

The elements comprising the alkaline earth and alkali groups are quite similar and have many properties in common. Their compounds exhibit only one stable oxidation state; they do not form amphoteric hydroxides, being distinctly basic; and they do not readily form complex ions with NH_3 or OH^-. The separation of the alkaline earth elements depends almost entirely on differences in the solubilities of their salts, which show a regular gradation through the periodic table. The salts of the alkali metals are almost all readily soluble. Table 27-1 will be useful in interpreting the analytical procedure.

This experiment includes the two alkaline earth ions Ca^{2+} and Ba^{2+}, the alkali ion K^+, and also NH_4^+ (whose salts are almost all soluble). We must test for NH_4^+ with a separate portion of the original unknown, since we have used it as a reagent.

TABLE 27-1
Solubilities of alkaline earth salts (g/100 g H_2O, at room temperature.)

	Mg^{2+}	Ca^{2+}	Sr^{2+}	Ba^{2+}
OH^-	0.001	0.16	1.74	3.89
CO_3^{2-}	0.09	0.0015	0.001	0.0018
SO_4^{2-}	35.5	0.2	0.01	0.00024
CrO_4^{2-}	138.	18.6	0.12	0.00037
$C_2O_4^{2-}$	0.015	0.0007	0.005	0.01

Precipitation with Ammonium Carbonate

The group reagent usually used to precipitate alkaline earth ions is $(NH_4)_2CO_3$. Since this is the salt of both a weak base and weak acid, it hydrolyzes in solution to a marked extent:

$$NH_4^+ + CO_3^{2-} \rightleftharpoons HCO_3^- + NH_3$$

For this reason we add excess NH_3 to reverse the hydrolysis and increase the concentration of CO_3^{2-}, to attain more complete precipitation.

The Separation of Barium and Calcium Ions

The analytical procedure below takes advantage of the above solubility differences. Note that, after dissolution of the carbonate precipitate with acetic acid, barium can be effectively separated as the chromate, $BaCrO_4$, and then further confirmed by transformation to the still less soluble sulfate, $BaSO_4$. The $BaCrO_4$ dissolves in HCl because H_2CrO_4 is a somewhat weak acid and forms dichromate ion, $Cr_2O_7^{2-}$. With barium ion removed, calcium ion is easily identified, by precipitation with oxalate ion in NH_3 solution, as CaC_2O_4.

EXPERIMENTAL PROCEDURE

Chemicals: 6 F $HC_2H_3O_2$, 3 F $NH_4C_2H_3O_2$, 0.1 F NH_4Cl, 3 F $(NH_4)_2CO_3$, 0.1 F $BaCl_2$, 0.1 F $CaCl_2$, 0.1 F KNO_3, 1 F K_2CrO_4, 1 F $K_2C_2O_4$, 0.1 F Na_2SO_4.

A. Typical Reactions of the Ions

Instead of performing a series of separate, prescribed tests to learn the properties of the ions, proceed at once to the study of the Procedure for the Analysis below, and at the same time complete the flow chart by filling in the proper formulas of reagents, precipitates or residues, ions in solution, etc. Compare these procedures with the data in the chart of the solubilities of alkaline earth salts, Table 27-1.

You may wish to try, on your own initiative, various test tube experiments, such as the separation of certain ions by a specific reagent, the subsequent precipitation identification tests, and the K^+ flame test in the presence of Na^+ (Figure 27-1).

B. The Analysis of a Known Solution for Ba^{2+}, Ca^{2+}, K^+, and NH_4^+

Prepare a known solution containing the ions of the group, and analyze it according to the following procedure; at the same time complete the flow chart (part B of the report form) in accord with your observations. Also complete the analysis summary for this solution (part B, report form), outlining the successive steps, reagents, observations, and conclusions.

Procedure for the Analysis

Precipitation of Ba^{2+} and Ca^{2+}. Step 11.[1] Start with a 3-ml sample of the solution to be tested. (If this is a general unknown, evaporate the filtrate from Step 10, Experiment 26, to a volume of 3 ml). If your 3-ml sample is not already basic (litmus test), add 6 F NH_3 until it is basic, and then add, with mixing, an additional 5–10 drops 3 F $(NH_4)_2$ CO_3. Let the mixture stand 3–5 minutes to complete precipitation, and then centrifuge it. Test for completeness of precipitation with further drop-by-drop additions of 3 F $(NH_4)_2CO_3$, until no further precipitation occurs. Centrifuge again if necessary. Separate the solution and save it for the K^+ test, Step 14.

Test for Ba^{2+}. Step 12. To the above precipitate add 5 drops of 6 F $HC_2H_3O_2$, warm as needed to dissolve the precipitate, add 1 ml of water, and buffer the solution by adding 5 drops of 3 F $NH_4C_2H_3O_2$. Add 3 drops of 1 F K_2CrO_4, mix, then centrifuge. If the solution is not yellow, add more 1 F K_2CrO_4, drop by drop, but avoid over 1–3 drops in excess. Wash any precipitate. Yellow $BaCrO_4$ indicates Ba^{2+}. (Save the solution for the Ca^{2+} test, Step 13.) This may be confirmed by adding 2–3 drops of 6 F HCl to dissolve the yellow precipitate, and then by adding 1 ml of 0.1 F Na_2SO_4. Centrifuge, if necessary, to see whether the precipitate is white in the yellow solution. White $BaSO_4$ proves the presence of Ba^{2+}.

Test for Ca^{2+}. Step 13. To the preceding yellow solution from which the $BaCrO_4$ was removed, add 10 drops of 1 F $K_2C_2O_4$, and then add 6 F NH_3 to

---

[1]Directions for Steps 1–5 are given in Experiment 25, and directions for Steps 6–10 in Experiment 26.

make the solution basic. Let stand 10 minutes if a precipitate does not appear before that time. White CaC_2O_4 proves the presence of Ca^{2+}. Further confirmation may be made by decanting the yellow solution, adding to the residue a few drops of 6 F HCl and 1 ml of water, to dissolve it, then adding another drop of 1 F $K_2C_2O_4$ and again making it basic with 6 F NH_3. The white precipitate of CaC_2O_4 will reappear.

Test for K^+. Step 14. Concentrate the solution from Step 11 (after removal of the Ba^{2+} and Ca^{2+}) by evaporation in a porcelain dish *almost* to dryness. Test the moist residue for potassium by the flame test: Clean a nichrome wire (See Figure 27-1A) by repeatedly heating it, plunging it into 5 ml of concentrated HCl in a small test tube, and reheating it in the *hottest* Bunsen flame to brilliant incandescence until there is only a minimum coloration to the flame. (Be sure the cleaned wire touches nothing except your test solution after this.) *Reduce* the flame to about 1-inch in height, touch the wire to the moist salt or solution to be tested, and heat it to incandescence in the colorless flame. Because the violet potassium flame coloration lasts only a few seconds (potassium salts are quite volatile), observe it at once. You may find it helpful to observe the color of the flame produced by a solution known to contain K^+. If sodium salts are present, their yellow flame masks the pale-violet potassium flame completely, and the use of two layers of blue cobalt glass is essential to absorb the sodium radiation so that you can observe the potassium flame (see Figure 27-1B). Calcium salts, if not completely removed, give an orange-red flame.

NOTE: The potassium flame test may be carried out, instead, on a sample of the original solution evaporated almost to dryness and tested as above. The other metal ions present, especially if the blue cobalt glass is used, probably will not interfere with the test.

Test for NH_4^+. Step 15. This test must always be carried out on the *original solution* or *sample,* since ammonium salts are used as reagents throughout the procedure.

Place 3 ml of the sample in an evaporating dish, and make it basic with 6 F NaOH. Cover the dish with a watch glass, on the under side of which is attached a moist strip of red litmus paper. Warm the solution *very gently* to liberate any ammonia present as the gas. (Avoid boiling, which would con-

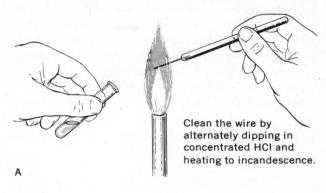

A colored flame indicates a contaminated wire.

Clean the wire by alternately dipping in concentrated HCl and heating to incandescence.

A

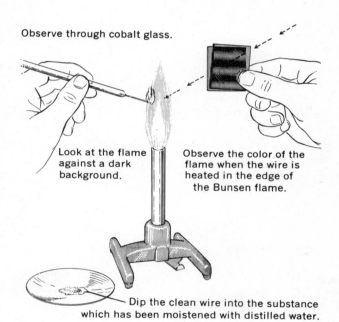

Observe through cobalt glass.

Look at the flame against a dark background.

Observe the color of the flame when the wire is heated in the edge of the Bunsen flame.

Dip the clean wire into the substance which has been moistened with distilled water.

B

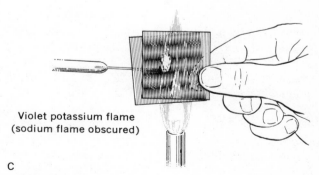

Violet potassium flame (sodium flame obscured)

C

FIGURE 27-1
The flame test for sodium and potassium: (A) shows the procedure for cleaning the wire; (B) and (C) illustrate the steps for observing the flame coloration.

taminate the litmus with spray droplets of the NaOH solution.) An even, unspotted blue color proves ammonium ion. The characteristic ammonia odor, observed soon after the solution is warmed, is also a positive test. (*Be very cautious* in bringing your nostrils close to a hot NaOH solution. If it is superheated, it may splatter in your face or eyes: Wear protective glasses.)

C. Analysis of Unknown Solutions

Obtain one or more unknown solutions from your instructor, and analyze them by the above procedure. *At the same time,* keep a record of each step, by completing the analysis summary in Part C of the report form, showing your actual observations, negative as well as positive, for each unknown.

REPORT

27

Identification of Alkaline
Earth and Alkali Metal Ions

NAME

SECTION   LOCKER

INSTRUCTOR   DATE

DATA

A. Typical Reactions of the Ions

NOTE: Indicate any experimental tests you have performed personally in order to be certain of the results.

1. Write net ionic equations for the reactions, *if any*, occurring when dilute solutions of the following are mixed. (Indicate if there is "no reaction".)

(a) NH_4Cl, KNO_3, $NaOH$, heated _____

(b) $CaCl_2$, $Ba(NO_3)_2$, $1\ F\ NH_3$ _____

(c) $CaCl_2$, $Ba(NO_3)_2$, $1\ F\ NaOH$ _____

(d) $Ca(NO_3)_2$, K_2CrO_4 _____

(e) Mixture (d) + $BaCl_2$ _____

(f) Mixture (e) + HCl _____

(g) Mixture (f) + Na_2SO_4 _____

(h) $BaCl_2$, KNO_3, $ZnCl_2$, H_2S _____

(i) KCl, $(NH_4)_2CO_3$, $Ba(NO_3)_2$ _____

(j) $CaCO_3(s)$, NH_4Cl, $6\ F\ HC_2H_3O_2$ _____

(k) Mixture (j) + $(NH_4)_2C_2O_4$, NH_3 _____

2. On the basis of periodic table relationships, what would you predict about the general solubility (indicate "soluble," "slightly soluble," or "insoluble") of each of the following:

(a) $Be(OH)_2$ _____   (c) $Ra(OH)_2$ _____

(b) $BeSO_4$ _____   (d) $RaSO_4$ _____

3. Why does the addition of HCl dissolve such insoluble salts as $BaCrO_4$ and CaC_2O_4?

B. Analysis of a Known Solution for Ba^{2+}, Ca^{2+}, K^+, NH_4^+

1. Complete the following flow chart in accord with your observations

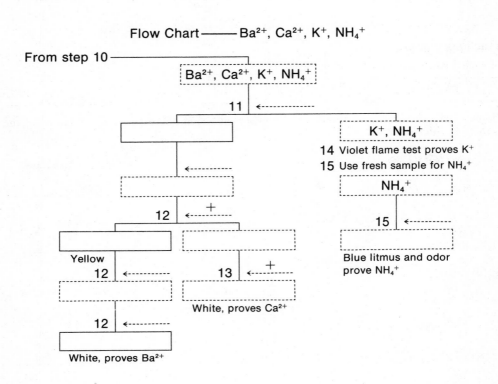

Flow Chart —— Ba^{2+}, Ca^{2+}, K^+, NH_4^+

From step 10

Ba^{2+}, Ca^{2+}, K^+, NH_4^+

11

K^+, NH_4^+

14 Violet flame test proves K^+
15 Use fresh sample for NH_4^+

NH_4^+

15

Blue litmus and odor prove NH_4^+

12 +

Yellow

12

White, proves Ca^{2+}

12

White, proves Ba^{2+}

13 +

2. Complete the following analysis summary for your known sample.

Appearance_____ Ions found_____

Step	Sample	Reagent(s)	Observations	Conclusions
11				
12				
13				
14				
15				
			Summary:	

REPORT 27	Sheet 2	NAME

C. Analysis of Unknown Solutions for Ba^{2+}, Ca^{2+}, K^{+}, NH_4^{+}

Unknown no.__________ Appearance:________________ Ions found:________________

Unknown no.__________ Appearance:________________ Ions found:________________

Analysis Summary: For your unknown samples, report below your actual observations, negative as well as positive, for each step of each analysis, and record above the ions found.

Step	Sample	Reagents	Observations	Conclusions

THE QUALITATIVE ANALYSIS OF
SOME COMMON ANIONS

PRE-STUDY

The principles that are employed in the identification of metal ions can also be applied to the analysis of anions. Thus the qualitative detection of anions in a sample depends upon the distinctive properties of particular ions and upon the possibility of controlling experimental conditions so that a separation or identification of a given ion may be attained. In this experiment we will explore ways to detect the presence of NO_3^-, Cl^-, Br^-, I^-, SO_4^{2-}, SO_3^{2-}, S^{2-}, PO_4^{3-}, and CO_3^{2-}.

Review thoroughly the following criteria upon which the identification of an anion generally depends. (These criteria are all summarized in appropriate tables in Appendix C.)

1. Does the anion form a *precipitate* with certain cations, such as Ag^+ or Ba^{2+}, in neutral or in acidic solutions? (See Table 11, Appendix C for solubility rules.)

2. Is the ion the *anion of a weak or a strong acid*? (See Table 9, Appendix C for information on the relative strength of acids.)

3. Is the ion the *anion of a volatile acid*?

4. Can the anion act as *an oxidizing agent, a reducing agent,* or *either*? What is its relative strength as an oxidizing or reducing agent? (See Appendix C, Table 13 for redox potentials.)

Review the descriptive chemistry of the anions we shall study in this experiment, in your text and in Experiment 16 on nitrogen, sulfur, and chlorine compounds, as you apply the above criteria to the typical reactions and tests for the following anions.

Typical Reactions of
Common Anions

Sulfate Ion. This ion is the anion of a *strong acid*; it forms characteristic precipitates with certain metal ions, as with barium ion:

$$Ba^{2+} + SO_4^{2-} \rightleftharpoons BaSO_4(s)$$

Other anions also form insoluble precipitates with barium ion:

$$Ba^{2+} + CO_3^{2-} \rightleftharpoons BaCO_3(s)$$
$$Ba^{2+} + SO_3^{2-} \rightleftharpoons BaSO_3(s)$$
$$3Ba^{2+} + 2PO_4^{3-} \rightleftharpoons Ba_3(PO_4)_2(s)$$

However, these are salts of *weak acids* and would all dissolve, or fail to precipitate, in an acid solution. In the presence of excess hydrogen ion, the concentration of the free anion is reduced to such a low value that the solubility equilibrium is reversed, and the precipitate dissolves or fails to form in the first place. We may therefore use barium ion in an acid solution as a test reagent for the presence of sulfate ion.

Sulfite Ion, Sulfide Ion, and Carbonate Ion. In each of these anions of *weak, volatile acids*, the free acid is formed by the addition of a strong acid:

$$SO_3^{2-} + 2H^+ \rightleftharpoons H_2SO_3(aq) \rightleftharpoons SO_2(g) + H_2O$$
$$S^{2-} + 2H^+ \rightleftharpoons H_2S(g)$$
$$CO_3^{2-} + 2H^+ \rightleftharpoons H_2CO_3(aq) \rightleftharpoons CO_2(g) + H_2O$$

The odors of sulfur dioxide (SO_2) and hydrogen sulfide (H_2S) are usually sufficient identification. Sulfite ion (SO_3^{2-}) may be oxidized to sulfate ion (SO_4^{2-}) by bromine water (Br_2) or hydrogen peroxide (H_2O_2), and then tested as the sulfate ion is tested. Sulfide ion (S^{2-}) may be further confirmed by placing it in contact with a lead salt solution on filter paper, which will result in the formation of a black deposit of lead sulfide (PbS).

In your study of the oxidation states of sulfur (Experiment 16), you learned that sulfite ion, which is intermediate in its oxidation state, can be reduced to free sulfur by sufide ion, which in turn is oxidized to sulfur. The equation is

$$SO_3^{2-} + 2S^{2-} + 6H^+ \rightarrow 3S(s) + 3H_2O$$

These ions, therefore, are not likely to be found together in the same solution, particularly if it is acid. A solution of sulfite is somewhat unstable because of its ease of oxidation by atmospheric oxygen:

$$2SO_3^{2-} + O_2(g) \rightarrow 2SO_4^{2-}$$

In the carbonate ion test, the evolution of a colorless, almost odorless gas when an acid is added to the sample is an indication, but not proof, that the gas is carbon dioxide. The sample must be reprecipitated as an insoluble carbonate salt, such as calcium carbonate ($CaCO_3$). When the slightly soluble carbon dioxide gas is bubbled into a neutral solution *containing calcium ion*, the weak acid which is formed, carbonic acid (H_2CO_3),

$$CO_2(g) + H_2O \rightleftharpoons H_2CO_3(aq) \rightleftharpoons 2H^+ + CO_3^{2-}$$

does not furnish a high enough concentration of carbonate ion to precipitate calcium carbonate. What would be the effect on the carbonate ion concentration if the carbon dioxide gas were bubbled into an alkaline solution instead of into the neutral one? This is the reason for using limewater, $Ca(OH)_2$, although the hydroxide is only slightly soluble, rather than some soluble calcium salt, in the carbonate ion test. The overall net ionic equation is

$$CO_2(g) + Ca^{2+} + 2OH^- \rightleftharpoons CaCO_3(s) + H_2O$$

Sulfite ion, if also present in the unknown, will interfere with the carbonate ion test, since it will liberate sulfur dioxide gas (SO_2) along with the carbon dioxide gas when the sample is acidified, and will form a calcium sulfite ($CaSO_3$) precipitate when the gases are absorbed in the limewater. (Sulfide ion, S^{2-}, would not interfere, as CaS is soluble.) To overcome this interference, it is necessary first to oxidize any sulfite ion to sulfate ion, and thus prevent the formation of any sulfur dioxide gas.

Chloride Ion, Bromide Ion, and Iodide Ion. These ions are, like the sulfate ion, the anions of *strong acids*. Their salts with silver ion are insoluble, whereas the silver salts of most other anions, which are also insoluble in neutral solution, will dissolve in acid solution. Silver sulfide (Ag_2S), however, is so very insoluble that the presence of hydrogen ion does not reduce the sulfide ion concentration sufficiently to dissolve it.

We may test for the presence of *chloride ion* in the mixed precipitate of AgCl, AgBr, and AgI by dissolving the AgCl from the still more insoluble AgBr and AgI by adding ammonia solution, thus forming the complex ion $Ag(NH_3)_2^+$ and Cl^-. When the filtrate is again acidified, the AgCl is reprecipitated:

$$AgCl(s) + 2NH_3(aq) \rightleftharpoons Ag(NH_3)_2^+ + Cl^-$$
$$Ag(NH_3)_2^+ + Cl^- + 2H^+ \rightleftharpoons AgCl(s) + 2NH_4^+$$

It is thus possible, by carefully controlling the NH_3 and Ag^+ concentrations, to redissolve the AgCl without appreciably redissolving the more insoluble AgBr and AgI. Furthermore, AgCl is white, AgBr is cream-colored, and AgI is light yellow.

The distinctive differences that enable us to separate and test for *bromide ion* and *iodide ion* in the presence of chloride ion depend on differences in their ease of oxidation. Iodide ion is easily oxidized to iodine (I_2) by adding ferric ion (Fe^{3+}), which does not affect the others. After adding carbon tetrachloride (CCl_4) to identify (purple color) and remove the iodine, we may next oxidize the bromide ion with potassium permanganate or chlorine water solutions, absorb the bromine (Br_2) formed in carbon tetrachloride, and identify it by the brown color produced.

Phosphate Ion. This is the anion of a moderately weak, nonvolatile acid—phosphoric acid (H_3PO_4). It is tested for by the typical method of forming a characteristic precipitate. The reagent used is ammonium molybdate solution, to which an excess of ammonium ion is added in order to shift the equilibrium further to the right. An acid solution is necessary. The equation is

$$3NH_4^+ + 12MoO_4^{2-} + H_3PO_4(aq) + 21H^+ \rightleftharpoons$$
$$(NH_4)_3PO_4 \cdot 12MoO_3(s) + 12H_2O$$

The yellow precipitate is a mixed salt, ammonium phosphomolybdate. Sulfide ion interferes with this

test but may be removed first by acidifying the solution with HCl and boiling it.

Nitrate Ion. In testing for the nitrate ion we cannot use a precipitation method, since all nitrates are soluble. Instead, two other facts are utilized. First, nitrate ion is a good oxidizing agent in an acid solution when a reducing agent such as ferrous ion (Fe^{2+}) is added:

$$4H^+ + NO_3^- + 3Fe^{2+} \rightarrow$$
$$3Fe^{3+} + NO + 2H_2O$$

Second, the nitric oxide, NO, reacts rapidly with the excess ferrous ion present to form a brown complex ion:

$$NO + Fe^{2+} \rightarrow Fe(NO)^{2+}$$

It is essential that an excess of ferrous ion be used, or the test will fail.

Summary. Table 28-1 may be of assistance to you in summarizing the behavior of the above negative ions with the metallic ions: silver ion, barium ion, calcium ion, and lead ion.

TABLE 28-1
The behavior of some negative ions with the metallic ions Ag^+, Ba^{2+}, Ca^{2+}, Pb^{2+}.

Negative ions	Metallic ions			
	Ag^+	Ba^{2+}	Ca^{2+}	Pb^{2+}
NO_3^-	Soluble	Soluble	Soluble	Soluble
Cl^-, Br^-, I^-	AgCl, white; AgBr, cream; AgI, yellow. All insoluble in HNO_3. AgCl soluble, AgBr slightly soluble, and AgI insoluble, in NH_3	Soluble	Soluble	$PbCl_2$ and $PbBr_2$, white, soluble in hot water. PbI_2, yellow, slightly soluble in hot water
SO_4^{2-}	Moderately soluble	$BaSO_4$, white, insoluble in HNO_3	$CaSO_4$, white, slightly soluble	$PbSO_4$, white, insoluble in HNO_3
SO_3^{2-}	Ag_2SO_3, white, soluble in NH_3 and in HNO_3	$BaSO_3$, white, soluble in HNO_3	$CaSO_3$, white, soluble in HNO_3	$PbSO_3$, white, soluble in HNO_3
S^{2-}	Ag_2S, black, soluble in hot, conc. HNO_3	Soluble	Soluble	PbS, black, soluble in HNO_3
PO_4^{3-}	Ag_3PO_4, yellow, soluble in NH_3 and in HNO_3	$Ba_3(PO_4)_2$, white, soluble in HNO_3	$Ca_3(PO_4)_2$, white, soluble in HNO_3	$Pb_3(PO_4)_2$, white, soluble in HNO_3
CO_3^{2-}	Ag_2CO_3, white, soluble in NH_3 and in HNO_3	$BaCO_3$, white, soluble in HNO_3	$CaCO_3$, white, soluble in HNO_3	$PbCO_3$, white, soluble in HNO_3

EXPERIMENTAL PROCEDURE

Chemicals: 0.5 F $(NH_4)_2MoO_4$, 0.1 F $BaCl_2$, saturated Br_2 water, saturated Cl_2 water, $CCl_4(l)$, 0.02 F $Ca(OH)_2$, 0.1 F $FeCl_3$, $FeSO_4 \cdot 7H_2O$, 3% H_2O_2, $Pb(C_2H_3O_2)_2$ paper, 0.1 F KBr, 0.1 F KI, 0.1 F KNO_3, 0.03 F $AgC_2H_3O_2$ (sat.), 0.1 F $AgNO_3$, 1 F Na_2CO_3, 0.1 F $NaCl$, 0.1 F NaH_2PO_4, 0.1 F Na_2SO_4, 0.1 F Na_2SO_3 (fresh), 0.1 F Na_2S (fresh).

First: Familiarize yourself with the following test procedures, using 2-ml samples of dilute solutions of each ion to be tested. Repeat any unsatisfactory results, after first improving your technique.

Second: Answer the review questions and statements in the experiment report before analyzing the unknowns.

Third: Obtain one or more unknowns, and perform analyses for each of the ions, using a fresh 2-ml sample for each test.

1. Test for Sulfide Ion. To about 2 ml of the test solution add a slight excess of 6 F HCl. Note any odor of H_2S, and, as a more sensitive test, place a piece of moistened lead acetate paper over the mouth of the test tube. Heat the test tube gently. A darkening of the paper indicates S^{2-} in the original solution. (If S^{2-} is found to be present, SO_3^{2-} cannot be present—why not?—and the SO_3^{2-} test may then be omitted. Note also the modification of the PO_4^{3-} test when S^{2-} is present.)

2. Test for Sulfate Ion. To 2 ml of the test solution add 6 F HCl by drops until the solution is slightly acid. Then add 1 ml of 0.1 F $BaCl_2$ solution, or more as needed to complete the precipitation. A white precipitate of $BaSO_4$ proves the presence of SO_4^{2-}. (Save the solution for the sulfite test.)

3. Test for Sulfite Ion. If the solution from the previous sulfate ion test had a sharp odor of SO_2 when it was made acid with HCl, then SO_3^{2-} is present. If you are in doubt, filter or centrifuge the solution to obtain a clear filtrate, add a drop or more of 0.1 F $BaCl_2$ to be sure all SO_4^{2-} is precipitated, and if necessary add more $BaCl_2$; then refilter or recentrifuge. To the clear solution add 1–2 ml of bromine water to oxidize any SO_3^{2-} to SO_4^{2-}. A second white precipitate of $BaSO_4$ now proves the presence of SO_3^{2-}.

4. Test for Carbonate Ion. Fit a 15-cm test tube with a one-hole rubber stopper and bent delivery tube (See Figure 3-1). Place about 3 ml of the test solution in this test tube. If sulfite ion (SO_3^{2-}) is present in the unknown, add 1 ml of 3% H_2O_2 to oxidize it to sulfate ion. Now insert the delivery tube into some clear limewater, $Ca(OH)_2$, in another test tube. When ready, remove the stopper just enough to add a little 6 F HCl to the test solution. Immediately close the stopper again, and heat the tube gently to boiling to drive any CO_2 gas into the limewater. Be careful not to let any of the boiling liquid escape through the delivery tube into the limewater. A white precipitate in the limewater indicates CO_3^{2-} or HCO_3^- in the test solution.

5. Test for Chloride Ion. To a 2-ml portion of the test solution add a few drops of 6 F HNO_3, as needed, to make the solution slightly acid. (Test with litmus paper.) Any sulfide ion present may be removed by boiling the solution a moment. The free sulfur formed in the oxidation of S^{2-} by HNO_3 does not interfere. Add 1 ml of 0.1 F $AgNO_3$. (No precipitate here proves the absence of Cl^-, Br^-, or I^-.) Centrifuge the mixture. Test the clear filtrate with 1 drop of 0.1 F $AgNO_3$, for complete precipitation. If necessary, centrifuge again. Discard the filtrate. Wash the precipitate with distilled water to remove excess acid and silver ion. To this precipitate add 3 ml of distilled water, 4 drops of 6 F NH_3, and $\frac{1}{2}$ ml of 0.1 F $AgNO_3$. (The proportions are important, as we wish to dissolve only the AgCl from any mixture of AgCl, AgBr, and AgI.) $Ag(NH_3)_2^+$ and Cl^- will form. Shake the mixture well, and centrifuge. Transfer the clear solution to a clean test tube, and acidify with 6 F HNO_3. A white precipitate of AgCl confirms Cl^-.

6. Test for Iodide Ion. To 2 ml of the test solution add 6 F HCl to make the solution acid. If S^{2-} or SO_3^{2-} is present, boil the solution to remove the ion. Add 1 ml of 0.1 F $FeCl_3$ to oxidize any I^- to I_2. (Br^- is not oxidized by Fe^{3+}.) Add 1 ml of carbon tetrachloride (CCl_4), and agitate the mixture. A purple color indicates I^-. (Save the mixture for the Br^- test.)

7. Test for Bromide Ion. *If no I^- was present in the above mixture,* add 2 ml of chlorine water, and agitate it. A brown color in the CCl_4 layer indicates Br^-. *If I^- was present,* separate, by means of a medicine dropper, as much as possible of the preceding iodide test solution above the CCl_4 layer that contains the I_2, and place it in a clean test tube. Again extract any remaining I_2 by adding 1 ml of CCl_4, agitating the mixture, and separating the solution. The solution may be boiled a moment to remove any remaining trace of I_2. Then add

2 ml of chlorine water and 1 ml of CCl_4, and agitate the mixture. A brown color indicates Br^-.

8. Test for Phosphate Ion. First mix about 1 ml of 0.5 F $(NH_4)_2MoO_4$ reagent with 1 ml of 6 F HNO_3. (If a white precipitate forms, dissolve it by making the solution basic with NH_3, then re-acidify with HNO_3.) If S^{2-} has been found in the unknown, first make a 2-ml portion of the test sample distinctly acid with HCl, boil it a moment to remove all H_2S, then add this (or a 2-ml sample of the original unknown if no S^{2-} is present) to the clear molybdate solution. A yellow precipitate of $(NH_4)_3PO_4 \cdot 12MoO_3$, appearing at once or after warming a few minutes to about 40°C, indicates the presence of PO_4^{3-}.

9. Test for Nitrate Ion. *If Br^- and I^- are absent from the unknown,* use 2 ml of test solution acidified with 3 F H_2SO_4. Add 1 ml of freshly prepared saturated $FeSO_4$. Incline the test tube at about a 45° angle, and pour about 1 ml of concentrated H_2SO_4 slowly down the side of the test tube. Be careful to avoid undue mixing. A brown ring of $Fe(NO)^{2+}$ at the interface of the two liquids indicates NO_3^-. A faint test may be observed more easily by holding the test tube against white paper and looking toward the light (Figure 28-1).

If Br^- and I^- are present in the unknown, free Br_2 or I_2 may form at the interface with the concentrated H_2SO_4 and invalidate the test. If this happens, add 4 ml of a saturated solution of silver acetate to 2 ml of the test solution, to precipitate AgBr or AgI. Add a drop or two of 6 F HCl to precipitate

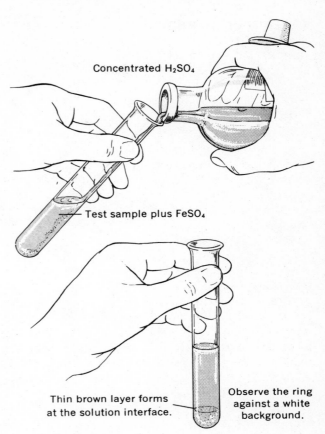

Concentrated H_2SO_4

Test sample plus $FeSO_4$

Thin brown layer forms at the solution interface.

Observe the ring against a white background.

FIGURE 28-1
The proper technique for the nitrate ion test.

any excess silver ion as AgCl. Decant the liquid into a test tube and centrifuge. Treat the clear centrifugate with $FeSO_4$ and concentrated H_2SO_4, as directed in the previous paragraph.

REPORT 28	The Qualitative Analysis of Some Common Anions	NAME	
		SECTION	LOCKER
		INSTRUCTOR	DATE

REVIEW QUESTIONS

1. In the tests for Cl^- and SO_4^{2-}, explain fully why the addition of acid will dissolve other insoluble silver salts and barium salts, respectively, which would interfere with the tests if the solutions were neutral. (See Table 28-1).

2. Suppose a white precipitate is obtained on adding $BaCl_2$ reagent to a neutral unknown solution. What might the precipitate be? Write as many formulas as you can of possible substances, considering both positive and negative ions as given in the chart.

3. A positive test for SO_4^{2-} is to be expected whenever you have SO_3^{2-} or S^{2-} in an unknown solution, even when no SO_4^{2-} has been placed in the solution. Why?

4. Why cannot both S^{2-} and SO_3^{2-} be present in the same solution? Explain, and write the equation.

5. Why is an excess of Fe^{2+} necessary in the test for NO_3^-? Explain, and write the equation.

6. Why do Br^- or I^-, if present, interfere with the NO_3^- test at the brown-ring interface of the solutions? Explain, and write the equation.

7. Both CO_3^{2-} and HCO_3^- respond to the usual carbonate ion test. Give specific procedures by which you could distinguish between these ions in an unknown. (*Hint:* Consider the *p*H of the solution.)

REPORT OF UNKNOWN SOLUTIONS

Write in the correct formula for each ion found, for each unknown. (The ions, if added at all in these unknowns, are present in more than traces.)

Unknown No.:__________ Ions found:__________

Unknown No.:__________ Ions found:__________

SUPPLEMENTARY DRILL

1. Give the formulas of all new molecules or ions formed when solutions of the following are mixed. (Indicate "no reaction" where there is none.)

(a) Cl^-, Br^-, I^-, Fe^{3+} _____

(b) Fe^{2+}, Br_2 water _____

(c) Cl^-, Br^-, I^-, Cl_2 water _____

(d) $CO_2(g)$, $Ca(OH)_2$ _____

(e) $SO_2(g)$, $CaCl_2$, HCl _____

(f) SO_3^{2-}, H_2O_2 (acid) _____

(g) S^{2-}, SO_3^{2-} (acid) _____

(h) $Pb(C_2H_3O_2)_2$, $H_2S(g)$ _____

(i) Ag^+, Cl^-, NH_3 (excess) _____

(j) Ag^+, I^-, NH_3 (excess) _____

2. An unknown solution (I) is to be tested for the ions, or their derivatives,[1] listed at the right. The following tests are performed on *separate* portions of the unknown.

(a) Litmus paper turns blue when touched with a drop of the solution.

(b) When a test portion is acidified with HCl, and lead acetate paper is held in the mouth of the test tube, the paper turns black.

(c) A test portion is acidified with HCl, boiled, and cooled, then Cl_2 water and CCl_4 are added. After the mixture is shaken, the CCl_4 layer is a clear, light reddish-brown color.

(d) Addition of dilute H_2SO_4 in excess results in a white precipitate.

On the basis of these tests, considered as a whole, mark, in column (I), a plus sign (+) for each ion or its derivative which is definitely present in the original solution, a minus sign (−) if the ion is definitely absent, or a question mark (?) if there is no evidence to prove the ion present or absent. (NOTE: Take account of the insolubility of certain salts. For example, if an acid solution is known to contain Pb^{2+}, then SO_4^{2-}, S^{2-}, and so on cannot be present.)

3. A second unknown solution (II) is to be tested for the same list of ions given in Problem 2. The following tests are performed.

(a) The solution is basic to litmus paper (blue color), and smells of ammonia.

(b) The addition of excess HNO_3 results in the formation of a pure white precipitate.

(c) If the mixture from (b) is filtered, and ammonium molybdate reagent is added to the clear filtrate, a yellow precipitate results.

On the basis of these tests, considered as a whole, mark, in column (II), a plus sign, a minus sign, or a question mark for each of the ions in the list. (Again, take account of the insolubility of certain salts.)

	(I)	(II)
Na^+	_____	_____
NH_4^+	_____	_____
Ag^+	_____	_____
Pb^{2+}	_____	_____
Ca^{2+}	_____	_____
Ba^{2+}	_____	_____
Cl^-	_____	_____
Br^-	_____	_____
I^-	_____	_____
SO_4^{2-}	_____	_____
SO_3^{2-}	_____	_____
S^{2-}	_____	_____
NO_3^-	_____	_____
CO_3^{2-}	_____	_____
PO_4^{3-}	_____	_____

---

[1]"Derivatives" are, for example, PO_4^{3-} or H_3PO_4, Ag^+ or $Ag(NH_3)_2^+$, NH_4^+ or NH_3, and so on.

IDENTIFICATION OF SOME CATIONS AND ANIONS IN A GENERAL UNKNOWN

PRE-STUDY

General Cation Analysis

In this experiment, we shall develop a general procedure for the analysis of cations and anions in an unknown. You can analyze the ten cations considered in Experiments 25 through 27 by starting with a single 4-ml sample of the dissolved unknown, and successively carrying out the separations and tests for each group of ions, as follows: (1) first remove and analyze the insoluble chlorides, as in Experiment 25; (2) treat the filtrate from this chloride precipitation as in Experiment 26, and test for the ions of that group; (3) after removal of $ZnS(s)$ from the NH_3 filtrate, then analyze the filtrate for alkaline earth and alkali cations, as in Experiment 27. Use a separate 3-ml portion of the original unknown for the NH_4^+ test.

Analysis for Both Cations and Anions

If both cations and anions are to be determined, it does not matter which are tested for first. For the anion tests, follow the procedures of Experiment 28. Always use the results of either analysis to shorten the procedures for the other. Thus, if the salt is soluble in water or in HNO_3, and if SO_4^{2-} is found, you know that Pb^{2+} and Ba^{2+} must be absent (why?), and tests for them may be omitted. Likewise, if Ag^+ is found in a neutral or acid solution, it is not necessary to test for the halide ions, or for S^{2-}.

Preliminary Work

First prepare, on a separate sheet of paper, a single *combined flow chart* showing how the several groups of cations may be separated and analyzed.

EXPERIMENTAL PROCEDURE

Chemicals: All reagents and solutions for Experiments 25–28.

A. Analysis of Unknown Solutions

Obtain your unknown sample(s), and record the code identification numbers. For the cation analyses, follow the combined flow chart you have prepared, and the procedures for analysis in Experiments 25, 26 and 27. At the same time, prepare an analysis summary showing all reagents, observations, and conclusions for each step of the analysis, as was done for each of the earlier experiments. Indicate negative as well as positive test results.

If anions are to be tested for, use a separate 2-ml portion of the unknown for each anion test, according to the procedures of Experiment 28.

B. Analysis of Unknown Solid Salts

An unknown solid salt or mixture of salts may be provided for analysis. This will be soluble in water, in 6 F HNO_3, or in 6 F NH_3. Test a very small portion of the unknown to determine the best solvent to use, and then prepare a solution of about 3 g of the solid to 35 ml of solution. If 6 F HNO_3 or 6 F NH_3 is needed, first treat the unknown with 2 ml (or more, if needed) of the acid or base, and then dilute to 35 ml with water.

Test 2-ml portions of the salt solution for negative ions according to the procedures of Experiment 28. (If HNO_3 was used as the solvent, test a water extract for NO_3^-, since all nitrates are soluble.) For the NH_4^+ test, treat a bit of the solid directly with 6 F NaOH in an evaporating dish, and follow the test directions. For the other positive metal ions, use a single 4-ml portion of the prepared solution, as outlined above. (Utilize the results of the anion tests to shorten your procedures where possible.)

Prepare an analysis summary showing all reagents, observations and conclusions, as before, including negative as well as positive results.

Identification of Some Cations and Anions in a General Unknown

NAME

SECTION   LOCKER

INSTRUCTOR   DATE

ANALYSIS OF UNKNOWN SOLUTIONS AND SOLID SALTS

NOTE: Include your combined Flow Chart with this report. Prepare additional Analysis Summary forms as needed.

Complete the following analysis summary for your unknown samples.

Unknown No._____ Appearance:_____ Ions found:_____

Unknown No._____ Appearance:_____ Ions found:_____

Unknown No._____ Appearance:_____ Ions found:_____

Step	Sample	Reagent	Observations	Conclusions

322

APPLICATION OF PRINCIPLES

1. From the following salts, list those which, considered individually, would behave in the manner indicated for each case below: $CaCO_3$, $(NH_4)_2CO_3$, $Zn_3(PO_4)_2$, Ag_2CO_3, $Ca(OH)_2$, Hg_2Cl_2, $FeCl_3$, $AgCl$, $AlCl_3$, $BaSO_4$, $Ba(NO_3)_2$, $Pb(NO_3)_2$.

   (a) Dissolves readily in water alone_____

   (b) Dissolves readily in water, but forms a precipitate with excess NH_3_____

   (c) Is insoluble in water, but dissolves when excess HNO_3 is added_____

   (d) Is insoluble in water, but dissolves when excess NH_3 is added_____

   (e) Will not dissolve in water, or when either HNO_3 or NH_3 is added_____

2. An unknown solution is to be tested for the ions, or their derivatives,[1] listed at the right. On the basis of the following tests, performed on *separate* portions of the unknown and considered as a whole, mark in column I a plus sign (+) if the ion is definitely present, a minus sign (−) if the ion is definitely absent, or a question mark (?) if there is no evidence to prove the ion present or absent.

   (a) Phenolphthalein gives a red color.

   (b) When a test portion is warmed, red litmus held at the mouth of the test tube turns blue.

   (c) Acidifying a test portion with HNO_3 causes vigorous effervescence and the formation of a white precipitate, which dissolves when excess NH_3 is added to the mixture.

   (d) Addition of $Na_2SO_4(aq)$ causes no visible effect.

3. A second unknown solution is to be tested for the same list of ions at the right. On the basis of the following tests, considered as a whole, mark in column II, a plus sign, a minus sign, or a question mark, for each ion in the list, as in Problem 2.

   (a) Methyl orange gives a red color.

   (b) Addition of excess Cl^- to one sample, and of SO_4^{2-} to another sample, gives a white precipitate in each sample.

   (c) The Cl^- precipitate from (b), treated with $NH_3(aq)$, turns black.

   (d) A sample is treated with excess Cl^- and filtered, and the filtrate is made basic with $NH_3(aq)$, giving an orange-red precipitate.

	(I)	(II)
Ag^+	___	___
Hg_2^{2+}	___	___
Al^{3+}	___	___
Fe^{3+}	___	___
Ba^{2+}	___	___
K^+	___	___
NH_4^+	___	___
Cl^-	___	___
SO_4^{2-}	___	___
NO_3^-	___	___
CO_3^{2-}	___	___
PO_4^{3-}	___	___

---

[1]By derivative is meant, for example, PO_4^{3-} or H_3PO_4, Ag^+ or $Ag(NH_3)_2^+$, NH_4^+ or NH_3, etc.

QUALITATIVE ANALYSIS OF UNLABELED SOLUTIONS: THE NINE-SOLUTION PROBLEM

PRE-STUDY

In this experiment we assume that you have been introduced to some of the techniques of qualitative analysis described in Experiments 25 through 29. The separation and identification of ions is generally accomplished by making use of the relative differences in solubilities of various salts and their tendencies to form complex ions. The strategy of developing a qualitative analysis scheme rests first on observing the behavior of cations and anions with various reagents, and then on attempting to construct a logical scheme of analysis.

In this experiment a different kind of problem will be posed. You will be given nine different unknown solutions, labeled 1, 2, 3, and so on. The composition of these solutions, not in order, is

6 F NaOH	0.5 F Na$_2$S	0.2 F BaCl$_2$
3 F NH$_4$Cl	0.3 F Na$_2$SO$_4$	0.2 F KIO$_3$
1 F NaBr	0.5 F KClO$_3$	0.1 F AgNO$_3$

Your task will be to identify the contents of each numbered bottle, using no other reagents, indicators, or instruments. One or more unknown solutions must serve as the reagent(s) for identifying another unknown solution. This task may seem an impossible one, since all the solutions are unknown at the beginning, and it will indeed be difficult to make all of the identifications correctly if you do not prepare beforehand.

In solving this chemical "crossword" puzzle you will need to make use of your knowledge of the solubilities of salts and the colors of their precipitates, the possibility of complex ion formation, the interactions of acids and bases and the possible evolution of volatile acids or bases, and the possibility of oxidation-reduction reactions.

It is essential that you do some preliminary preparation before coming to the laboratory. Prepare your pre-study work sheet (Tables 1 and 2 of the report form), showing in tabular form the chemical behavior to be expected of these solutions when roughly equal volumes are mixed.

For some ions, certain types of reactions can be excluded at the outset. For example, Na^+, K^+, and Ba^{2+} are not likely to be oxidized because they have only one stable oxidation state. Also, they are not likely to be reduced in aqueous solution. In contrast, such ions as Br^-, I^-, or S^{2-} can be readily oxidized by moderately strong oxidizing agents.

In working out the predictions of the chemical reactions that you would expect between various ions, you may find it helpful to consult Tables 9 through 13 of Appendix C. Your instructor may also suggest other reference books on inorganic chemistry or qualitative analysis that will provide information not given in your textbook. For example, you can consult the *Handbook of Chemistry and Physics*[1] to determine the color and solubility of various salts. If the solubility of a salt is between 1 and 10 g per 100 ml, record the figure. If the solubility is below this range, you can consider the compound insoluble and assume that it is likely to form a precipitate when its constituent ions are mixed.

EXPERIMENTAL PROCEDURE

Special Supplies: Nine 30-ml bottles and droppers.

Chemicals: 6 *F* NaOH, 3 *F* NH$_4$Cl, 1 *F* NaBr, 0.5 *F* Na$_2$S, 0.3 *F* Na$_2$SO$_4$, 0.5 *F* KClO$_3$, 0.2 *F* BaCl$_2$, 0.2 *F* KIO$_3$, 0.1 *F* AgNO$_3$.

Prepare beforehand as directed in the pre-study and fill in Tables 1 and 2 of the report form. Follow the directions of the instructor in receiving your unknown solutions. You might receive them in test tubes or in dropping bottles, or the nine solutions might be furnished to the entire laboratory section in stock bottles, from which you should fill nine 30-ml bottles equipped with droppers.

---

[1]*Handbook of Chemistry and Physics,* published annually by The Chemical Rubber Company, Cleveland, Ohio 44114.

In mixing solutions, place 10–20 drops of a solution in a small test tube and add drop by drop a like quantity of another unknown solution. Since the order of addition may change the nature of some of the reactions, you might try adding the reagents in reverse order. Record your observations in brief form in Table 3 of the report form, and give a complete description of them in Table 4.

On the basis of your observations, identify as many of the solutions as you can. With these known reagents now at your disposal, devise a logical scheme to identify any remaining unknowns. For example, you may wish to mix three different solutions, or to mix a drop or two of one reagent with a ml of another. If you plan thoughtfully, you will probably not find it necessary to study all 36 of the possible binary combinations of the unknowns.

Your laboratory report should include the results of your pre-study in Tables 1 and 2, a record of your observations in Tables 3 and 4, and the identification of the unknowns on the answer sheet. In addition to the accuracy of your identifications, part of your grade will be based on the reasoning you adopt in your approach to the problem.

Bibliography

MacWood, G. E., Lassettre, E. N., and Breen, G., "A Laboratory Experiment in General Chemistry," *J. Chem. Educ.* **17**, 520 (1940).

Ricketts, J. A., "Laboratory Exercise Emphasizing Deductive Reasoning," *J. Chem. Educ.* **37**, 311 (1960). (This article describes a procedure that you might employ for working out a qualitative analysis scheme based on your own observations on the individual ions that might be present in an unknown.)

Zuehlke, R. W., "The Case of the Unlabeled Bottles," *J. Chem. Educ.* **43**, 601 (1966). (The author presents a variation of the nine-solution problem.)

NAME		
SECTION		LOCKER
INSTRUCTOR		DATE

PREPARATION AND OBSERVATIONS

Table 1

In Table 1 below, write in the net ionic reactions that you expect after solutions of each pair of reagents have been mixed. Classify each reaction as either a precipitation (P), complex ion formation (CIF), acid-base (neutralization) (AB), or oxidation-reduction (OR) reaction. Note also any color changes that you expect and indicate whether you anticipate the evolution of any gas with a characteristic odor. If you expect no reaction, write "N.R."

Solution	Expected reaction	Classification
$NaOH + NH_4Cl$		
$NaOH + NaBr$		
$NaOH + Na_2S$		
$NaOH + Na_2SO_4$		
$NaOH + KClO_3$		
$NaOH + BaCl_2$		
$NaOH + KIO_3$		
$NaOH + AgNO_3$		
$NH_4Cl + NaBr$		
$NH_4Cl + Na_2S$		
$NH_4Cl + Na_2SO_4$		
$NH_4Cl + KClO_3$		
$NH_4Cl + BaCl_2$		
$NH_4Cl + KIO_3$		
$NH_4Cl + AgNO_3$		

TABLE 1 (continued)

Solution	Expected reaction	Classification
$NaBr + Na_2S$		
$NaBr + Na_2SO_4$		
$NaBr + KClO_3$		
$NaBr + BaCl_2$		
$NaBr + KIO_3$		
$NaBr + AgNO_3$		
$Na_2S + Na_2SO_4$		
$Na_2S + KClO_3$		
$Na_2S + BaCl_2$		
$Na_2S + KIO_3$		
$Na_2S + AgNO_3$		
$Na_2SO_4 + KClO_3$		
$Na_2SO_4 + BaCl_2$		
$Na_2SO_4 + KIO_3$		
$Na_2SO_4 + AgNO_3$		
$KClO_3 + BaCl_2$		
$KClO_3 + KIO_3$		
$KClO_3 + AgNO_3$		
$BaCl_2 + KIO_3$		
$BaCl_2 + AgNO_3$		
$KIO_3 + AgNO_3$		

Tables 2 and 3

Before you go to the laboratory, fill in the squares of Table 2 with a brief description of the results you expect based on your pre-study of the chemical reactions. For example, if you expect a yellow precipitate you might record "yel. ppt.", or if you expect evolution of NH_3 gas, you might write "$NH_3 \uparrow$".

As you begin the process of identification in the laboratory, record in the squares of Table 3 a brief description of the results obtained from the mixing of the unknown solutions, penciling in the identification beside the number when you think you have correctly identified an unknown. (In Table 4, you will record a fuller description of those observations.) Remember that you may want to mix three or more solutions or to change the order in which you add the reagents before making positive identifications of some unknowns.

TABLE 2	NH_4Cl	NaBr	Na_2S	Na_2SO_4	$KClO_3$	$BaCl_2$	KIO_3	$AgNO_3$
NaOH →								
NH_4Cl →								

TABLE 3								
1 ↓			NaBr →					
2 ↓			Na_2S →					
3 ↓			Na_2SO_4 →					
4 ↓				$KClO_3$ →				
5 ↓					$BaCl_2$ →			
6 ↓						KIO_3 →		
7 ↓								
8 ↓								
9								

Table 4

In the table below, record a complete description of the results obtained from mixing two or more unknown solutions.

Solutions mixed	Observations	Conclusions

REPORT 30	Sheet 3	NAME

ANSWER SHEET

Record below your identification of the contents of each of the nine unlabeled solutions. If you are not confident of an identification, leave it blank.

Solution number	Identity
1	______________________
2	______________________
3	______________________
4	______________________
5	______________________
6	______________________
7	______________________
8	______________________
9	______________________

In the space below, indicate by chemical equations or a flow sheet the reasoning you have employed in making the identifications.

THE USE OF PAPER CHROMATOGRAPHY IN THE SEPARATION OF IRON(III), COBALT(II), NICKEL(II), AND COPPER(II) IONS

PRE-STUDY

In Experiments 25 through 29 the separation and identification of a number of cations and anions are described. These separations are accomplished by selective precipitation and by the selective formation of complex ions with the use of a variety of reagents. It would be ideal if each cation or anion formed a precipitate or colored complex with just one unique reagent. Then no prior separation would be necessary and the addition of the proper reagent would provide a test for the presence of the sought for ion. However, this ideal situation has not yet been attained, and consequently simpler and more efficient methods of separation have been sought to replace the somewhat tedious methods in which precipitation is employed.

One of these methods is extraction, in which a metal ion (usually in the form of a complex ion) is extracted into a solvent that is immiscible with water. For example, the compound dithizone forms many soluble complexes with metal ions that can be extracted into chloroform. However, this procedure is often not too selective, and therefore it is possible that several ions will be extracted at the same time.

Perhaps the most powerful method of separation that has been discovered to date is the method of *chromatography,* a procedure by which one separates the components of a mixture by passing them over a selectively absorbing medium. In general, a two-phase system—comprising a stationary phase and a moving phase—is employed. In *column chromatography,* small particles of the stationary phase are packed in a tube, and the moving phase travels between the particles confined in the tube. In *thin-layer chromatography,* the stationary phase is a thin layer of fine particles spread on a glass plate; in *paper chromatography,* it is the paper itself.

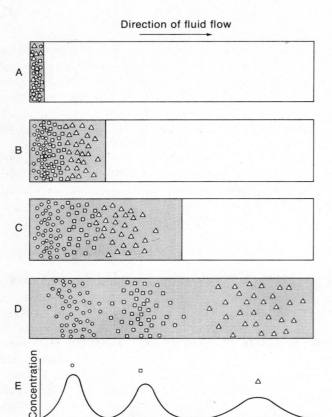

Direction of fluid flow →

FIGURE 31-1
Schematic illustration of separation by the differential migration mode. Parts (A) through (D) illustrate the separation of a three-component mixture by a differential migration process. Note that at the beginning of a separation (A), three components are clustered together at the point of sample application. As the sample migrates across the system—(B) through (D)—the three sample components are gradually separated, the fastest moving component spending the most time in the moving fluid phase. (E) displays a smoothed plot of the concentration of each substance as a function of the distance from the origin for the distribution shown in (D). [After Barry L. Karger et al., *Introduction to Separation Science*, Wiley. Copyright © 1973.]

The separation process in chromatography depends on the fact that the solute species have different affinities for the stationary and moving phases. Solutes that have a greater affinity for the moving phase will spend more time in the moving phase and therefore will move along faster than solutes that spend more time in the stationary phase. This process is illustrated in Figure 31-1, which represents the separation of a sample mixture as it is caused to migrate through a porous medium (such as paper) by fluid flow.

Several different kinds of chromatography exist. Which kind is employed depends on the nature of the stationary phase and on whether the moving fluid phase is a liquid or a gas. Figure 31-2 shows the relationships between the various kinds of chromatography.

There are several kinds of interactions that may give rise to differential mobilities of the solutes. In *partition* chromatography, the stationary phase is generally a liquid that is held either on the surface or within the structure of a porous inert solid. The partition of the solute between these two phases is determined largely by the relative solubilities of the solute in the stationary liquid phase and the moving phase.

In *adsorption* chromatography, the immobile phase is a solid capable of adsorbing a compound on its surface. Alumina and silica gel are often used as adsorbents. Those solutes that are adsorbed most strongly will move most slowly.

When the mobile phase is a gas, the technique is called *gas chromatography,* whether the principle of separation is based on partition or adsorption. Although gas chromatography is widely used, it applies only to substances that are appreciably volatile, mainly low-molecular-weight organic

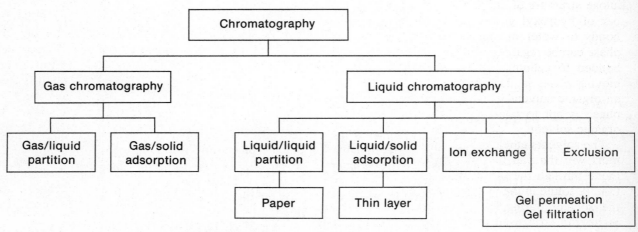

FIGURE 31-2
The interrelation of different kinds of chromatography.

molecules, and it is not useful for separating ionic salts or large polymers.

When the mobile phase is a liquid, the family of techniques is called *liquid chromatography,* but most of the individual techniques are named according to the predominant principle employed — such as *adsorption chromatography, ion-exchange chromatography, exclusion chromatography,* and so on.

In *ion-exchange* chromatography, the stationary phase is a polymer matrix containing fixed negative or positive charges. These fixed charges are neutralized by mobile ions of opposite charge that can be exchanged for other similarly charged ions in the liquid mobile phase.

In *exclusion* (gel) chromatography gels with a porous matrix are employed. Large molecules that cannot get into the solvent filled pores are excluded and move along with the mobile liquid phase. Small molecules that can penetrate into the pores move along more slowly. The molecules are therefore separated according to their sizes. This method is widely used for separating both synthetic polymers and natural polymers, such as proteins.

In this experiment, we will employ paper chromatography, one of the simplest forms to use experimentally, but a rather difficult one to describe by a quantitative theory because it may involve a combination of adsorption and partition, and even ion exchange, if the cellulose molecules contain any free carboxyl (—COOH) groups. The exact nature of the processes depends on the solutes that are used, and the nature of the solvent being used as the mobile phase, as well as any pretreatment of the paper.

In most paper chromatography, partition is the major factor in the separation of solutes. The cellulose structure of the paper contains a large number of hydroxyl groups that can form hydrogen bonds to water molecules so that the stationary phase can be regarded as a layer of water hydrogen bonded to cellulose. If the solvent is water, the moving phase is also aqueous, but if a mixture of an organic solvent is used with water, the moving phase is apt to contain a high proportion of the organic solvent.

The solvents employed in paper chromatography must wet the paper so that the mobile phase will move through the paper fibers by capillary attraction. A solute in the mobile phase moves along with the solvent during the chromatographic *development* (the term used to describe the process that occurs as the solute moves along and is partitioned between the two phases). As it moves, it undergoes many successive distributions between the mobile and stationary phases, the fraction of time it spends in the mobile phase determining how fast it moves along. If it spends all of its time in the moving phase it will move along with the solvent front. If it spends nearly all of its time in the stationary phase it will stay near the point of application.

After the chromatogram has been developed, and the solutes on the paper located, the movement of the solute on the paper is mathematically expressed by the R_f value, where

$$R_f = \frac{\text{distance traveled by the solute}}{\text{distance traveled by the solvent front}}$$

If all conditions could be maintained constant, R_f values would be constant. However, variations in temperature, the composition of the solvent phase, or changes in the paper may alter the R_f value. The R_f value is useful mainly for expressing the relative mobility of two or more solutes in a particular chromatographic system. The absolute R_f values may change from day to day, but their values in relation to each other remain nearly constant.

Paper chromatography can be performed on strips or sheets of paper, so that the solvent is allowed to flow either upward or downward. Horizontal development may also be used on filter paper circles, causing the solvent to flow radially outward from the center. Horizontal movement of the solvent may be assisted by spinning the paper in a specially designed centrifuge. One can separate a very complex mixture in a single sheet of paper with the use of two different solvent systems, developing first in one direction, and then turning the paper by 90° and developing in the second direction.

Development is accomplished in the following way. First apply the solution (containing the solutes to be separated) to the dry filter paper. Use a micropipet or glass capillary to put a very small drop (microliters) of the solution on the paper. Then allow the spots to dry, and place the filter paper in a tightly closed container containing the developing solvent whose vapor saturates the atmosphere of the container. The paper is supported so that the developing solvent moves through the spots that have been applied. The point of application must not be immersed in the developing solvent; if it is, the solute spots will be spread out and greatly diluted.

Allow the solvent flow to continue for a definite time or until the solvent front (the limit of wetness) reaches a specified point on the paper. Then remove the paper, mark the solvent front with a pencil, and dry the paper, fixing the spots in their position at the end of development.

If the solutes are colored, they can be readily seen on the paper. Colorless substances are located by various means. Often the paper is exposed to the vapor of a reagent, or it can be dipped or sprayed with a reagent that reacts with the colorless substances to form a visible spot. The resulting spots and their measured R_f values are compared with the colors and R_f values of standard known materials run under the same conditions. If possible, known and unknown compounds are run on the same paper. If the known and unknown compound have the same R_f values in different solvent systems and give the same reactions with various color-forming (chromogenic) reagents, it is probable that they are identical.

Obtain the distances used in calculating R_f values as follows: for the "distance traveled by the solute," measure from the point of application to the center of density of the spot; for that "traveled by the solvent front," measure from the point of application to the limit of movement of the solvent front (which must be marked immediately after the paper is removed from the developing chamber before the solvent evaporates).

Separation of Fe^{3+}, Co^{2+}, Ni^{2+}, and Cu^{2+}

In separating these transition metal ions, we make use of a mixed solvent system containing water, 2-butanone, and hydrochloric acid. We may envision the stationary phase as a layer of water adsorbed or hydrogen-bonded to the many hydroxyl groups contained in the cellulose molecules of the paper. The moving phase contains HCl and is rich in the less polar 2-butanone. The metal ions, ranging from water solvated cations to anionic chloro complexes, will be distributed between these two phases. The metal ions that do not form chloro complexes, or that form only weak ones, will spend most of their time in the predominantly aqueous stationary phase. The metal ions that form chloro complexes will migrate more rapidly because they will spend most of their time in the 2-butanone rich phase, either as neutral chloro complexes, MCl_2 or MCl_3 (where M represents a metal ion), or as protonated anionic chloro complexes, such as $HMCl_3$ or $HMCl_4$ (for example, $HFeCl_4$).

EXPERIMENTAL PROCEDURE

Special Supplies: Whatman No. 1 paper, cut in 11- by 10-cm sheets; thin polyethylene film, such as "Handi-wrap"; rubber bands; plastic mm rulers; pencils; glass capillaries about 0.5-mm ID; ovens or hair dryers are useful.

Chemicals: $FeCl_3$, $CoCl_2$, $NiCl_2$, $CuCl_2$ (nitrate salts may also be used); concentrated NH_3 solution, 3 F Na_2S, 7 F HCl, 2-butanone,[1] 6 F H_2SO_4. You will be provided with four metal-ion solutions—one each of Fe(III), Co(II), Ni(II), and Cu(II)—containing 5 g/liter of the ion dissolved in 1 F HCl; you will also be given a solution mixture of all four of the metal ions, 5 g/liter of each dissolved in 1 F HCl. Finally, you will also be provided with three unknown solutions containing one or more of the metal ions at similar concentrations.

Take an 11- by 10-cm piece of Whatman No. 1 paper and, with a pencil, draw a line parallel to the 10-cm dimension and 2 cm from the edge. Then pleat the paper with three folds into four equal sections. The folds should be parallel to the 11-cm edge and the pleated sheet should be able to stand upright, as shown in Figure 31-3.

Prepare a fresh developing solvent by putting 10.0 ml of 7.0 F HCl and 35 ml of 2-butanone in the bottom of a clean dry 600-ml beaker, and cover the beaker with plastic wrap held in place with a rubber band.

Lay the pleated paper down on a clean paper towel and carefully spot a tiny drop of each of the

---

[1]Acetone may be used in place of 2-butanone, with only a slight sacrifice in the separation. If acetone is used, the 7 F HCl is replaced with 6 F HCl.

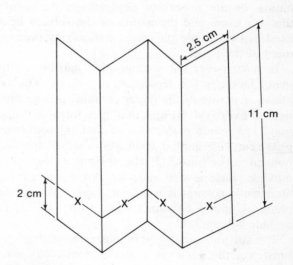

FIGURE 31-3
Preparation of the paper for paper chromatography. A pencil line is drawn 2 cm from the bottom edge, and the paper is pleated so that it will stand upright. The sample spots are placed at the "x" marks.

four metal-ion solutions on the pencil line in the middle of the section as shown in Figure 31-3. Use a glass capillary of about a 0.5-mm internal diameter to spot the solutions. Each spot should be no larger than 3 mm. In chromatogram no. 1 each section will contain a spot of only one metal ion. Prepare a second pleated sheet just like the first one. Mark this one with a pencil as chromatogram no. 2. In the first section put a tiny drop of the solution mixture containing all four metal ions. In the other three sections place a drop of each of the three unknown solutions.

You can expedite drying of the spots by gently waving the paper in the air. After they have dried, put the folded papers in the beaker containing the developing solvent, placing the spots at the bottom. (The surface of the solution must be below the level of the spots.) It is possible to put both papers in the same beaker, if you are careful. They should not touch each other or the walls of the beaker. Immediately seal the beaker with plastic wrap and allow the solvent to rise to within 1 cm of the top of the paper (about 20–30 minutes). Observe the development frequently so that you do not let the solvent front rise all the way to the top of the paper. Also note the colors of the spots as they move along the paper.

When development is finished, remove the papers from the beaker and immediately mark the solvent fronts with a soft pencil before the solvent evaporates. Then dry the paper thoroughly. You can do so in a hood by gently waving the paper, or more quickly if you put them in an oven,[2] or if you use a hair dryer to blow a warm stream of air over the paper. (The laboratory hot-air blower is not recommended, because it will scorch the paper unless it is used very judiciously.)

After the paper has dried, note the colors and locations of any visible spots. Then place the papers in a liter beaker containing a 100-ml beaker half full of concentrated NH_3 solution and seal with plastic wrap and a rubber band. Within 3–5 min-

utes, visible changes should be observable. Note the color and location of any spots that change or become visible. (The presence of a heavy deposit of NH_4Cl or fuming indicates that the paper has not been dried sufficiently so that most of the HCl is not removed.)

Finally, remove the papers from the beaker of NH_3 vapor, gently wave in the hood to remove most of the NH_3, and place the papers in a liter beaker containing H_2S gas. This gas can be generated from an H_2S generator or cylinder, or you can obtain it by rapidly pouring 2 ml of 6 F H_2SO_4 into a 100-ml beaker containing 50 ml of 3 F Na_2S solution placed in the liter beaker. Then seal the liter beaker immediately with plastic wrap and a rubber band. Formation of brown to black spots should take place within 1–2 minutes, and each of the four metal ions should produce a visible spot. Locate each of the spots and note its color.

Locate the densest part of the spot for each metal ion, and draw a pencil line through it. Measure the distance from the pencil line at the point of application to the densest part of the spot, and the distance from the point of application to the solvent front. Calculate the R_f values for each spot, and record on your data sheet.

Using the measured R_f values for the known solutions and the unknown solutions and the characteristic colors and intensities of the spots, identify and report the cations that are present in each of the three unknown solutions.

If time permits, run another set of chromatograms to check the reproducibility of the R_f values. Each time you repeat the experiment, make up a fresh developing solvent mixture.

You might also wish to investigate the effect of varying the HCl concentration, keeping the amount of water and 2-butanone constant.

Bibliography

Skovlin, D. O., "The Paper Chromatographic Separation of the Ions of Elements 26 through 30," *J. Chem. Educ.* **48**, 274 (1971).

---

[2]The temperature should be set at 60°C. If it is 100°C, the paper is likely to char and turn dark.

**The Use of
Paper Chromatography**

NAME	
SECTION	LOCKER
INSTRUCTOR	DATE

DATA AND RESULTS

1. Record your paper chromatography data below.

Distance of solvent front from point of application: chromatogram no. 1, (mm) _____ ; chromatogram no. 2, (mm) _____ .

Sample	Color with solvent	Color with NH_3	Color with H_2S	Zone distance from point of application (mm)
Fe^{3+}				
Co^{2+}				
Ni^{2+}				
Cu^{2+}				
Mixture containing Fe^{3+} Co^{2+} Ni^{2+} Cu^{2+}				
Unknown no. ____				
Unknown no. ____				
Unknown no. ____				

2. Give your paper chromatograph results below.

	Fe^{3+}	Co^{2+}	Ni^{2+}	Cu^{2+}	Mixture	Unknown no. ____	Unknown no. ____	Unknown no. ____
Average R_f values of zones					Fe^{3+} Co^{2+} Ni^{2+} Cu^{2+}			

Unknown no. ____ ; cations present: _____

Unknown no. ____ ; cations present: _____

Unknown no. ____ ; cations present: _____

338

EXERCISES

1. Write an equation which represents the formation of the tetrachloroiron(III) complex, $FeCl_4^-$, from the aquo complex.

2. Write an equation which explains the formation of the blue color when $Cu(II)$ is exposed to the vapors of NH_3.

3. Write equations for the reactions of $Fe(III)$, $Co(II)$, $Ni(II)$, and $Cu(II)$ ions with H_2S.

4. What would you expect to happen to the R_f values of $Fe(III)$, $Co(II)$, and $Cu(II)$ if you decreased the HCl concentration? Why?

5. Why is it important to have clean hands when handling chromatograms?

6. Why is a pencil used to mark the points of application rather than a pen?

SECTION **VIII**

**OXIDATION-REDUCTION:
ELECTRON TRANSFER REACTIONS**

COMMON OXIDIZING AND REDUCING AGENTS. THE BALANCING OF OXIDATION-REDUCTION EQUATIONS

OXIDATION STATES FOR COMMON REAGENTS

Much useful information about the behavior of oxidizing and reducing agents, under various conditions, can be summarized in the form of charts. Such charts for some common elements are presented in Table F-1, and are repeated again in later experiments where these elements are studied in greater detail. The charts, with their comments on the behavior of the various compounds, will help you to predict the probable changes in oxidation state in a particular reaction. Note that the oxidation state is given just before each formula in the chart. Let us comment briefly on the charts for sulfur and oxygen compounds.

It should be obvious from the chart of sulfur compounds that since H_2S represents the lowest possible oxidation state it can act *only* as a reducing agent in oxidation-reduction processes. In such processes it can be oxidized to free sulfur or to some higher state, such as a sulfate, its extent of oxidation depending on the conditions and on the strength of the oxidizing agent. Sulfuric acid (H_2SO_4), representing the highest oxidation state, can act *only* as an oxidizing agent, its reduction products being any of the lower states of sulfur. However, since sulfurous acid (H_2SO_3) represents an intermediate state of sulfur, it can act either as a reducing agent with substances that can take on electrons, such as chlorate ion (ClO_3^-), or as an oxidizing agent with substances that can lose electrons, such as a metal like zinc.

Hydrogen peroxide and *the peroxides* are important oxidizing agents, both commercially and in the laboratory. Note, from the chart, that when hydrogen peroxide acts as an oxidizing agent it is *reduced to water* or, in basic solution, to hydroxide ion. It is *oxidized to free oxygen* only when it acts as a reducing agent, in the presence of stronger oxidizing agents. The instability of hydrogen peroxide, especially in the presence of certain catalysts, is

TABLE F-1
Charts of oxidation states for common reagents

Element			Oxidation state	Formulas	Comments
SULFUR COMPOUNDS	o	r	+6 +4 ave +2 0 −1 −2	SO_3, H_2SO_4, SO_4^{2-} SO_2, H_2SO_3, SO_3^{2-} $S_2O_3^{2-}$ S S_2^{2-} or S_x^{2-} H_2S, S^{2-}	Concentrated acid is a strong oxidizing agent. Active either as oxidizing or reducing agent. Thiosulfate ion. Decomposes to S and H_2SO_3 in acid solution. Oxidized to $S_4O_6^{2-}$ (tetrathionate ion) by free I_2. Polysulfide ion. Decomposes to S and H_2S in acid solution. Strong reducing agent, usually oxidized to S.
OXYGEN COMPOUNDS, PEROXIDES	o	r	 0 −1 −2	*Acidic* *Basic* O_2 H_2O_2 HO_2^- H_2O OH^-	 Active as an oxidizing or as a reducing agent.

CHLORINE COMPOUNDS	o	r	+7
			
			+5
			+4
			+3
			+1
			0
			−1
				(Cl_2O_7), $HClO_4$, ClO_4^-
				
				   $HClO_3$, ClO_3^-
				ClO_2
				   $HClO_2$, ClO_2^-
				Cl_2O, $HClO$, ClO^-
				Cl_2
				Cl^-
					Cl_2O_7 is unstable. $HClO_4$ is a strong oxidizing agent. Reduced to Cl^-.
					Strong oxidizing agent. Reduced to Cl^-.
					Unstable, explosive.
					Good oxidizing agent. Reduced to Cl^-.
					Good oxidizing agent. Reduced to Cl^-.
					Good oxidizing agent. Reduced to Cl^-.

NITROGEN COMPOUNDS	o	r	+5
			
			+4
			+3
			
			+2
			+1
			0
			−3
				N_2O_5, HNO_3, NO_3^-
				
				NO_2, (N_2O_4)
				$(N_2O_3$, $HNO_2)$, NO_2^-
				
				NO
				N_2O
				N_2
				NH_3, NH_4^+
					Strong oxidizing agent, usually reduced to NO_2 and NO;
					  largely to NO_2 in concentrated acid, and to NO in dilute acid.
					  With strong reducing agent may go to NH_3.
					A heavy brown gas.
					N_2O_3 and HNO_2 are unstable; nitrites are fairly stable. Active as
					  oxidizing or as reducing agent.
					Oxidized by the air to NO_2.
					Supports combustion quite vigorously.
					
					Good reducing agents.

CHROMIUM COMPOUNDS	o	r	+6
			
			+3
			
			+2
			
			0
				Acidic *Basic*
				$Cr_2O_7^{2-}$ CrO_4^{2-}
				
				Cr^{3+} $Cr(OH)_4^-$
				
				Cr^{2+}
				
				Cr
					Strong oxidizing agents. Dichromate ion is orange, chromate ion
					  is yellow.
					Amphoteric. Chromic ion is green to violet. Chromic hydroxide
					  complex ion is green.
					An uncommon ion, because it is such a strong reducing agent
					  that it reduces water to hydrogen gas.
					The metal.

MANGANESE COMPOUNDS	o	r	+7
			
			+6
			
			+4
			+3
			+2
			
			0
				MnO_4^-
				
				MnO_4^{2-}
				
				MnO_2, $MnO(OH)_2$
				Mn^{3+}, $Mn(OH)_3$
				Mn^{2+}, $Mn(OH)_2$
				
				Mn
					Permanganate ion, purple. Strong oxidizing agent, reduced to
					  Mn^{2+} in acid solution or to MnO_2 (sometimes to MnO_4^{2-}) in
					  neutral or basic solution.
					Manganate ion, green. Stable only in base. Easily reduced to
					  manganese dioxide.
					Brown as precipitated from solution.
					Mn^{3+} is unstable, gives Mn^{2+} and MnO_2.
					Colorless in solution, pale pink as solid manganese(II) salts.
					  $Mn(OH)_2$ is oxidized by air to $Mn(OH)_3$.
					The metal.

due to this ability of one molecule to oxidize another molecule of the same substance (auto-oxidation-reduction).[1] The half-reactions corresponding to these statements are

$$2H^+ + H_2O_2 + 2e^- \rightarrow 2H_2O \quad \text{(reduction)}$$
$$H_2O_2 \rightarrow 2H^+ + O_2 + 2e^- \quad \text{(oxidation)}$$

$$2H_2O_2 \rightarrow 2H_2O + O_2$$
$$\text{(auto-oxidation-reduction)}$$

Listed below are some important categories of common oxidizing and reducing agents.

Some Common Oxidizing Agents. 1. The halogens and oxygen—reduced to their negative ions, such as the following:

$$F_2 - \text{reduced to } F^-$$
$$Cl_2 - \text{reduced to } Cl^-$$
$$O_2 - \text{reduced to } O^{2-}$$

2. Ions in which the metal ion has a stable lower oxidation state such as the following:

$$Ce^{4+} - \text{reduced to } Ce^{3+}$$
$$Mn^{3+} - \text{reduced to } Mn^{2+}$$
$$Co^{3+} - \text{reduced to } Co^{2+}$$

3. Oxygen containing complex ions, where the central atom is in a high oxidation state, for example:

$$MnO_4^- - \text{reduced to } Mn^{2+}$$
$$ClO_3^- - \text{reduced to } Cl^-$$
$$Cr_2O_7^{2-} - \text{reduced to } Cr^{3+}$$
$$NO_3^- - \text{reduced to } NO$$

Some Common Reducing Agents. 1. The metals—oxidized to their positive ions—such as these:

$$Sn - \text{oxidized to } Sn^{2+}$$
$$Zn - \text{oxidized to } Zn^{2+}$$

2. Ions in which the metal has another higher oxidation state, such as the following:

$$Sn^{2+} - \text{oxidized to } Sn^{4+}$$
$$Fe^{2+} - \text{oxidized to } Fe^{3+}$$
$$Hg_2^{2+} - \text{oxidized to } Hg^{2+}$$

3. Carbon and organic compounds—may be oxidized to other organic compounds or to CO_2 and H_2O; for example:

C (coke, much used in industry)—oxidized to CO or to CO_2

CH_3CH_2OH (alcohol)—may be oxidized to CH_3COOH (acetic acid); thus,

$$CH_3CH_2OH + H_2O \rightarrow CH_3COOH + 4H^+ + 4e^-$$

HCHO (formaldehyde)—oxidized to HCOOH (formic acid); thus,

$$HCHO + H_2O \rightarrow HCOOH + 2H^+ + 2e^-$$

THE BALANCING OF OXIDATION-REDUCTION EQUATIONS

In balancing any oxidation-reduction reaction, you must first know all of the reactants and products. If you do not, you will not be able to balance it correctly. Once the reactants and products are known, balance the reaction by keeping in mind that, in an oxidation-reduction reaction which is essentially an electron-transfer reaction, relative amounts of the reactants must be taken in such a way that all the electrons supplied by the oxidation process are used by the reduction process.[2] There are several methods for doing this, each differing in the mechanics of operation, but all based on the same principle.

THE HALF-REACTION METHOD

Separate half-reactions, or electron reactions, are first written for the oxidation and for the reduction processes. In developing these, we can first determine the number of electrons required from the change in oxidation number, then insert H^+ (or OH^- if the solution is basic) to balance the charges, and finally add H_2O to balance the atoms.

The reverse process is sometimes used. First, balance the atoms in the half-reaction by inserting H^+ and H_2O as needed, then insert as many electrons as needed to balance the charges. Study the following examples.

---

[1] Auto-oxidation-reduction reactions are also called *disproportionation* reactions.

[2] In one class of oxidation-reduction reactions atom transfer rather than electron transfer takes place. Such reactions will not be considered here.

Example 1. Let us consider the oxidation of sodium sulfite (Na_2SO_3) by potassium dichromate ($K_2Cr_2O_7$) in an acid solution. Referring to the charts of oxidation states for sulfur compounds and for chromium compounds (Table F-1), we find that sulfite ion will be oxidized to sulfate ion, and that the dichromate ion will be reduced to chromic ion (Cr^{3+}), as will be evidenced by the green color of the solution. We may write first a partial equation including only the sulfite ion (SO_3^{2-}) and its oxidation product:

Step (1) $$SO_3^{2-} \rightarrow SO_4^{2-}$$

We need another oxygen atom on the left; this is supplied by water, and we write the hydrogen as $2H^+$ on the right. (Note that we keep the oxidation states of oxygen -2 and of hydrogen $+1$ on both sides of the equation, since they are not the substances oxidized and reduced.) Our partial equation then becomes

Step (2) $$H_2O + SO_3^{2-} \rightarrow SO_4^{2-} + 2H^+$$

We still need to balance the charges so they are the same on both sides of the equation. To do so we add 2 electrons on the right,

Step (3)
$$H_2O + SO_3^{2-} \rightarrow SO_4^{2-} + 2H^+ + 2e^- \quad (a)$$

thereby completing the oxidation half-reaction, and demonstrating the fact that in an oxidation process, electrons are lost.

The reduction of dichromate ion to chromic ion may be expressed similarly. The several steps are as follows.

Step (1) $$Cr_2O_7^{2-} \rightarrow 2Cr^{3+}$$

Then balance the hydrogen and oxygen by inserting $14H^+$ to react with the 7 oxygen atoms to form $7H_2O$,

Step (2) $$Cr_2O_7^{2-} + 14H^+ \rightarrow 2Cr^{3+} + 7H_2O$$

Finally, we balance the charges. As above written, we have 12 positive charges on the left and 6 positive charges on the right, or a net charge of $6+$ on the left. We therefore need 6 electrons on the left to complete the reduction half-reaction:

Step (3) $$Cr_2O_7^{2-} + 14H^+ + 6e^- \rightarrow$$
$$2Cr^{3+} + 7H_2O \quad (b)$$

This emphasizes the fact that in a reduction process electrons are gained. Note that the requirement of 6 electrons corresponds to the change in oxidation state of chromium from $+6$ to $+3$, so the two chromium atoms decrease by a total of 6 charges. However, it is not necessary to assume these oxidation states in order to balance the reduction half-reaction properly.

Finally, we may combine (a) and (b) in such a way that we balance the electrons gained against those lost, since free electrons never appear in the final equation. To do this, we multiply (a) by 3, and add algebraically to (b):

$$3H_2O + 3SO_3^{2-} \rightarrow 3SO_4^{2-} + 6H^+ + 6e^-$$
$$(a \times 3)$$
$$\underline{Cr_2O_7^{2-} + 14H^+ + 6e^- \rightarrow 2Cr^{3+} + 7H_2O \quad (b)}$$
$$Cr_2O_7^{2-} + 3SO_3^{2-} + 8H^+ \rightarrow$$
$$2Cr^{3+} + 3SO_4^{2-} + 4H_2O$$

Always check your results to make certain that *both* atoms and charges are balanced in the equation.

Example 2. The spontaneous decomposition of *aqua regia* results from the slow oxidation of chloride ion by nitrate ion in a strongly acid solution. This time we shall give only the completed half-reactions involved. See if you can develop them, on a piece of scratch paper, by either or both of the above techniques. For the half-reactions, we have

$$2Cl^- \rightarrow Cl_2 + 2e^- \qquad \text{(oxidation)} \quad (a)$$

$$4H^+ + NO_3^- + 3e^- \rightarrow NO + 2H_2O$$
$$\text{(reduction)} \quad (b)$$

Now we multiply (a) by 3 and (b) by 2 to give the same number of electrons in each case, and then add:

$$6Cl^- \rightarrow 3Cl_2 + 6e^- \qquad (a \times 3)$$
$$\underline{8H^+ + 2NO_3^- + 6e^- \rightarrow 2NO + 4H_2O \quad (b \times 2)}$$
$$8H^+ + 2NO_3^- + 6Cl^- \rightarrow$$
$$2NO + 3Cl_2 + 4H_2O^3$$

---

[3]Nitric oxide, NO, will combine with free chlorine, Cl_2, to give nitrosyl chloride, so the final equation may be written to include $2NOCl + 2Cl_2$, instead of $2NO + 3Cl_2$.

Example 3. A case in basic solution. If the same reaction studied in Example 1 is carried out in basic solution, we proceed as follows. We write the reducing agent and its oxidized form as before:

Step (1) $\qquad\qquad SO_3^{2-} \rightarrow SO_4^{2-}$

But now the reaction must be balanced in terms of H_2O and OH^- (since H^+ is not available), and so we add $2OH^-$ on the left (twice as much as the required oxygen) and then water on the right to balance the atoms:[4]

Step (2) $\qquad 2OH^- + SO_3^{2-} \rightarrow SO_4^{2-} + H_2O$

Finally we add electrons to balance the charges, as before:

Step (3)
$$2OH^- + SO_3^{2-} \rightarrow SO_4^{2-} + H_2O + 2e^- \quad (a)$$

For the reduction of dichromate ion, note in the chart on chromium compounds that in basic solution this will be present as chromate ion (CrO_4^{2-}), and when reduced to the trivalent state it will be present in basic solution as the hydroxide complex ion ($Cr(OH)_4^-$). We therefore write

Step (1) $\qquad\qquad CrO_4^{2-} \rightarrow Cr(OH)_4^-$

The oxygen now balances, but there are four extra hydrogen atoms on the right; by adding $4OH^-$ on the right and $4H_2O$ on the left, we achieve a balance of atoms:

Step (2)
$$4H_2O + CrO_4^{2-} \rightarrow Cr(OH)_4^- + 4OH^-$$

Finally, we add electrons to balance the charges:

Step (3) $\quad 3e^- + 4H_2O + CrO_4^{2-} \rightarrow$
$$Cr(OH)_4^- + 4OH^- \quad (b)$$

---

[4]For reactions in which it is easy to assign definite oxidation states, it may be simpler for you to note, as in this case, that sulfur changes from oxidation state +4 in SO_3^{2-} to +6 in SO_4^{2-}, and therefore $2e^-$ are needed on the right side to give for Step (2) $SO_3^{2-} \rightarrow SO_4^{2-} + 2e^-$. Finally, you can balance the charges by adding $2OH^-$ on the left and H_2O on the right to balance the atoms.

We may now combine the oxidation half-reaction (a) with the reduction half-reaction (b) to eliminate electrons:

$$6OH^- + 3SO_3^{2-} \rightarrow 3SO_4^{2-} + 3H_2O + 6e^-$$
$$(a \times 3)$$

$$6e^- + 8H_2O + 2CrO_4^{2-} \rightarrow$$
$$2Cr(OH)_4^- + 8OH^- \quad (b \times 2)$$

$$\overline{}$$

$$3SO_3^{2-} + 2CrO_4^{2-} + 5H_2O \rightarrow$$
$$3SO_4^{2-} + 2Cr(OH)_4^- + 2OH^-$$

Example 4. A Reaction involving an *organic compound.* The oxidation of an alcohol to a carboxylic acid by $K_2Cr_2O_7$ in acid solution shows how the method can be applied to organic compounds. We write the reducing agent and its oxidized form:

Step (1) $\qquad CH_3CH_2OH \rightarrow CH_3COOH$

We balance the half-reaction for oxygen by adding H_2O to the left side, and for hydrogen by adding H^+ to the right hand side so that all of the atoms are balanced:

Step (2)
$$CH_3CH_2OH + H_2O \rightarrow CH_3COOH + 4H^+$$

Finally we add electrons to balance the charges:

Step (3) $\quad CH_3CH_2OH + H_2O \rightarrow$
$$CH_3COOH + 4H^+ + 4e^- \quad (a)$$

The half-reaction for the reduction of dichromate ion is balanced as in Example 1 to give:

$$Cr_2O_7^{2-} + 14H^+ + 6e^- \rightarrow 2Cr^{3+} + 7H_2O \quad (b)$$

To obtain the overall reaction, we multiply (a) by 3, and add algebraically to (b) $\times$ 2:

$$3CH_3CH_2OH + 3H_2O \rightarrow$$
$$3CH_3COOH + 12H^+ + 12e^- \quad (a \times 3)$$
$$2Cr_2O_7^{2-} + 28H^+ + 12e^- \rightarrow$$
$$4Cr^{3+} + 14H_2O \quad (b \times 2)$$

$$\overline{}$$

$$2Cr_2O_7^{2-} + 3CH_3CH_2OH + 16H^+ \rightarrow$$
$$4Cr^{3+} + 3CH_3COOH + 11H_2O$$

THE OXIDATION NUMBER METHOD

This technique differs from the preceding one principally in that both the oxidizing and the reducing agents, and their respective products, are written as a single preliminary equation. The relative amounts of each are then computed from the changes in oxidation numbers. The total charge on each side of the equation is then balanced by the addition of H^+ (or OH^- in a basic solution), and the atoms are balanced by the addition of H_2O. Study the following example.

Example 1. Consider again the oxidation of sodium sulfite by potassium dichromate in an acid solution.

Write the principal substances used and produced:

Step (1) $SO_3^{2-} + Cr_2O_7^{2-} \rightarrow SO_4^{2-} + Cr^{3+}$

Note the change in oxidation number, that is, the number of electrons lost and gained. These changes may be indicated by brackets:

Step (2)

$$\overset{+4}{SO_3^{2-}} + \overset{+6}{Cr_2O_7^{2-}} \rightarrow \overset{+6}{SO_4^{2-}} + \overset{+3}{Cr^{3+}}$$

loses $2e^-$

gains $3e^-$ per Cr

Insert stoichiometric coefficients in the equation which will make the number of electrons lost equal the number gained. Because S loses $2e^-$ and each Cr gains $3e^-$, we need 3S to 2Cr, or

Step (3)
$$3SO_3^{2-} + Cr_2O_7^{2-} \rightarrow 3SO_4^{2-} + 2Cr^{3+}$$

Note the total net charges on each side of the equation — in this case 8 negative charges on the left and 0 charges on the right. Then add H^+ to balance the charges, and insert H_2O to balance the atoms. Here we need $8H^+$ on the left and $4H_2O$ on the right. The final equation (balanced) is then

Step (4) $3SO_3^{2-} + Cr_2O_7^{2-} + 8H^+ \rightarrow$

$$3SO_4^{2-} + 2Cr^{3+} + 4H_2O$$

Example 2. If this same reaction were being carried out in basic solution, the chromium would be present as chromate ion (CrO_4^{2-}) before the reaction and as chromic hydroxide complex ion ($Cr(OH)_4^-$) afterward. (See the chart for chromium compounds in the table). The steps in balancing the equations are

Step (1)
$$SO_3^{2-} + CrO_4^{2-} \rightarrow SO_4^{2-} + Cr(OH)_4^-$$

Step (2)

$$\overset{+4}{SO_3^{2-}} + \overset{+6}{CrO_4^{2-}} \rightarrow \overset{+6}{SO_4^{2-}} + \overset{+3}{Cr(OH)_4^-}$$

loses $2e^-$

gains $3e^-$

Step (3)
$$3SO_3^{2-} + 2CrO_4^{2-} \rightarrow 3SO_4^{2-} + 2Cr(OH)_4^-$$

Since we have 10 negative charges on the left and 8 negative charges on the right, we need 2 OH^- on the right, and $5H_2O$ on the left:

Step (4) $3SO_3^{2-} + 2CrO_4^{2-} + 5H_2O \rightarrow$

$$3SO_4^{2-} + 2Cr(OH)_4^- + 2OH^-$$

REPORT

F

Common Oxidizing and Reducing Agents.
The Balancing of
Oxidation-Reduction Equations

NAME

SECTION   LOCKER

INSTRUCTOR   DATE

APPLICATION OF PRINCIPLES

1. Show the change in oxidation number (give number of electrons gained or lost *per atom*; e.g., $3e^-$ gained) in the following:

   (a) $NO_2^- \rightarrow NO_3^-$ _____

   (c) $MnO_4^- \rightarrow MnO_2$ _____

   (e) $NH_3 \rightarrow NH_4^+$ _____

   (b) $SO_2 \rightarrow S_2^{2-}$ _____

   (d) $KClO_2 \rightarrow KCl$ _____

   (f) $HCOOH \rightarrow HCHO$ _____

2. Give the formula of a product (derived from the first-named substance) that may be formed in the following reactions. (Note in the example that any lower oxidation state compound is possible; but not any higher one. Some are more probable than others.)

   Example: H_2SO_3 is treated with a reducing agent S, S_2^{2-}, H_2S _____

   (a) $HClO_2$ is treated with a reducing agent   _____

   (b) H_2SO_3 is treated with an oxidizing agent   _____

   (c) $SnCl_4$ is treated with zinc dust   _____

   (d) $Cr_2O_7^{2-}$ is treated with $SnCl_2$ _____

   (e) $KMnO_4$ is treated with $FeSO_4$ _____

   (f) MnO_2 is treated with concentrated HCl   _____

3. Write the half-reaction equation for the oxidation of

   (a) NO_2^- to NO_3^- (acidic)   _____

   (b) H_2S to SO_4^{2-} (acidic)   _____

   (c) NH_4^+ to NO_3^- (acidic)   _____

   (d) H_2O_2 to $O_2(g)$ (basic)   _____

   (e) $Cr(OH)_4^-$ to CrO_4^{2-} (basic)   _____

   (f) ClO^- to ClO_4^- (acidic)   _____

   (g) $HCOOH$ to CO_2 (acidic)   _____

4. Write the half-reaction equation for the reduction of

(a) SO_3^{2-} to H_2S (acidic) _____

(b) MnO_4^{2-} to MnO_2 (basic) _____

(c) HO_2^- to OH^- (basic) _____

(d) HCOOH to CH_3OH (acidic) _____

(e) ClO_3^- to Cl^- (acidic) _____

(f) $Cr_2O_7^{2-}$ to Cr^{3+} (acidic) _____

(g) CH_3NO_2 to CH_3NH_2 (acidic) _____

5. Given the reactants and products, write balanced net ionic equations for the following reactions. (Supply H_2O, H^+, or OH^- as needed.)

(a) Iron filings are added to $FeCl_3$ solution.

$$\underline{\quad}Fe + \underline{\quad}Fe^{3+} \rightarrow \underline{\quad}Fe^{2+}$$

(b) Bismuth metal is dissolved in hot concentrated HNO_3, and a brown gas is given off.

$$\underline{\quad}Bi + \underline{\quad}NO_3^- \rightarrow \underline{\quad}Bi^{3+} + \underline{\quad}NO_2(g)$$

(c) A mixture of Na_2S, NaClO, and NaOH solutions is warmed, giving a suspended precipitate.

$$\underline{\quad}S^{2-} + \underline{\quad}ClO^- \rightarrow \underline{\quad}S + \underline{\quad}Cl^-$$

(d) SO_2 gas is bubbled into $K_2Cr_2O_7$ solution (acidic).

$$\underline{\quad}SO_2 + \underline{\quad}Cr_2O_7^{2-} \rightarrow \underline{\quad}Cr^{3+} + \underline{\quad}SO_4^{2-}$$

6. Predict the products and write balanced net ionic equations for the following reactions.

(a) $SnCl_2$ is added to $KMnO_4$ solution (acidic).

(b) Zinc dust is treated with dilute HNO_3, forming NH_4^+.

(c) Oxalate in CaC_2O_4 is oxidized to CO_2 by $KMnO_4$ in acidic solution.

OXIDATION-REDUCTION.
ELECTRON TRANSFER REACTIONS

PRE-STUDY

Fundamental Definitions

Oxidation-reduction always involves a change in the valence or oxidation number of *two* species in a given reaction. The species oxidized attains a higher positive, or less negative, charge, while the species reduced becomes less positive, or more negative, in charge. This may be shown graphically as in Figure 32-1.

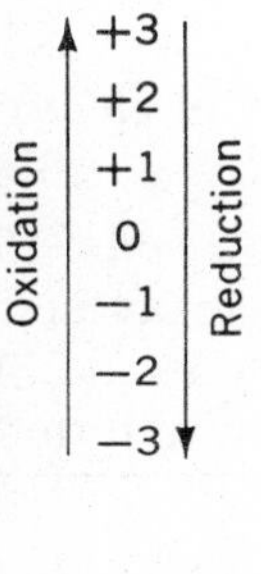

Basically, this change in electrical charge is synonymous with the *transfer of one or more electrons* to or toward one element, which is thereby reduced, and away from the other element, which is thereby oxidized. *Example:* In the reaction of metallic copper with silver ion (Experiment 4), the silver ion attracts electrons away from the neutral copper atom:

$$Cu(s) \rightarrow Cu^{2+} + 2e^-$$

(oxidation — loss of electrons)

$$2Ag^+ + 2e^- \rightarrow 2Ag(s)$$

(reduction — gain of electrons)

$$Cu(s) + 2Ag^+ \rightarrow Cu^{2+} + 2Ag(s)$$

(involves both oxidation and reduction)

FIGURE 32-1
When an oxidation number changes, both oxidation and reduction have taken place.

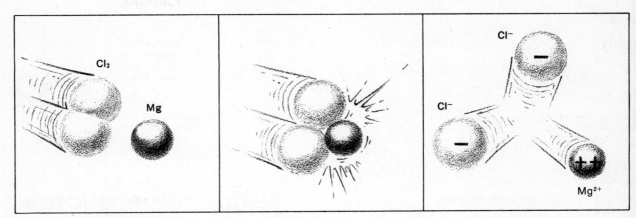

FIGURE 32-2
Oxidation-reduction and electron transfer. When chlorine, $Cl_2(g)$, reacts with magnesium metal, two electrons are transferred from one magnesium atom to two chlorine atoms.

The Oxidation State of an Element

In *compounds containing simple ions,* the oxidation number or oxidation state is identical with the ionic valence of the element concerned. Thus, the oxidation number is +2 for iron in Fe^{2+} or $FeCl_2$; it is +1 for silver, and −2 for oxygen, in silver oxide, Ag_2O. Any element in the *free state,* such as zinc metal, Zn, or hydrogen gas, H_2, has an oxidation number of zero.

In compounds in which *polar covalent bonds* are present, the valence electrons are assigned completely to the more electronegative element. Thus, in ammonia NH_3, nitrogen is assigned the oxidation number −3, and hydrogen +1. In compounds of three or more elements, hydrogen, oxygen,[1] and the alkali and alkaline earth metals are assigned their usual oxidation numbers. The third element is then given an oxidation number such that the algebraic sum of the charges of all the elements in the neutral molecule will be zero. *Example:* In H_3PO_4 there are three positive charges due to the three hydrogen atoms, and eight negative charges due to the four oxygen atoms. The phosphorus must then have an oxidation state of +5. Again, in $K_2Cr_2O_7$, the sum of +2 (due to potassium) and −14 (due to oxygen) is −12, which is balanced by the opposite charges due to *two* chromium atoms, so the oxidation state of chromium is +6.

In *complex ions,* the charge on the ion as a whole must also be considered. Example: In permanganate ion, MnO_4^-, there are eight negative charges due to oxygen, and thus the oxidation state of manganese must be +7, to be consistent with the total ionic charge of −1.

It should always be kept in mind that oxidation states are assigned by formal rules or conventions and do not represent the actual distribution of electronic charge in the species. A permanganate ion is not an Mn^{7+} ion surrounded by four O^{2-} ions; the Mn—O bonds have significant covalent character.

Let us now examine a more complicated reaction—the oxidation of hydrochloric acid by potassium permanganate, for which the net ionic equation is

$$16H^+ + 10Cl^- + 2MnO_4^- \rightarrow$$

$$2Mn^{2+} + 5Cl_2(g) + 8H_2O$$

Here, the $10Cl^-$ are oxidized to $5Cl_2$, with the corresponding transfer of 10 electrons:

$$10Cl^- \rightarrow 5Cl_2(g) + 10e^-$$

$$\text{(oxidation—loss of electrons)}$$

The $2MnO_4^-$ (manganese oxidation state +7) are reduced by these $10e^-$ to $2Mn^{2+}$ (manganese oxidation state +2). The eight oxygen atoms in $2MnO_4^-$ require $16H^+$ from the acid solution to form $8H_2O$

---

[1] Peroxides constitute an exception to the usual valence of −2 for oxygen. In these, the two covalently linked oxygen atoms each have an oxidation state of −1.

no oxidation-reduction), so the reduction "half-reaction" is:

$$2MnO_4^- + 16H^+ + 10e^- \rightarrow 2Mn^{2+} + 8H_2O$$

(reduction—gain of electrons)

The $10Cl^-$ lose the same number of electrons that the $2MnO_4^-$ gain. These quantities are therefore *equivalent* from the standpoint of oxidation-reduction.

The Relative Strength of Oxidizing and Reducing Agents

Oxidation-reduction processes involve a *relative competition* of substances for electrons. The stronger oxidizing agents are those substances with greater affinity for additional electrons; the stronger reducing agents are those substances with the least attraction for electrons which they already possess. Thus silver ion is a stronger oxidizing agent than is cupric ion because the reaction

$$2Ag^+ + Cu(s) \rightarrow Cu^{2+} + 2Ag(s)$$

takes place as indicated, but not appreciably in the reverse direction. That is, silver ion has a strong enough attraction for electrons to take them away from copper atoms. Similarly, copper metal is a stronger reducing agent than is silver metal, because copper releases its electrons more easily. It is possible to arrange the various oxidizing and reducing agents as "redox couples" in a series, according to their relative tendencies to gain or lose electrons. In this experiment we shall explore such relative tendencies for a limited number of reactions.

Fluorine is one of the strongest oxidizing agents known and lithium is one of the strongest reducing agents: therefore these two redox couples would be at opposite ends of the series, with most other redox couples falling somewhere in between.

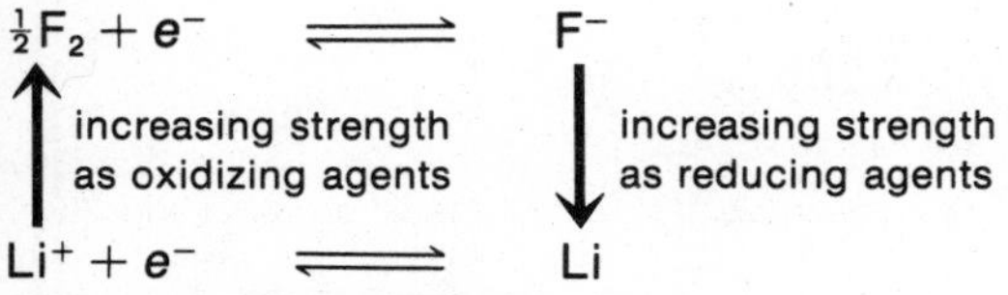

FIGURE 32-3
Schematic diagram illustrating the most powerful and the weakest oxidizing and reducing agents.

EXPERIMENTAL PROCEDURE

NOTE: Before beginning this experiment, complete the preliminary exercise in your experiment report form, indicating as directed there the substances oxidized and reduced in the several equations given and the change in oxidation state, if any.

Chemicals: Metal strips (about 5 mm × 15 mm) of Cu, Pb, Ag, and Zn; Br_2 water (saturated solution), $CCl_4(l)$, Cl_2 water (saturated solution), Cu (turnings), 0.1 F $Cu(NO_3)_2$, 0.1 F $FeCl_3$, 0.05 F I_2, 0.1 F $Pb(NO_3)_2$, 0.1 F KBr, 0.1 F $K_3Fe(CN)_6$, 0.1 F KI, $KMnO_4(s)$, 0.1 F $AgNO_3$, 0.1 F $Zn(NO_3)_2$.

In this experiment, we shall explore qualitatively the relative position of a limited number of oxidation-reduction couples in the potential series. We shall start with only three metals and their ions, and then expand our determinations until ten couples in all have been considered.

1. A Simple Potential Series for the Metals. (a) *Copper, Zinc, Lead, and their Ions.* Explore the behavior of small pieces of Cu, Zn, and Pb metals, each with 3 ml of 0.1 F solutions of the ions of the other metals—that is, Cu with Zn^{2+} and with Pb^{2+}, Zn with Cu^{2+} and with Pb^{2+}, and Pb with Cu^{2+} and with Zn^{2+}—to determine which metal is the strongest and which the weakest reducing agent. Which ion is the strongest and which the weakest oxidizing agent? Write equations wherever reactions occur. Note that the oxidized form of one redox couple is always mixed with the reduced form of the other couple, or vice versa. Prepare a table as directed in the report form, under part 1 of the experimental data.

(b) *The Silver Ion–Silver Couple.* Recall your data from Experiment 4, or try again the reaction of Cu with Ag^+, and also other combinations of Ag or Ag^+ with the metal ions or metals of (a) preceding, until you are able to include the Ag^+–Ag couple in your potential series.

(c) *The Hydrogen Ion–Hydrogen Couple.* Recall from previous experience, or test as needed, to prove the ability of 6 F HCl to dissolve the above four metals. (Nitric acid cannot be used to test the activity of H^+, because nitrate ion is a stronger oxidizing agent than H^+ and would react first and confuse the results. Lead metal reacts too slowly with HCl, even when hot, to give definite evidence. Electric cell-potential measurements, discussed in Experiment 33, prove lead to be a slightly stronger reducing agent than hydrogen.) Now place the H^+–H_2 couple in its proper place in your potential series.

2. The Oxidizing Power of the Halogens. First, if not familiar with the colors of the free halogens in carbon tetrachloride, add a little Br_2 water to 3 ml of H_2O and 1 ml of CCl_4.[2] Mix these liquids, and observe. Repeat, using 0.05 F I_2 instead of Br_2 water. (Cl_2 in CCl_4 is colorless.) Now explore the behavior of 3-ml samples of 0.1 F solutions of each of the halide ions Br^-, Cl^-, and I^- (with 1 ml of CCl_4 added to each), when each is treated with a little of the other free halogen—that is, Br^- with Cl_2 and with I_2, Cl^- with Br_2 and with I_2, and I^- with Cl_2 and with Br_2—to determine which halogen is the strongest and which the weakest oxidizing agent. Which ion is the strongest and which the weakest reducing agent? Write equations. Now arrange the three halogens in a separate potential series, as directed in the report form.

3. The Iron(II) Ion–Iron(III) Ion Couple. Determine whether Fe^{3+} ion is a stronger or a weaker oxidizing agent than I_2 or Br_2 by adding 1 ml of 0.1 F $FeCl_3$ to 2 ml each of 0.1 F KBr and 0.1 F KI. Add CCl_4 to each, mix, and note the formation of any free halogen. You can test for any reduction of Fe^{3+} to Fe^{2+} by adding a little potassium ferricyanide solution, $K_3Fe(CN)_6$. If Fe^{2+} is present, the deep blue precipitate of $Fe_3(Fe(CN)_6)_2$ will form. Write equations for any reactions in which iron(III) is reduced; note that it does not go to metallic iron. Place the Fe^{3+}–Fe^{2+} couple in its proper place in your potential series of the halogens.

4. The Permanganate Ion–Manganese(II) Ion Couple. Add a few $KMnO_4$ crystals to 3 ml of 6 F HCl. Warm the mixture, if necessary, and cautiously note the odor (see p. 350 for this equation). Place the MnO_4^-–Mn^{2+} couple in your potential series.

5. The Reaction of the Halogens with Metals. Test the ability of the halogens to oxidize metals by adding 10 ml of Br_2 water to one of the less active metals—some Cu turnings in a 15-cm test tube. Shake the mixture for several minutes and, if necessary, boil it for a moment to remove the last of the free bromine. Test for the formation of any bromide ion by adding a few drops of 0.1 F $AgNO_3$. Also test for the formation of any cupric ion by adding 1 ml of 6 F NH_3. (Blue $Cu(NH_3)_4^{2+}$ indicates copper; any $AgBr(s)$ present will not interfere with the test.) What would you conclude from

the result as to the ability of other metals such as Fe, Pb, and Zn to be oxidized by bromine?

The less active metal silver can also be oxidized by the halogens. If the halogen is I_2, the silver is oxidized only because of the formation of $AgI(s)$ which keeps the Ag^+ concentration so low that Ag^+ is then a weaker oxidizing agent than is I_2. When Ag^+ is not precipitated, other experimental data beyond our present study show that I_2 and also Fe^{2+} are weaker oxidizing agents than 1 M Ag^+, but stronger than 1 M Cu^{2+}. Br_2, however, is a stronger oxidizing agent than 1 M Ag^+.

Summary of Data

Your experimental observations, together with the preceding data, will enable you to combine your two separate potential series into one general oxidation-reduction potential series of ten couples, which shows the relative tendencies of the various elements and ions to lose electrons. Note that each couple is written so that the change from left to right represents gain of electrons (reduction). On that table, in your report form, designate clearly which are the *reducing agents*, which the *oxidizing agents*, the end of each column that is the strongest (S), and the end that is the weakest (W).

Such a table can be expanded to include many more oxidation-reduction couples and is useful in predicting the course of many reactions. (See Table 13 in Appendix C.) In reading the table, note that *any oxidizing agent (on the left) has the possibility of reacting with any reducing agent (on the right) which is stronger than its own reduction product—that is, which is lower in the series* (see Figure 32-4). The table makes no prediction, however, of the *rate* of a given reaction—some reactions are too slow to be practical. Again, the *concentration* of the ions in a solution has a definite effect on the tendency for reaction, in accordance with the Le Châtelier principle. This factor is considered in Experiment 33.

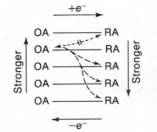

FIGURE 32-4
Tendencies for reactions between oxidizing and reducing agents. The dashed arrows indicate which reducing agent can react with a given oxidizing agent.

---

[2] CCl_4 is toxic and volatile. Use appropriate precautions.

Oxidation-Reduction. Electron Transfer Reactions

PRELIMINARY EXERCISE

As a review, in the following equations for familiar reactions, underline the reducing agent once, and the oxidizing agent twice. At the right, indicate the number of electrons gained or lost *per atom* for each element concerned. If there is no change, write in "no valence change."

Reactions	Element oxidized	Electrons lost per atom	Element reduced	Electrons gained per atom
$2Al(s) + 3Cl_2(g) \rightarrow 2Al^{3+} + 6Cl^-$				
$Ag^+ + Cl^- \rightarrow AgCl(s)$				
$Mg(s) + 2H^+ \rightarrow Mg^{2+} + H_2(g)$				
$Cu^{2+} + H_2S(g) \rightarrow CuS(s) + 2H^+$				
$Ba(s) + 2H_2O \rightarrow Ba^{2+} + 2OH^- + H_2(g)$				
$3Na_2O_2(s) + Cr_2O_3(s) + H_2O \rightarrow 6Na^+ + 2CrO_4^{2-} + 2OH^-$				
$H_2O_2(aq) \rightarrow H_2O + \frac{1}{2}O_2(g)$				
$CO_2(g) + H_2O \rightarrow H_2CO_3(aq)$				
$CO_2(g) + C(s) \rightarrow 2CO(g)$				
$HCl(g) + NH_3(g) \rightarrow NH_4Cl(s)$				
$2ZnS(s) + 3O_2(g) \rightarrow 2ZnO(s) + 2SO_2(g)$				
$4H^+ + 2Cl^- + MnO_2(s) \rightarrow Mn^{2+} + Cl_2(g) + 2H_2O$				
$10H^+ + SO_4^{2-} + 8I^- \rightarrow 4I_2(s) + H_2S(g) + 4H_2O$				

EXPERIMENTAL DATA

I. A Simple Potential Series for the Metals

a) *Copper, Zinc, Lead, and their Ions.* Write net ionic equations for any reactions taking place between the metals and metal ions listed below (indicate any cases of no action).

Copper and zinc ion _____

Zinc and cupric ion _____

Lead and cupric ion _____

Lead and zinc ion _____

Copper and lead ion _____

Zinc and lead ion _____

Which metal is the stronger reducing agent, copper or zinc? _____

lead or copper? _____

lead or zinc? _____

In the space at the right, construct an oxidation-reduction potential series for lead, copper, and zinc, and their ions. Arrange the three metals in a column at the right hand side, with the strongest reducing agent at the bottom and the weakest at the top. To the left of each metal symbol, make a dash; to the left of the dash write the symbol for the oxidized form of the metal—for example, Cu^{2+}——Cu. You now have a brief "redox potential series."

Which is the strongest oxidizing agent: lead ion, cupric ion, or zinc ion?

_____

Explain how this table summarizes your observations.

Suggest a specific reason why many lead pipes, frequently used for draining laboratory sinks, develop leaks.

(b–c) *The Ag^+–Ag and H^+–H_2 Couples*. Describe your observations (either in this experiment or in earlier experiments) which enabled you to place the Ag^+–Ag and the H^+–H_2 couples in the potential series. Also, write net ionic equations for the evidence thus obtained.

Ag+–Ag Couple:

H+–H₂ Couple:

In the space at the right, construct a potential series for lead, copper, zinc, silver, and hydrogen, and their ions, in accordance with your experimental results.

The strongest oxidizing agent in this series is _____

The strongest reducing agent in this series is _____

2. The Oxidizing Power of Halogens

List your observations, and write net ionic equations for the reactions of the free halogens with the halide ions, in accordance with your experimental data.

	Observations	Net ionic equations
Chlorine and bromide ion		
Chlorine and iodide ion		
Bromine and iodide ion		
Iodine and bromide ion		

At the right, construct a potential series for the common halogens and their ions, placing the strongest *reducing* agent at the *bottom* and on the *right* of the series; and the strongest oxidizing agent at the *top* and on the *left*.

The strongest oxidizing agent in this series is _____

The strongest reducing agent in this series is _____

If fluorine were included in the series, the F_2–F^- couple would be located _____

3. The Iron(III) Ion–Iron(II) Ion Couple

Write net ionic equations for the reaction of Fe^{3+} with the halide ions, in accordance with your experimental results.

Iron(III) ion with iodide ion _____

Iron(III) ion with bromide ion _____

The Fe^{3+}–Fe^{2+} couple should therefore be placed between the_____couple

and the_____couple.

4. The Permanganate Ion–Manganese(II) Ion Couple

What was the observed result on mixing MnO_4^- with Cl^- in acid solution?

_____

The equation for this reaction is given in part 3. Rewrite it here.

_____

In this reaction, the oxidizing agent is_____; the reducing agent

is_____.

In the space at the right, construct a potential series, including the three halogens, the Fe^{3+}–Fe^{2+} couple, and the MnO_4^-–Mn^{2+} couple, in accordance with your observations.

5. The Reaction of the Halogens with Metals

Write a net ionic equation showing the results of your experiments on the reaction of bromine water with copper.

_____

What is the experimental evidence for the above reaction products?

_____

6. Summary of Data

In the space to the right, construct an oxidation-reduction potential series for all ten couples studied in this experiment. Write "Oxidizing agents" and "Reducing agents" along the proper sides of the table. Indicate the position of the strongest and weakest oxidizing agents and reducing agents by placing (S) and (W) beside the formulas for these substances.

APPLICATION OF PRINCIPLES

_____

NOTE: In answering these questions, refer to Table 13, Appendix C.

1. In the first parentheses following each formula, write "O" if the substance may be used as an oxidizing agent, then write the formula of the reduced form. In the second parentheses, write "R" if the substance may be used as a reducing agent, then write the oxidized form. (Note that some substances may be used for either, depending on the substance with which they react.)

Al ( )_____ ( )_____   HBr ( )_____ ( )_____ MnO_4^- ( )_____ ( )_____

Sn^{2+} ( )_____ ( )_____ Br_2 ( )_____ ( )_____ H_2SO_3 ( )_____ ( )_____

2. Indicate by "T" or "F" whether the following statements are true or false.

Manganese metal can dissolve in dilute HCl ( )
Tin metal will reduce Sn^{4+} to Sn^{2+} ( )
Mercury metal will dissolve in nitric acid, liberating H_2 gas ( )
Oxygen in moist air can oxidize Fe^{2+} to Fe^{3+} ( )
Copper metal will dissolve in HNO_3 but not in HCl ( )
Gold may be dissolved in 1 F HNO_3 ( )

3. Name a substance for each of the following descriptions:

It can oxidize Cd to Cd^{2+}, but cannot oxidize Pb to Pb^{2+}_____

It can reduce Br_2 to Br^-, but cannot reduce I_2 to I^-_____

It can oxidize Fe to Fe^{2+}, but cannot oxidize Fe^{2+} to Fe^{3+}_____

It can reduce Sn^{4+} to Sn^{2+}, but cannot reduce Sn^{2+} to Sn_____

4. Hypothetical elements, A, B, C, and D form the respective divalent ions A^{2+}, B^{2+}, C^{2+} and D^{2+}. The following equations indicate reactions that can, or cannot, occur. Use these data to arrange the ion–metal couples into a short redox potential series.

$B^{2+} + D \rightarrow D^{2+} + B$

$B^{2+} + A$ (will not react)

$D^{2+} + C \rightarrow C^{2+} + D$

STUDY ASSIGNMENT G

ELECTROLYSIS AND FARADAY'S LAW. A DEMONSTRATION

NOTE: This additional information and demonstration may be assigned by your instructor to provide an understanding of the role of oxidation and reduction in electrolysis, and further study of electrochemical cells.

In Study Assignment F we saw that, in an oxidation-reduction reaction, electrons are transferred from a reducing agent to an oxidizing agent when the two are combined. It is possible to construct a device that allows the oxidation-reduction reaction to be carried out in such a way that the electrons are transferred through a wire rather than by actual contact of the oxidizing agent with the reducing agent. When the electron flow in the external wire is spontaneous, the device is called an electrochemical cell (also called a galvanic or voltaic cell), and the flow of electrons in the external circuit constitutes an electric current that can produce useful electrical work.

The Daniell cell shown in Figure 33-1 of the next experiment is a characteristic example. It consists of a copper electrode immersed in a solution of copper sulfate and a zinc electrode immersed in a solution of zinc sulfate. The cell is represented by the following diagram

$$Zn|ZnSO_4||CuSO_4|Cu$$

in which a vertical bar (|) represents an electrode-to-solution contact, and a double slash (||) represents an ionic contact of two solutions, sometimes through an intermediate solution of chemically unreactive ions called a salt bridge. The chemical reactions that spontaneously take place in the Daniell cell are the following

$$Cu^{2+} + 2e^- \rightarrow Cu(s) \qquad \text{(cathode reaction)}$$

$$Zn(s) \rightarrow Zn^{2+} + 2e^- \qquad \text{(anode reaction)}$$

$$Zn(s) + Cu^{2+} \rightarrow Zn^{2+} + Cu(s) \qquad \text{(net cell reaction)}$$

so that when the cell is operated, zinc metal is consumed at the anode and copper is plated out at the cathode. The cell voltage can be calculated from the Nernst equation (see Experiment 33) if the concentrations of Cu^{2+} and Zn^{2+} in the cell are known.

ELECTROLYSIS

When the desired reaction does not occur spontaneously, an external battery or power supply can be used to drive the flow of electrons (the electric current in the wire) in the desired direction. This process is called electrolysis. For example, if we place two carbon or graphite electrodes (which act as inert conductors of electrons) in a concentrated $CuCl_2$ solution (Fig. G-1A) and connect an ammeter between the terminals, we will note that the ammeter reads zero current. This cell is not an electrochemical cell because there are no chemical reactions present that can spontaneously cause the flow of electric current between the two identical electrodes.

However if we now use a battery to force current through the cell as in Fig. G-1B, we can cause chemical reactions to take place at the two electrodes. This process is called *electrolysis*.[1] As in an electrochemical cell, oxidation will take place at the anode, reduction will take place at the cathode, and electrons will flow away from the anode and into the cathode of this *electrolytic cell* from the external voltage source. In the electrolysis of a concentrated $CuCl_2$ solution the reactions are

cathode: $Cu^{2+} + 2e^- \rightarrow Cu(s)$ (reduction)

anode: $2Cl^- \rightarrow Cl_2(g) + 2e^-$ (oxidation)

Note that the cathode is connected to the negative terminal of the external voltage source, and that the cathode of the electrolytic cell will therefore be negative with respect to the anode. Copper metal is deposited at the cathode, and bubbles of chlorine gas are evolved at the anode. This establishes an electrochemical cell, which is composed of the two half-reactions written above.

When the external battery is disconnected and the electrode terminals connected through the ammeter (Fig. G-1C), the spontaneous flow of current is in the opposite direction: the copper-plated electrode is now the anode and is the negative elec-

---

[1]Current flow in the solution inside the cell consists of the movement of ions of both positive and negative charge. The anions (negative ions) migrate toward the anode; the cations (positive ions) migrate toward the cathode. The external circuit is usually composed of metallic wires connected to the electrodes. In the wires, electrons are the charge carriers, and the flow of current in a wire consists entirely of a flow of negative charge. Since current is conventionally defined as a flow of positive charge, the conventional current flow is opposite to the electron flow. The movement of an electron in one direction in the wire is equivalent to the movement of a hypothetical positive charge in the opposite direction.

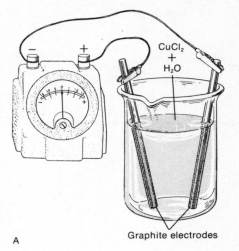

A

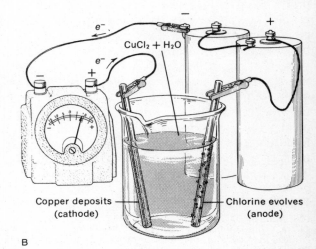

B

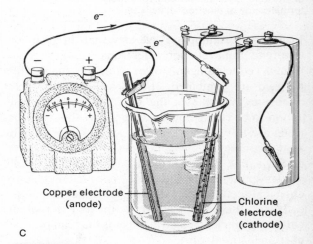

C

FIGURE G-1
Formation of an electrochemical cell by electrolysis. In (A) no current flows through the electrodes. The products of the electrolysis of a concentrated $CuCl_2$ solution in (B) will create an electrochemical cell that operates spontaneously in the reverse direction, as in (C).

trode, whereas the chlorine electrode is now the cathode.

cathode: $Cl_2(g) + 2e^- \rightarrow 2Cl^-$ (reduction)

anode: $Cu(s) \rightarrow Cu^{2+} + 2e^-$ (oxidation)

Always keep in mind that, in both the electrolytic cell and the electrochemical cell, *reduction* occurs at the *cathode* and oxidation occurs at the *anode*.[2] The *sign* of the electrode in the electrolytic cell is different from that in the electrochemical cell, because in the electrolytic cell (Fig. G-1B) we are forcing current to flow in the direction that is opposite that of the spontaneous flow of current in the electrochemical cell (Fig. G-1C).

This creation of a reverse electric cell by the products of the electrolysis results in a counter electromotive force, so that, if any electrolysis at all is to occur, the externally applied voltage must be large enough to overcome this reverse internal voltage of the cell produced. In addition to the voltage required to overcome the internal voltage of the cell (which is governed by the concentrations of the reactants and products of the half-cell reactions) it is necessary to have an extra driving force (called the *overpotential*) to cause electron transfer to take place at an appreciable rate. The overpotential depends upon the chemical composition and physical state of the electrode surface. Finally the internal resistance of the cell and the resistance of the external circuit, which act in accord with Ohm's law, also will tend to limit the electric current (amount of electrolysis) and will require extra voltage above the minimum required to overcome the internal voltage of the cell and the overpotential requirement.

Electrolytic Separation of Substances

In any process of electrolysis, electrons will act as the reducing agent, being conducted to the reducible species by the cathode. The substances most readily reduced are those which are themselves the strongest oxidizing agents. Thus, in a mixture of gold, copper and silver salts, Au(III) ion (the strongest oxidizing agent) is most readily reduced to gold, followed by silver, and then copper, as the voltage is successively increased. As long as the voltage is kept below the reduction potential for the remaining ions, *only* the more easily reduced ion can plate out,[3] or form a coherent layer of the metal on the cathode.

Likewise, at the anode (conducting electrons away from oxidizable species), the strongest reducing agent will be oxidized first. Thus, in a mixture of halide ions (Cl^-, Br^-, and I^-), iodine will be formed first, then bromine, finally chlorine. But each will be formed *only* as the voltage is increased enough to oxidize that corresponding ion.

In the commercial preparation of chlorine, a concentrated chloride ion solution must be used. At low concentrations the chloride ion is a weaker reducing agent than the water present; consequently, oxygen gas from the water, rather than chlorine gas, is liberated. At intermediate concentrations, or if the current density is too high to allow time for the chloride ion to migrate to the electrode, both oxygen and chlorine will be liberated.

Faraday's Law

The amount of chemical change that takes place at an electrode, in either an electrochemical or electrolytic cell, depends on the amount of electricity (or charge) that passes through the electrode. The quantity of electricity is usually expressed in coulombs, and can be determined experimentally by measuring the rate of flow of electricity (current in amperes, i.e., coulombs per second), and multiplying this figure by the time in seconds:

$$\text{coulombs} = \text{amperes} \times \text{seconds}$$

When a mole of silver (107.870 g) is deposited at a cathode, the half-cell equation

$$Ag^+ + e^- \rightarrow Ag(s)$$

indicates that the Avogadro number (6.0225×10^{23}) of individual silver ions must be reduced by the same number of electrons—"a mole of electrons." The charge on a mole of electrons is

$$6.023 \times 10^{23} \text{ electrons} \times 1.6021$$

$$\times 10^{-19} \frac{\text{coulomb}}{\text{electron}} = 96,487 \text{ coulombs}$$

---

[2]This statement may be regarded as a definition of anode and cathode. The anode is the electrode at which oxidation takes place; the cathode is the electrode at which reduction takes place.

[3]If a voltage higher than that necessary to plate out gold is applied, those of lesser activity, such as Ag^+ and Cu^{2+}, will also plate out along with the gold.

which is called a *faraday*. When copper is plated out at the cathode, the half-cell equation

$$Cu^{2+} + 2e^- \rightarrow Cu(s)$$

shows that 1 mole of electrons (1 faraday) will produce only $\frac{1}{2}$ mole of copper (63.54/2 g).

To summarize, when 1 faraday (96,487 coulombs of charge) is passed through a cell, chemical change equivalent to transfer of 1 mole of electrons (oxidation at the anode and reduction at the cathode) occurs at each electrode. This quantity of material is called 1 *electrochemical equivalent*.

OPTIONAL DEMONSTRATION OF ELECTROLYSIS

Your instructor may demonstrate the electrolysis of a potassium iodide solution, as follows:

Directions. Mix 2 drops of phenolphthalein with 15 ml of 0.1 *F* KI, and pour this solution into a 10-cm U-tube supported upright in a beaker. Use the copper wires connected to two or more dry cells in series as electrodes and dip these into the solution in the U-tube arms. After a moment observe the results.

Questions and Discussion. Explain the formation of the yellow color at one of the electrodes. Which one? Has this electrode acted as the anode or the cathode? Write the equation for this half-reaction. The red color at the other electrode indicates the formation of what substance? What gas is being liberated? The reason for this is as follows. At this electrode (cathode or anode?), the most easily reduced substance will be the one to react. This is hydrogen ion from the water, not potassium ion; therefore, instead of liberating potassium metal, the half-reaction here is

$$2e^- + 2H_2O \rightarrow H_2(g) + 2OH^-$$

As shown in the preceding discussion, the products of the electrolysis will act as an electrochemical cell, generating a counter electromotive force which tends to work against the applied voltage. From Table 13, Appendix C, we see that the standard electrode potential for the H_2O–H_2 (10^{-7} *M* H^+) couple, corresponding to the reverse of the half-reaction above, is +0.414 volts. From this, subtract algebraically the standard potential for the I_2–I^- couple, to get the minimum external voltage necessary to apply to cause any oxidation-reduction for this solution. Would a Daniell cell be *satisfactory* to electrolyze a potassium iodide solution? (The student should remember that standard electrode potentials are based on solutions of unit activity—approximately one molar unless otherwise stated. Since these conditions are not maintained during the electrolysis, particularly with respect to the hydroxide ion around the hydrogen electrode, the calculated voltages are only a rough approximation. In addition, overpotential[4] effects may be large if appreciable current is flowing in the cell and may lead to significant deviations from the predicted voltage.)

---

[4]For a fuller discussion of overpotential, see the article by J. O'M. Bockris, "Overpotential—A Lacuna in Scientific Knowledge," *J. Chem. Educ.* **48**, 352 (1971).

REPORT

G

Electrolysis and
Faraday's Law

NAME

SECTION   LOCKER

INSTRUCTOR   DATE

DEMONSTRATION OF ELECTROLYSIS

At the right, sketch a neat diagram for the electrolysis of
a potassium iodide solution. Indicate the polarity, the
cathode, the anode, the directions in which the various
ions move in the solution, and the products formed at
each electrode.

The yellow color at
one electrode is due
to the formation of_____

Is this electrode the
anode or the
cathode?_____

Write the equations for the half-reaction at the anode, for that at the cathode ($H_2(g)$ is liberated, see the directions in
the demonstration) and for the overall electrolysis reaction. From the $E°$ values for each half-reaction (Table 13, Appendix C), calculate the minimum external voltage needed to cause electrolysis of a 1 F KI solution.

$E°$ Values

Anode reaction_____   _____

Cathode reaction_____   _____

Overall reaction_____   _____

Would a Daniell cell (Zn and Cu^{2+}) be able to bring about this oxidation-reduction? Explain.

APPLICATION OF PRINCIPLES AND PROBLEMS

1. How many hours would it take to silverplate a cup to a thickness of approximately 0.01 cm, assuming a surface
   area of the cup of 300 cm², and 31.5 g of silver deposited, if a current of 1.00 ampere is used?

362

2. Magnesium metal is produced commercially by the electrolysis of fused magnesium chloride. Write the half-reaction equation for each electrode process.

Anode————————————————————————

Cathode————————————————————————

What weight of magnesium would be produced by a 100-ampere current operating for one hour?

————————————

What volume of chlorine gas, at standard conditions, would be produced at the same time?

————————————

3. For each of the following, indicate the substance(s) that will be liberated first at each electrode when electrolysis takes place. Assume low current density and inert electrodes, unless otherwise specified. Use Table 13, Appendix C, to determine relative oxidation or reduction tendencies, when applicable. (Assume negligible overpotential effects.)

	Anode	Cathode		Anode	Cathode
Fused NaCl	————	————	Dilute NaCl solution	————	————
Dilute $CuCl_2$ solution	————	————	Concentrated $SnCl_2$ and $FeBr_2$ solution	————	————
Dilute $CuSO_4$ solution (Cu electrodes)	————	————	Fused Al_2O_3 and Na_3AlF_6 (C electrodes — Hall process)	————	————

4. In the space at the right, make a schematic sketch of the voltaic cell: $Cd|Cd(NO_3)_2$ (1 F)$||AgNO_3$ (1 F)$|$ Ag. (A single vertical line means a phase boundary and a double vertical line indicates a solution boundary.) Indicate the following: (a) the polarity, (b) the name of each electrode, (c) the direction of electron flow in the wire outside the cell, (d) the directions in which the ions move in the solutions.

Write equations, with $E°$ values, for the spontaneous:

Oxidation
reaction ——————————— ————

Reduction
reaction ——————————— ————

Total cell
reaction ——————————— ————

5. What will happen to the voltage of the cell in Problem 4 if the following occurs:

A quantity of NaCl is stirred into the $Ag|AgNO_3$ half-cell? _____

A quantity of Na_2S is stirred into the $Cd|Cd(NO_3)_2$ half-cell to precipitate $CdS(s)$? _____

6. In the lead storage battery, the overall cell reaction on discharge is

$$Pb(s) + PbO_2(s) + 4H^+ + 2SO_4^{2-} \rightarrow 2PbSO_4(s) + 2H_2O$$

Write the half-cell reaction that occurs at each electrode, with the *proper E°* value for each (Table 13, Appendix C).

Pb electrode _____ _____

PbO$_2$ electrode _____ _____

Total cell voltage _____

If a current of 25.0 amperes is drawn from a car battery for 10.0 seconds in starting, calculate the following:

the number of coulombs used _____

the number of faradays used _____

the number of grams of $PbSO_4$ (303 g/mole) deposited on the Pb electrode _____

the number of grams of $PbSO_4$ (303 g/mole) deposited on the PbO_2 electrode _____

How many coulombs are required to recharge the battery to its original state? _____

What happens to the concentration of sulfuric acid during the following processes:

During discharge _____ During charge _____

OXIDATION-REDUCTION REACTIONS AND ELECTROCHEMICAL CELLS

PRE-STUDY

The Nature of Electrochemical Cells

We have observed that whenever an oxidation-reduction reaction occurs, there is a transfer of electrons from the substance oxidized to the substance reduced. Thus, when zinc is oxidized by cupric ion, the zinc atom loses two electrons, and cupric ion gains two electrons. We may express this as two separate half-reactions:

$$Zn(s) \rightarrow Zn^{2+} + 2e^- \quad \text{(oxidation)}$$
$$Cu^{2+} + 2e^- \rightarrow Cu(s) \quad \text{(reduction)}$$

The sum of these two half-reactions gives the total equation for the process:

$$Zn(s) + Cu^{2+} \rightarrow Zn^{2+} + Cu(s)$$

This total equation does not contain any free electrons, because all the electrons lost by the zinc are gained by cupric ion.

An *electrochemical cell,* (also called a voltaic or galvanic cell), is simply a device used for carrying out an oxidation-reduction reaction in such a way that the electrons are transferred through a wire rather than by actual contact of the oxidizing agent with the reducing agent (see Figure 33-1). The chemical reactions taking place at the two electrodes of the cell are the half-reactions described above. The oxidation half-reaction takes place at the anode, liberating electrons to the external circuit. The reduction half-reaction takes place at the cathode, consuming electrons from the external circuit. Electron flow in the external circuit is always from − to +. The *conventional* current flow is in the opposite direction, because the conventional current is assumed to be a flow of positive charge.

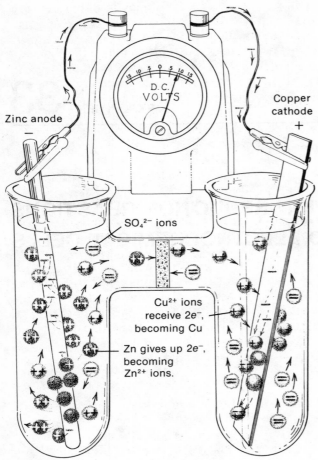

FIGURE 33-1
A simple electrochemical cell, which transforms the energy liberated by a chemical reaction into electrical energy. The electrical current in the solution consists of sulfate ions moving toward the left, and of zinc and copper(II) ions moving toward the right.

Cell Voltage

The *volt* is the unit of electrical potential, or driving force. The product of the voltage × the charge of the electron, e^-, is a measure of the work done when this unit electrical charge is transferred from one substance to another. *The voltage of a cell—sometimes called its electromotive force or potential—is thus a quantitative value expressing the tendency of the chemical reaction occurring in the cell to take place.* The magnitude of this voltage depends on the relative strengths of the oxidizing and reducing agents used. If the oxidizing agent has an affinity for electrons that is stronger than the tendency of the reducing agent to hold electrons, the electrical potential, or voltage, is correspondingly large.

Standard Electrode Potentials

The voltage of an electrochemical cell is the sum of separate voltages due to the oxidation half-reaction and the reduction half-reaction. Since the voltage due to a single couple cannot be directly measured, the voltage of all oxidation-reduction couples is measured with respect to a standard reference couple, the H^+-H_2 couple. This couple is arbitrarily assigned a potential of zero, so that the total cell voltage is ascribed to the other couple. For example, in a cell composed of the $Zn^{2+}-Zn$ half-reaction and the H^+-H_2 half-reaction, where all species are at unit activity,[1] the potential of the cell is measured to be −0.763 volts, with the zinc electrode being more negative than the hydrogen electrode. This value, −0.763 volts, is called the *standard electrode potential*[2] for the $Zn^{2+}-Zn$ couple (see Figure 33-2). All standard electrode potentials are the values of voltage obtained when all substances in solution are present at unit activity (approximately 1 molar), all gases are at 1 atmosphere pressure, and the temperature is at a fixed, convenient value, usually 25°C (see Table 13, Appendix C).

---

[1]The activity of an ion in solution is usually less than the molar concentration because of ionic interactions. It may be crudely thought of as the "effective concentration." A textbook of physical chemistry can be consulted for further details.

[2]We have adopted in this edition of the laboratory manual the reduction potential convention recommended by the IUPAC. In Table 13 of Appendix C, all half-reactions are written as reductions. In this convention the sign of the standard reduction potential will be the same as the sign of the electrode measured against the standard hydrogen electrode. For this reason it is preferred to the oxidation-potential convention, which has been widely used in the United States in the past.

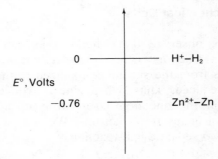

FIGURE 33-2
On the potential scale, the H^+-H_2 half-reaction is arbitrarily assigned the value zero. The zinc electrode has a voltage of −0.76 volts measured against the hydrogen electrode (See Figure 33-3). This value is assigned as the standard electrode potential of the $Zn^{2+}-Zn$ couple. This procedure is analogous to measuring the elevations from sea level (rather than from the center of the earth), with sea level being assigned zero in the scale of elevation.

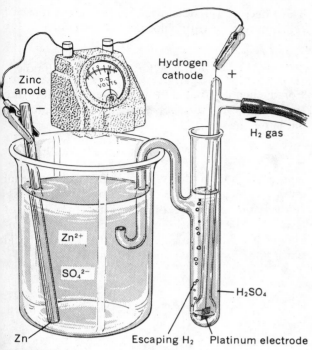

FIGURE 33-3
Hydrogen gas, adsorbed on the platinum electrode and in contact with 1 F H^+, acts as the reference electrode. When this electrode is coupled with a Zn electrode in contact with 1 F Zn^{2+} to form the cell $(Pt)H_2|H^+||Zn^{2+}|Zn$, the meter reads approximately 0.76 volts, with the zinc electrode being more negative than the hydrogen electrode. The cell voltage will not be the same as the theoretical $E°$ of the cell because of activity and junction potential effects.

To obtain the voltage for any given cell, calculate the *algebraic difference* of the potentials of the two oxidation-reduction couples concerned. For example, a cell may be made utilizing the strong reducing agent zinc, reacting with the strong oxidizing agent chlorine, according to the reaction

$$Zn(s) + Cl_2(g) \rightarrow Zn^{++} + 2Cl^-$$

A zinc rod is placed in 1 F zinc chloride, and a platinum electrode is placed in 1 F potassium chloride saturated with chlorine gas. A salt bridge connects the solutions. The platinum electrode is inactive, and it merely conducts electrons from the solution as chlorine is reduced to chloride ion. The corresponding half-reactions, and the voltages of the two couples, Zn^{2+}–Zn and Cl_2–Cl^-, as given in Table 13, Appendix C, are

Note that we had to subtract half-reaction (1) from half-reaction (2) in order to obtain the net chemical reaction. When we do so, we must also subtract $E_1°$ from $E_2°$ in order to obtain the standard potential of the cell. (This step is equivalent to reversing the direction of half-reaction (1) and the sign of $E_1°$ and adding.) A positive value for the standard cell potential indicates that the reaction can occur spontaneously, but does not indicate how fast it will occur.

It should also be noted that we had to multiply half-reaction (2) by a factor of 2 so that the electrons would cancel in the overall reaction. However we did not multiply $E_2°$ by a factor of 2. The reason for this is that the voltage of a half-reaction is an *intensive* quantity (like temperature or pressure); it does not depend on the absolute amounts of reactants present, but only on their activity (or concentration). This is in contrast to the work done when current is passed through the cell. The work is an *extensive* quantity proportional to the product of the moles of electrons transferred per mole of reaction times the cell voltage. (Work$_{electrical}$ = nFE, where F is the faraday.)

Electrode Potentials and the Principle of Le Châtelier

The Effect of Concentration. We have observed that the tendency for an oxidation-reduction reaction to take place is measured by the voltage created when the reaction takes place in an electrochemical cell. Thus, the electromotive force of 2.122 volts created by the cell in the preceding paragraph, for the reaction

$$Zn(s) + Cl_2(g) \rightleftharpoons Zn^{2+} + 2Cl^-$$

is indicative of the behavior of quite a strong reducing agent, Zn, with a strong oxidizing agent, Cl_2. There is also some tendency for the reverse reaction to occur. However, zinc ion (Zn^{2+}) is quite a weak oxidizing agent, and chloride ion (Cl^-) is a weak reducing agent; hence this reverse tendency, as indicated by the short reverse arrow in the equation, is not very great. A reaction in which the forward and the reverse processes have

(1)	$Zn^{2+} + 2e^- \rightleftharpoons Zn(s)$	$E_1° = -0.763$ v
(2)	$Cl_2(g) + 2e^- \rightleftharpoons 2Cl^-$	$E_2° = +1.359$ v
(3) = (2) − (1)	$Zn(s) + Cl_2(g) \rightleftharpoons Zn^{2+} + 2Cl^-$	$E_3° = E_2° - E_1° = 2.122$ v

attained a balance, and are taking place at equal but opposing rates, is said to be in *chemical equilibrium*.[3]

According to *Le Châtelier's principle,* and in accord with observed fact, an increase in the concentration (or pressure) of chlorine gas, $Cl_2(g)$, will increase the voltage of the cell. Conversely, an increase in the concentration of zinc ion, or of chloride ion, will favor the reverse process and therefore decrease the voltage.

The quantitative relationship between the voltage of a cell and the concentrations of the reactants and products, is expressed by the thermodynamic Nernst equation,

$$E = E° - \frac{0.059}{n} \log Q$$

where E is the measured voltage, $E°$ is the standard electrode potential as calculated from the half-reaction potentials as given in Table 13, Appendix C, n is the number of electrons gained by the oxidizing agent in the reaction equation, and Q is the product of the concentrations (molarity of solutes, gas pressure in atmospheres) of the reaction products, divided by the product of the concentrations of the reacting substances.[4]

Example. In the reaction above we found that

$$Zn(s) + Cl_2(g)(1 \text{ atm}) \rightarrow$$
$$Zn^{2+}(1 \text{ } M) + 2Cl^-(1 \text{ } M)$$

and $E° = 2.122$ volts. To calculate the corresponding voltage if the $Cl_2(g)$ were at 4.0 atm, the Zn^{2+} were 0.010 M, and the Cl^- remains at 1 M,

$$E = E° - \frac{0.059}{2} \log \frac{0.010}{4.0} = 2.122 + 0.077$$
$$= 2.199 \text{ volts}$$

Both the increase in the $Cl_2(g)$ pressure and the decrease in the Zn^{2+} concentration have a modest effect in increasing the voltage of the cell.

The Effect of Temperature.   Again, in the forward reaction of zinc with chlorine, heat is evolved:

$$Zn(s) + Cl_2(g) \rightarrow$$
$$Zn^{2+}(aq) + 2Cl^-(aq) + 115 \text{ kcal/mole}$$

For a chemical reaction at constant pressure the thermal energy change, $q_{chemical}$, is equal to the enthalpy change

$$q_{chemical} = \Delta H = \Delta U + P\Delta V = -115 \text{ kcal/mole}$$

where ΔH is the enthalpy change, ΔU is the internal energy change, P is the constant pressure, and ΔV is the volume change of the system per mole of reaction. The thermal energy change in an electrochemical cell, $q_{electrochemical}$, includes the electrical work done by the cell and is given by

$$q_{electrochemical} = \Delta U + \text{Work}_{total}$$
$$= \Delta U + P\Delta V + \text{Work}_{electrical}$$
$$= \Delta U + P\Delta V + nFE$$
$$= \Delta H + nFE$$

Therefore we see that

$$q_{chemical} = q_{electrochemical} - nFE.$$

Generally, if $q_{chemical}$ is exothermic, $q_{electrochemical}$ will be exothermic also. However, it is possible for the direct chemical reaction to be exothermic and the electrochemical reaction to be endothermic, because the thermal energy evolved in the electrochemical process is less than that evolved by the direct "chemical" process by the amount of electrical work done.[5] This is true, for example, for the reaction

$$Pb + Cu^{2+} \rightarrow Pb^{2+} + Cu$$

where $q_{chemical} = -15.0$ kcal/mole and $q_{electrochemical} = +6.4$ kcal/mole. In order to make a prediction on the basis of the Le Châtelier principle of the effect of changing the cell temperature, it is necessary to know $q_{electrochemical}$, the thermal energy change in the electrochemical cell. If $q_{electrochemical}$ is negative (thermal energy liberated in the cell), raising the temperature of the cell would tend to decrease the cell voltage, since this higher temperature would repress the evolution of thermal energy in the cell, and the forward tendency of the reaction would decrease.

---

[3]The experimental study of chemical equilibrium is considered in this manual beginning with Study Guide E and Experiment 19.

[4]In the expression for Q, the concentrations of each substance must be raised to a power equal to its coefficient in the reaction equation.

[5]We are indebted to Professor Grover C. Willis, California State University, Chico, for pointing out an error in the previous edition of this manual on the effect of temperature on an electrochemical cell.

EXPERIMENTAL PROCEDURE

Special Supplies: Voltmeter (1000 ohms/volt) for class use; porous cups of unglazed porcelain (1 × 3 inches), 10-cm U-tube; electrode strips 1 × 10 cm of Cu, Zn and Fe (or large iron nail), Pb and Ag strips for optional part; carbon (graphite) rods $\frac{5}{16}$ × 6 inches; very narrow strips or wire (1 mm × 8 cm) of Zn, Sn, and Cu; several iron nails (for part 4); agar-agar.

Chemicals: Br_2 water (saturated solution), 0.1 F $CuSO_4$, 0.1 F $FeSO_4$ (freshly prepared), 0.1 F $ZnSO_4$, 2 F Na_2S, $KBr(s)$ 0.1 F $AgNO_3$, 0.1 F $Pb(NO_3)_2$ (optional). If U-tube cells are used, concentrations of the preceding solutions must be 0.5 F to decrease the resistance of the circuit; also include 0.5 F NH_4NO_3.

NOTE: For the construction of the cells in this experiment, the use of a porous cup, nested in a 150-ml beaker, as illustrated in Figure 33-4 and as described in part 1, is preferred.

An alternate arrangement consists of two 150-ml beakers, each of which contains one of the metal electrode strips and 35 ml of the corresponding 0.5 F salt solution of that metal ion. The solutions, at equal levels in the beakers, are then connected by an inverted U-tube salt bridge. First fill a 10-cm U-tube with 0.5 F H_2SO_4 or a salt such as 0.5 F NH_4NO_3 that will not form a precipitate with the solutions being used. Place a cotton plug, *well* saturated with the solution, in each end of the U-tube, and invert this into the solutions, after connecting the electrodes to the 1000 ohm/volt meter.

For an even simpler arrangement, place one half-cell in a 150-ml beaker; put the other half-cell in a 1-cm o.d. × 10-cm long tube, whose end is closed with cellophane tubing (dialysis tubing) held in place with a ring of rubber or Tygon tubing. The dialysis tubing junction has low electrical resistance, but diffusion through the cellophane is slow.

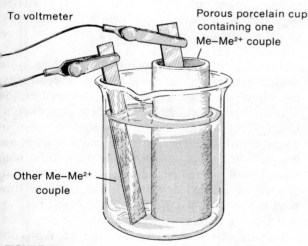

To voltmeter

Porous porcelain cup containing one Me–Me²⁺ couple

Other Me–Me²⁺ couple

FIGURE 33-4
Apparatus for a simple electrochemical cell, consisting of a 1-inch by 3-inch porous porcelain cup containing one Me–Me²⁺ couple, and a 150-ml beaker containing the other Me–Me²⁺ couple.

1. A Simple Daniell Cell: $Zn|ZnSO_4||CuSO_4|Cu$.[6]

Soak the porous cup (1 × 3 inches) in a dilute ion solution—your tap water supply probably is adequate—then place the cup in a 150-ml beaker. Put 0.1 F $CuSO_4$ in the cup and 0.1 F $ZnSO_4$ in the beaker outside the cup, to approximately equal heights of 3 cm, and then lift the cup out of the beaker. Place a copper electrode strip in the $CuSO_4$ solution and a zinc electrode strip in the $ZnSO_4$ solution. (Sandpaper the electrodes to clean them, if necessary.) This establishes the two separate equilibria:

$$Zn^{2+} + 2e^- \rightarrow Zn$$

$$Cu^{2+} + 2e^- \rightarrow Cu$$

Which of these two metals will give up its electrons more readily? To determine the answer, connect the metal electrodes by means of copper-wire connectors to the 1000-ohms/volt voltmeter,[7] *first* with the beaker and cup separated, *second* with the cup placed within the beaker. Read the meter carefully to the nearest 0.01 volt, noting which electrode carries the negative charge. (As further proof of this polarity, recall the comparative behavior of Zn metal in Cu^{2+} solution, and of Cu metal in Zn^{2+} solution, Experiment 32, part 1.) Explain fully all aspects of the operation of the Daniell cell, and complete a diagram of this cell in your report sheet. Be sure you understand the following. (a) What constitutes the electric current in the wire? (b) What constitutes the electric current in the solution? (c) Why must there be actual contact of the two solutions? (d) What are the chemical reactions at each electrode?

2. Standard Electrode Potentials.

In your report, rewrite the oxidation-reduction potential series for the couples studied in Experiment 32, and also, by referring to Table 13 in Appendix C, include for later comparison the Fe^{2+}–Fe and the Sn^{2+}–Sn couples. In addition, by referring to Table 13,

---

[6]In this conventional designation of a cell, a single vertical bar (|) indicates an electrode-to-solution contact, and a double bar (||) indicates ionic contact of two solutions, sometimes through a central inactive solution or salt bridge, to avoid the formation of precipitates.

[7]An ordinary voltmeter, which is rated at about 250 ohms/volt, will give a low reading because of the high internal resistance of cells constructed as suggested here. Voltmeters with a resistance of about 1000 ohms/volt, such as Weston Model 301, are quite satisfactory. These (one for each 10–15 students) may be located conveniently on a wall shelf for class use.

enter the "standard electrode potentials" of all the couples you have listed. Now calculate the accepted standard potential of your Daniell cell by algebraically subtracting (why do you subtract?) the voltage for the Zn^{2+}–Zn couple from that for the Cu^{2+}–Cu couple.[8]

For the experimental part of this section, set up cells similar to the Daniell cell above. Clean the electrodes if necessary with sandpaper. A large iron nail, cleaned by immersion in 6 F HCl in a test tube, may be used for the iron electrode.

(a) $Zn|ZnSO_4||FeSO_4|Fe$. Remove the cup containing the $CuSO_4|Cu$ half-cell, decant the solution into another beaker for later use, rinse the cup thoroughly, and return it to the $Zn|ZnSO_4$ half-cell beaker. Add 0.1 F $FeSO_4$ (freshly prepared) in the cup to the height of the $ZnSO_4$ solution, insert the iron electrode and connect it, and measure the potential of the cell. (Do not take the reading until the meter has attained a steady value.)

(b) $Fe|FeSO_4||CuSO_4|Cu$. Determine, as you did in (a), the potential of this cell, using the original $CuSO_4|Cu$ half-cell. (Again save the $CuSO_4$ solution.)

(c) *Electric Cells with Nonmetals.* Nonmetallic elements may be used as oxidizing agents in cell reactions by employing an inert electrode. Place 15–25 ml of saturated Br_2 water in a 150-ml beaker, add a few crystals of KBr and dissolve these to provide Br^-, and insert a carbon (graphite) rod (about $\frac{5}{16}$ by 6 inches) to serve as an inert electrode. Use this, together with one or more (as directed by your instructor) of the porous-cup half-cells previously constructed—$Zn|ZnSO_4$, $Fe|FeSO_4$, or $Cu|CuSO_4$—and measure the potential of these cells.

(d) *Additional Cell Measurements* (optional). Your instructor may demonstrate additional cell measurements, using such half-cells as $Pb|Pb(NO_3)_2$ or $Ag|AgNO_3$ with any of the half-cells previously used. If you attempt to make such measurements yourself, be sure to avoid the formation of insoluble precipitates such as $PbSO_4$ or $AgBr$ at the ionic solution contacts. Instead of the porous-cup and beaker arrangement, a salt bridge containing an electrolyte that can form only soluble salts may be advisable.

---

[8]In this cell we have used 0.1 F solutions (0.5 F in the U-tube cell) rather than the 1 F solutions on which the standard electrode potentials are based. This does not affect the equilibrium potential in this reaction, since the concentrations of both reactant and product are affected alike. See the Pre-study section, the discussion on electrode potentials and the principle of Le Châtelier.

3. The Effect of Concentration.

In order to observe how changes in concentration affect the voltage, again set up the $Zn|ZnSO_4||CuSO_4|Cu$ cell, preferably with the $CuSO_4|Cu$ half-cell in the outer beaker, to facilitate later cleaning of the porous cup. Connect the electrodes to the voltmeter to check the voltage. Then add, while stirring, a slight excess of 6 F NH_3 solution to the $CuSO_4$ solution until the deep blue $Cu(NH_3)_4^{2+}$ complex ion is obtained, and the concentration of Cu^{2+} is thus reduced. Read the voltage. Further reduce the concentration of Cu^{2+} by adding, while stirring, an excess (several milliliters) of 2 F Na_2S. Again read the voltage. Interpret the changes you observe, in terms of the Le Châtelier principle.

4. The Corrosion of Iron.

This familiar phenomenon is closely related to oxidation-reduction because it occurs in electrochemical cells and in electrolysis. Two alternate procedures are suggested for your study, which may be set up 24 hours in advance as demonstrations by selected students or by your instructor.

First Method. Select three bright nails. Around one of them, wind a small, narrow strip of tin; around another, a copper wire; and around the third, a small, narrow strip of zinc. Space the windings 1–2 mm apart, and twist them tightly. Place them in separate test tubes and partially cover them with water. Let them stand overnight. In the report form, explain your observations, commenting on the electrochemical cell action and the relative reduction potentials of the metals used.

Compare the protective action with the corrosion that has taken place in a "tin" can (iron coated with tin), such as that used in food preservation and the corrosion in a galvanized water pipe (iron coated with zinc).

Alternate Method. (This method provides more explicit evidence about the nature of the processes and about the effect of stress in the metal on the oxidation potential.) Prepare a 1% agar-agar solution by first soaking 2 g of powdered agar-agar in 25–50 ml of distilled water, preferably for several hours, then adding this to hot distilled water, with stirring and continued heating, to prepare about 200 ml of dispersed agar solution. Add to this with mixing, about 10 drops of 0.1 F $K_3Fe(CN)$ and 5 drops of phenolphthalein indicator.

Prepare three bright nails as follows. Bend one to a sharp angle with pliers, wrap the second with copper wire, and the third with a narrow strip of zinc, as in the first method. Place these nails, and a fourth straight clean nail, on the bottom of a large petri dish or in two beakers, so that each is well separated from the others. Now pour the warm, still fluid agar solution over these nails to a depth of about 5 mm, and let them stand overnight.

Make simple sketches in your report to indicate the changes you observe in the nails and the surrounding solution, and explain these in terms of electric cell action and the potentials of the metals concerned. The $K_3Fe(CN)_6$ reagent will detect any formation of Fe^{2+} by the blue color of $Fe_3[Fe(CN)_6]_2$, or of Zn^{2+} by the white $Zn_3[Fe(CN)_6]_2$ precipitate. The phenolphthalein will detect any reduction of atmospheric oxygen to form OH^- in the solution according to the half-reaction

$$O_2(g) + 2H_2O + 4e^- \rightarrow 4OH^-$$

Bibliography

Lawrence, R. M., and Bowman, W. H. "Electrochemical Cells for Space Power," *J. Chem. Educ.* **48**, 359 (1971).

Robbins, O., Jr. "The Proper Definition of Standard Electromotive Force," *J. Chem. Educ.* **48**, 737 (1971).

1. A Simple Daniell Cell

Complete the diagram at the right by writing in the formulas for the composition of the electrodes and of the ions in the solutions. Place these formulas near the appropriate arrows leading into the solutions. Also indicate which electrode is the anode and which electrode is the cathode at the top.

What constitutes an electric current in a wire?

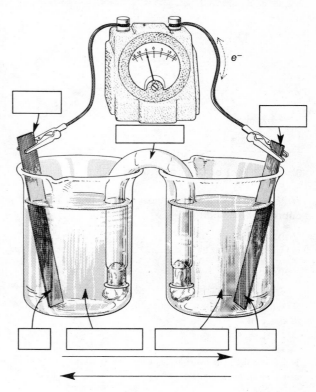

_____

Indicate the direction of these particles by drawing an arrow over each wire.

What constitutes an electric current in a solution?

_____

Indicate the direction of movement of these particles in the solution by placing their formulas on the proper arrow below the sketch.

Why must there be actual contact of the two solutions?

Write the half-reaction taking place at the cathode:

_____

Write the half-reaction taking place at the anode:

_____

Write the total cell reaction:

_____

Give the experimental voltage for this cell. _____

_____

2. Standard Electrode Potentials

In the left-hand column on the right, rewrite the potential series for the couples studied in the experimental procedure of Experiment 32. In the right-hand column, enter the standard electrode potential for each, as obtained from Table 13 in Appendix C. Calculate the potential for your Daniell cell, as directed in the procedure for this experiment.[1]

Potential Series

Couple	Volts

---

[1]The values in Table 13 are obtained by measurements in a reversible manner, with very small currents, using a potentiometer, so the internal resistance of the cells does not tend to decrease the voltage reading, as occurs when a voltmeter is used.

In the spaces below, write the equation for each cell reaction measured in part 2 (a–d). Indicate which element is the positive electrode and which the negative electrode, and give the experimental and calculated voltages in each case.

Cell reaction	Positive electrode	Negative electrode	Experimental voltage	Calculated voltage

3. The Effect of Concentration

Note your observed voltages, and write net ionic equations for the reactions, first of NH_3, and then of Na_2S, with cupric sulfate solution.

Voltage with
NH_3 added _____   _____

Voltage with
Na_2S added _____   _____

Explain the reasons for the effects on the cell voltage you observed when these reagents were added to cupric sulfat in the cell.

How would you adjust the concentrations of the Cu^{2+} and of the Zn^{2+} in the Daniell cell in order to obtain the maxi mum voltage possible for this cell?

4. The Corrosion of Iron

Summarize, by comment, by suitable sketches, and by equations, your observations on the rusting of iron exposed moisture and to air when in contact with each of the following metals—tin, copper, and zinc.

Utilizing these observations, and the potential data, explain why a "tin" can (iron plated with tin) rusts more eas than a galvanized pipe (iron coated with zinc), when both are exposed to weathering.

Why are cans of tin plate, rather than of galvanized iron, used for preserving fruits and other food products?

REDOX TITRATIONS.
THE OXIDIZING CAPACITY
OF A HOUSEHOLD CLEANSER

PRE-STUDY

Oxidation-reduction reactions, like acid-base reactions, are widely used as the basis for the analytical determination of substances by titration.[1] In a volumetric titration, a known volume of the *titrant,* usually contained in a buret, is added to the substance being determined, called the *titrand.* The conditions needed for redox titrations are the same as for any titration: the reaction between the titrant and the titrand must be rapid, stoichiometric, and quantitative: that is, the equilibrium must greatly favor the products. In addition to these basic requirements the titrant solution must be stable, and there must be some means of determining its concentration accurately. Finally, some means of detecting the endpoint of the titration reaction must be available.

---

[1]Techniques of volumetric analysis are described in the Section "Basic Laboratory Operations," in the Introduction.

Oxidizing and Reducing Agents

Oxidizing agents available in pure form, such as $K_2Cr_2O_7$, may be weighed out directly to form a titrant solution of known concentration. Potassium permanganate and cerium(IV) salts are often used as oxidizing agents, but they are not pure enough to be weighed out directly, and ordinarily one must titrate them against reducing substances of known high purity (such as As_2O_3) to determine their concentrations accurately. This process is called *standardization.*

The redox potentials of several commonly used oxidizing agents are shown below

	$E°$, volts
$Ce^{4+} + e^- \rightleftharpoons Ce^{3+}$	$+1.61$
$MnO_4^- + 8H^+ + 5e^- \rightleftharpoons Mn^{2+} + 2H_2O$	$+1.51$
$Cr_2O_7^{2-} + 14H^+ + 6e^- \rightleftharpoons 2Cr^{3+} + 7H_2O$	$+1.31$
$I_2(aq) + 2e^- \rightleftharpoons 2I^-$	$+0.54$

Cerium(IV) is one of the strongest oxidizing agents available for use as a titrant, and its solutions are very stable. However it is more expensive than either potassium permanganate, $KMnO_4$, or potassium dichromate, $K_2Cr_2O_7$. Iodine is a rather weak oxidizing agent and is most often used in indirect procedures in which iodide ion is first oxidized to iodine, and then the iodine is titrated with sodium thiosulfate solution.

Most strong reducing agents are easily oxidized by the oxygen in the air. Therefore they are not often employed as titrants. However they are commonly employed in analysis to reduce a substance to a lower oxidation state before it is titrated with a standard solution of an oxidizing agent. Reducing agents of moderate strength, whose solutions are stable, also exist. One of these is sodium thiosulfate, $Na_2S_2O_3$, which is most often used to titrate I_2.

Endpoint Detection in Redox Titrations

The most general methods of detecting the endpoints in a redox titration are like those used in acid-base titrations (see Experiment 23). When both the oxidized and reduced species are soluble in solution (as $MnO_4^- $-$Mn^{2+}$, for example) an inert platinum or gold electrode will respond to the redox potential of the system provided that the electron transfer processes at the electrode surface are reasonably fast so that the equilibrium potential is established. Such an inert electrode may be used in conjunction with a suitable reference electrode in much the same way that the glass electrode is used in acid-base titrations. At the endpoint of the titration an abrupt change, of about several hundred millivolts, usually takes place in the redox potential of the system, and a voltmeter can be used to detect the change. Such a titration is called a *potentiometric redox titration*.

Redox indicators are available that function as acid-base indicators except that they respond to changes in the redox potential of the system rather than to changes in the hydrogen ion concentration. For example, the redox indicator *ferroin* is red in its reduced form, at potentials of less than +1.12 volts. In titrations in which strong oxidizing agents, such as cerium(IV), are used, the potential jumps abruptly at the endpoint to potentials of greater than +1.12 volts, and the indicator changes to its oxidized form, which is blue in color.

In some titrations, a titrant or the substance titrated, acts as its own indicator. For example, the permanganate ion, MnO_4^-, is a very deep reddish-purple color, whereas the reduced product of its reaction with reducing agents in acid solution is the colorless manganese(II) ion, Mn^{2+}. Solutions titrated with standard MnO_4^- are usually colorless until the endpoint, then turning pink when no reducing species are left to react with the added MnO_4^-. Conversely, in the titration of iodine solutions with standard thiosulfate, $S_2O_3^{2-}$, the solutions are brown or yellow until the endpoint is reached, then becoming colorless when all of the iodine is titrated. (Also, in these titrations, starch—which forms an intense blue with I_2 and I^-—is added just before the endpoint, making the endpoint sharper and easier to detect).

Redox Titration of the Oxidizing Agents in a Household Cleanser

In this experiment we will make use of two oxidation reduction reactions to determine the oxidizing capacity of a household cleanser. To a solution containing the sample of cleanser will be added an unmeasured excess of KI, in acid solution. Oxidizing agents contained in the cleanser, such as sodium hypochlorite, oxidize the iodide ion to iodine according to the following reaction

	$E°$, volts
$HClO(aq) + H^+ + 2e^- \rightleftharpoons Cl^- + H_2O$	+1.49
$2I^- \rightleftharpoons I_2 + 2e^-$	−0.54
$HClO(aq) + H^+ + 2I^- \rightleftharpoons$ $I_2 + Cl^- + H_2O$	+0.95

The iodine produced in the solution will then be determined by titration with a standardized thiosulfate solution, which reduces iodine stoichiometrically according to the following reaction

	$E°$, volts
$2S_2O_3^{2-} \rightleftharpoons S_4O_6^{2-} + 2e^-$	−0.09
$I_2 + 2e^- \rightleftharpoons 2I^-$	+0.54
$2S_2O_3^{2-} + I_2 \rightleftharpoons 2I^- + S_4O_6^{2-}$	+0.45

Even though the method is indirect, only one standard solution is required.

Establishing the Endpoint in the Iodine-Thiosulfate Titration

Starch forms an intense blue compound with traces of iodine in the presence of I^-.[2] Water is also involved in the complex which is essentially a hydrated colloidal particle with iodide and iodine adsorbed upon it. The blue color is reversible, and may be used as a sensitive indicator for traces of iodine in solution.

$$\text{Starch} + I_2 + I^- \rightleftharpoons \underset{\text{blue}}{(\text{Starch}, I_2, I^- \text{ complex})}$$

Iodine in concentrations as low as $10^{-6}\ M$ may be detected, provided that the concentration of iodide ion is $10^{-3}\ M$ or greater. Most iodine-containing solutions are titrated with standard thiosulfate to the disappearance of the blue color at the endpoint. In such titrations, the starch indicator should not be added until just before the endpoint, when the iodine concentration is low. If it is added too early in the titration, the formation of the blue is not as easily reversible. In practice, therefore, withhold the starch indicator until the last tinge of yellow due to excess iodine has almost disappeared; then add the starch, and quickly complete the titration.

Iodide ion is oxidizable by air according to the reaction

$$4I^- + O_2 + 4H^+ \rightleftharpoons 2I_2 + 2H_2O$$

Although this reaction is slow in neutral solution, the rate increases with acid concentration. If an iodimetric titration must be carried out in acid solution, exclude the air if possible, and perform the titration quickly.

EXPERIMENTAL PROCEDURE

Special Supplies: several brands of household cleansers (such as Comet, Bab-O, and Ajax); two 50-ml burets (or one buret and one 25-ml pipet).

Chemicals: $KI(s)$, 1 F HCl, 0.0100 F KIO_3 (potassium iodate); 0.05 F $Na_2S_2O_3$ (sodium thiosulfate); starch indicator (prepare 1 liter by making a paste of 1 g of soluble starch and 10 mg of HgI_2 as a preservative in about 30 ml of water; pour into 1 liter of boiling water, and heat until the solution is clear; cool and store in a stoppered bottle).

---

[2]In Experiment 39, Figure 39-6 shows the structure of the starch–I_2–I^- complex.

1. Standardization of the Sodium Thiosulfate Solution. If the approximately 0.05 F $Na_2S_2O_3$ solution has not been standardized, it may be titrated against a solution made up by dissolving pure dry potassium iodate, KIO_3, to a concentration of 0.0100 F. KIO_3 reacts with excess KI in acid solution according to the following reaction

$$IO_3^- + 5I^- + 6H^+ \rightleftharpoons 3I_2 + 3H_2O$$

To perform the standardization, dispense 25 ml of 0.0100 F KIO_3 solution by buret or pipet into a 250-ml Erlenmeyer flask, add 25 ml of distilled water and 2 g of solid KI. Swirl the contents of the flask until the KI dissolves, then add 20 ml of 1 F HCl, and mix by swirling. A deep brown color should appear, indicating the presence of iodine. Titrate immediately with 0.05 F $Na_2S_2O_3$ solution contained in a buret until the brown fades to a pale yellow, and then add 5 ml of starch indicator solution and continue the titration until the deep blue color of the starch indicator disappears. The change from blue to a colorless solution is very sharp. Record the volume of sodium thiosulfate solution used in the titration. Repeat the procedure in a duplicate titration. In the report form calculate the accurate concentration of the $Na_2S_2O_3$ solution.

2. Titration of a Sample of Household Cleanser. The oxidizing agent in a household cleanser is determined by placing 50 ml of distilled water in a 250-ml Erlenmeyer flask and adding 2 g of KI, swirling the contents of the flask until the KI dissolves. Add an accurately weighed sample (about 10 g) of cleanser to the flask, and swirl to dissolve the soluble material; a residue of insoluble polishing agent contained in the cleanser will remain. Then add 20 ml of 1 F HCl and proceed with the titration, using 0.05 F thiosulfate solution as in the standardization procedure. Record the volume of thiosulfate solution used in the titration. If time permits, make a duplicate titration. Then, if time permits, repeat the procedure, using a second brand of cleanser, and duplicate it.

From the concentration and volume of the added sodium thiosulfate solution used to titrate the different brands of cleanser, calculate the weight of oxidizing agent present, assuming it to be sodium hypochlorite, NaOCl. Express the final result as the percent by weight of sodium hypochlorite: g NaOCl/g cleanser $\times$ 100.

If you have the time to compare two cleansers, remember that the effectiveness of a cleanser is influenced by several factors. First the ability of a cleanser to remove food stains will be related to its oxidizing (bleaching) power. Second, its ability to remove stains from smooth surfaces is also assisted by the abrasive action of its polishing agent, which is usually pumice. Finally detergents are also added to cleansers to provide foaming action and to emulsify greases and dirt. In your comparison, you must consider all three of these factors, as well as the cost per unit weight of the cleanser.

DATA AND CALCULATIONS

In all of the calculations below, be sure to show the units of each quantity and the proper number of significant figures.

1. Standardization of the Sodium Thiosulfate Solution

Note the volume of KIO_3 solution taken. _____ ml

Give the concentration of KIO_3 solution. _____ moles/liter

Give the volumes of $Na_2S_2O_3$ solution: Sample 1 _____ ml; Sample 2 _____ ml

Give the average of samples 1 and 2. _____ ml

Calculations: Using the average value of the volume for the two duplicate titrations, calculate the accurate concentration of the sodium thiosulfate solution.

2. Determination of the Oxidizing Capacity of a Household Cleanser

(a) Record the following data for the titration of a household cleanser.

Brand name of cleanser_____.

Sample weight of cleanser: Sample 1 _____ g; Sample 2 _____ g.

Volume of thiosulfate solution to titrate the sample: Sample 1 _____ ml; Sample 2 _____ ml.

Calculations: Calculate the oxidizing capacity of the cleanser as g NaOCl/g cleanser $\times$ 100 (% NaOCl by weight), and average the results for the two samples of the cleanser.

(b) If you have titrated a sample of another cleanser, record your data below.

Brand name of cleanser_____.

Sample weight of cleanser: Sample 1 _____g; Sample 2 _____g.

Volume of thiosulfate solution to titrate the sample: Sample 1 _____ml; Sample 2 _____ml.

Calculations: Calculate the oxidizing capacity of the cleanser as g NaOCl/g cleanser × 100 (% NaOCl by weight), and average the results for the two samples of the cleanser.

EXERCISES

1. Some cleansers may contain bromate salts as oxidizing agents that will react with iodide ion under the conditions we are using according to the following reaction:

$$BrO_3^- + 6H^+ + 6I^- \rightleftharpoons 3I_2 + Br^- + 3H_2O$$

What percent by weight of $KBrO_3$ would a cleanser have to contain in order to produce an amount of iodine equivalent to that produced by an equal weight of cleanser containing 0.5% NaOCl by weight?

---

2. Many household bleaching solutions are essentially solutions of sodium hypochlorite, NaOCl. Describe the procedure you would employ to analyze the concentration of sodium hypochlorite, which is approximately 5% by weight in a household bleaching solution, if you were using 0.05 F sodium thiosulfate to titrate the iodine produced by reaction of the NaOCl with excess KI.

ORGANIC CHEMISTRY
AND BIOCHEMISTRY

INTRODUCTORY ORGANIC CHEMISTRY

PRE-STUDY

The distinctive character of organic compounds and their reactions is associated with the unique position of carbon in the periodic table. Carbon, the smallest group IV element, has four valence electrons. Its intermediate electronegative position causes it to form covalent bonds with other elements—hydrogen, oxygen, nitrogen, and, less frequently, sulfur, phosphorus, and the halogens. The very great number of carbon compounds arises from the tendency of carbon atoms to form strong covalent bonds with one another. These may be linked as open chains or as rings.

Hydrocarbons

These organic compounds contain only carbon and hydrogen and may be classified according to behavior into two principal types. *Saturated* hydrocarbons have all four carbon valences satisfied by other carbon or hydrogen atoms. *Unsaturated* hydrocarbons contain double or triple bonds between carbon atoms; they contain a smaller proportion of hydrogen and are more reactive than saturated hydrocarbons. Table 35-1 contains some examples of various types of hydrocarbons. Listed in order are the classification, the generic formula, a specific example, and its "graphic" structural formula.

Nearly all of the alkanes are found in our increasingly scarce deposits of natural gas and petroleum. The series of compounds is called a homologous series because, as can be seen from an inspection of the generic formula in the table, the members of the series differ by the number of $-CH_2-$ groups present. The first four alkanes—methane (CH_4), ethane (CH_3CH_3), propane ($CH_3CH_2CH_3$), and butane ($CH_3CH_2CH_2CH_3$)—are gases and are commonly used as fuel sources for stoves and furnaces in the home, and in industry for the generation of electrical power. These compounds having the carbon atoms arranged in a straight line are called normal or *n*-hydrocarbons. Because the structural formulas become cumbersome as the length of the compounds increases, they are frequently shortened to specify only the

TABLE 35-1
Examples of various types of hydrocarbons

Values of n	Classification	Generic formula	Name	Structural formula
1, 2, 3, . . .	Alkanes (saturated)	C_nH_{2n+2}	Ethane	$\begin{array}{c} H\ \ \ H \\ |\ \ \ \ | \\ H-C-C-H \\ |\ \ \ \ | \\ H\ \ \ H \end{array}$
2, 3, 4, . . .	Alkenes (unsaturated)	C_nH_{2n}	Ethene (ethylene)	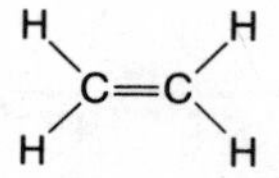 $\begin{array}{c} H\ \ \ \ \ \ \ \ H \\ \backslash\ \ \ \ / \\ C=C \\ /\ \ \ \ \backslash \\ H\ \ \ \ \ \ \ \ H \end{array}$
2, 3, 4, . . .	Alkynes (unsaturated)	C_nH_{2n-2}	Ethyne (acetylene)	$H-C\equiv C-H$

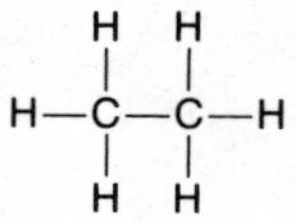

number of carbon and hydrogen atoms. Thus, an alternative way to write the structure of the butane molecule shown above is $n\text{-}C_4H_{10}$.

The next twelve members of the alkane series, C_5H_{12} to $C_{16}H_{34}$, are liquids at 20°C under normal pressures. These compounds include octane, $n\text{-}C_8H_{18}$. An isomer[1] of this compound, isooctane, is the octane standard against which all gasoline mixtures are rated for use in the internal combustion engine. Gasoline is a mixture of hydrocarbons having about 5 to 10 carbon atoms. Kerosene is a mixture of somewhat larger hydrocarbons that are distilled from the crude oil between 150–300°C.

All hydrocarbons larger than $C_{16}H_{34}$ are solids. They include the paraffins used to make candles and the tars used in asphalt paving.

The ultimate alkane is man-made. Polyethylene contains as many as 800 methylene ($-CH_2-$) groups linked in one giant molecule.

The benzene molecule, C_6H_6, is representative of another series of hydrocarbons. As shown below, the carbon atoms in benzene are arranged in a hexa-

gon with alternating double bonds. Although these bonds are shown as alternating double and single bonds, the concept of resonance describes them as being identical. Compounds like benzene are called aromatic compounds and have very different chemical reactivities in comparison with other unsaturated compounds. You will contrast the reactivity of benzene with an alkyne in part 1 of the experimental procedure.

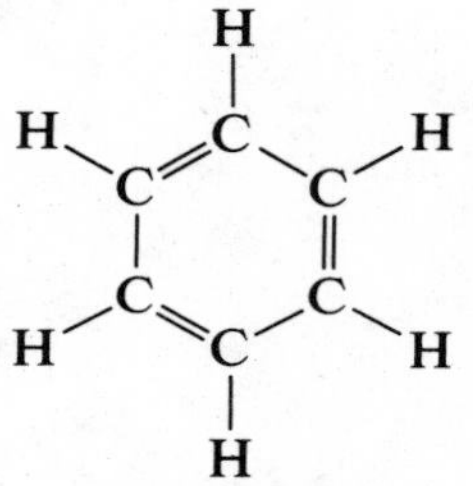

Benzene

Oxygen Derivatives

Successive stages of oxidation, beginning with a hydrocarbon, result in some important classes of compounds, each with a particular grouping of carbon, oxygen, and hydrogen atoms. In the flow diagram in Figure 35-1 each oxidation step (represented here simply by $\xrightarrow{o}$) increases the oxidation number of a given carbon atom by two.

[1]The structure of isooctane is
$$CH_3-\underset{\underset{CH_3}{|}}{\overset{\overset{H}{|}}{C}}-CH_2-\underset{\underset{CH_3}{|}}{\overset{\overset{CH_3}{|}}{C}}-CH_3$$

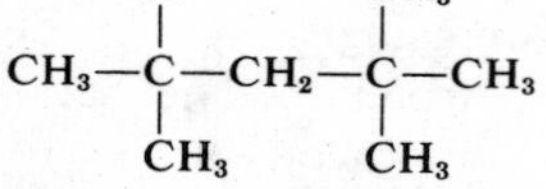

This compound has the same chemical formula, C_8H_{18}, as n-octane but the atoms are bonded together differently. Such compounds are called structural isomers.

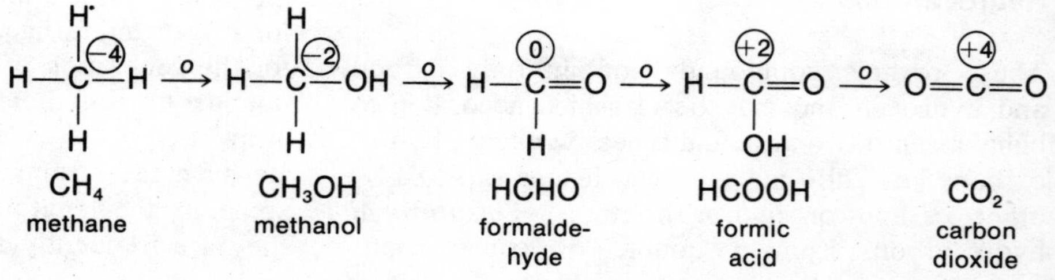

FIGURE 35-1
Flow chart showing successive stages of oxidation.

The successive addition of oxygen atoms to a hydrocarbon results in the formation of functional groups. These groups impart a characteristic reactivity to the molecule and result in the progressive increase in the formal oxidation state of the carbon atom bonded to the oxygen atom(s). Thus all *alcohols* consist of a hydrocarbon chain with an —OH group in place of an H. They behave somewhat like water, H—OH. All *aldehydes* contain the —CHO group, the hydrogen atom of which is easily oxidized; thus all aldehydes are good reducing agents. The carboxyl group, —COOH, imparts *acid* properties, because the hydrogen atom is not firmly held by the oxygen atom and therefore dissociates in water.

Nitrogen Derivatives

Nitrogen (group V) is less electronegative than oxygen (group VI). Therefore, if —NH_2 (the amine group) replaces the —OH group of the quite neutral alcohol, a basic "ammonia-like" compound, CH_3NH_2, methyl *amine*, results. If the —NH_2 group replaces the —OH of the —COOH group, acid properties are much reduced, and a more nearly neutral substance, for example, CH_3CONH_2, *acetamide*, results.

Summary

The letter R is used to represent the inactive carbon-hydrogen radical to which the functional group is attached. Common group formulas for the important classes of organic compounds are shown in Table 35-2. A common example of each compound is also given in the table, together with its structural formula.

EXPERIMENTAL PROCEDURE

Chemicals: calcium carbide, $CaC_2(s)$; benzene, $C_6H_6(l)$; naphthalene, $C_{10}H_8(s)$; chloroform, $CHCl_3(l)$; bromine, Br_2, dissolved in chloroform; methyl alcohol, $CH_3OH(l)$; amyl alcohol, $CH_3CH_2CH_2CH_2CH_2OH(l)$; ethyl alcohol, $CH_3CH_2OH(l)$; formaldehyde, HCHO, dissolved to give a 40% solution called formalin; acetone, $CH_3COCH_3(l)$; copper, Cu wire; sodium carbonate, $Na_2CO_3(s)$ (also called soda ash); calcium hydroxide, $Ca(OH)_2$, as a saturated aqueous solution (also called limewater); sodium acetate trihydrate, $NaC_2H_3O_2 \cdot 3H_2O(s)$; salicylic acid (or 2-hydroxy benzoic acid), $C_7H_6O_3(s)$; potassium dichromate, $K_2Cr_2O_7(s)$; silver nitrate, 0.1 F $AgNO_3$; acetic acid, 17 F CH_3COOH (also called glacial acetic acid).

TABLE 35-2
Important classes of organic compounds

Class of compound	Group formula	Example
Alcohol	R—OH	Methanol or methyl alcohol CH_3OH
Ether	R—O—R	Dimethyl ether CH_3OCH_3
Aldehyde	R—CH=O	Formaldehyde HCHO
Ketone	R—CO—R	Acetone CH_3COCH_3
Acid	R—CO—OH	Acetic acid CH_3COOH
Ester	R—CO—OR	Ethyl acetate $CH_3CO_2CH_2CH_3$
Amine	R—NH_2	Butyl amine $CH_3CH_2CH_2CH_2NH_2$
Amide	R—CO—NH_2	Acetamide CH_3CONH_2
Amino acid	R—CH(NH_2)—CO—OH	Glycine $CH_2(NH_2)COOH$

1. Properties of Some Hydrocarbons. (a) *Saturated Hydrocarbons.* Using the natural-gas supply in the laboratory as a source of methane, bubble gas slowly through 3 ml of Br_2 in $CHCl_3(l)$ for several minutes. Record your results so that you will be able to compare them with those of the similar treatment of acetylene, part (b). Invert a test tube filled with water in a beaker of water, and then bubble in natural gas to fill the test tube by displacement. Ignite the gas in the test tube, and observe the flame for comparison with that observed in the similar treatment of acetylene, part (b). At once add 1 ml of the saturated $Ca(OH)_2$ solution, stopper the test tube, and shake it with the products of combustion. From the observed results, write an equation for the combustion of methane in an excess of oxygen.

(b) *Unsaturated Hydrocarbons, Open-chain Type.* Acetylene, C_2H_2, has the structure H—C≡C—H. Prepare this gas, using a small generator, similar to that shown in Figure 13-2, Experiment 13. In this procedure, however, the dry test tube should contain a small lump of calcium carbide, $CaC_2(s)$. When ready, generate the gas by adding water, drop by drop, as needed, to the $CaC_2(s)$. Collect a 15-cm test tube of acetylene by displacement of water, and then bubble acetylene through a 3-ml portion of Br_2 in $CHCl_3(l)$ by connecting a rubber tube to the test tube containing the acetylene, leading it into the chloroform solution and turning the C_2H_2 tube upright. Collect another sample of acetylene in a test tube and ignite it (exercise caution) by bringing it near a Bunsen flame. Compare the character of the flame, and also compare the results of adding Br_2 in $CHCl_3(l)$ with the behavior of the saturated hydrocarbons in part (a).

(c) *Aromatic Hydrocarbons: Benzene, Naphthalene.* Test 3 ml of benzene, $C_6H_6(l)$, with 1 ml of Br_2 in $CHCl_3(l)$. Also ignite about five drops of benzene on a watch glass and note the type of flame produced.

Repeat the above tests, using a few crystals (volume of half a pea) of naphthalene. Look up the structure of the naphthalene molecule.

2. Alcohols, R—OH. (a) *Solubility.* Compare the solubility of 1-ml samples of methyl alcohol, CH_3OH, and of amyl alcohol, $C_5H_{11}OH$, in a polar solvent by adding to each 1 ml of water. Repeat, using 1 ml of the solvent, chloroform, $CHCl_3$, instead of water. Correlate the polar character of the solvent, and the lengths of the carbon chain, with the observed solubilities. Predict the solubility of glycerol, a trihydroxy alcohol, with the formula $C_3H_5(OH)_3$, in water, and in a hydrocarbon. Do the same for the alcohol, $C_{16}H_{33}OH$.

(b) *Reaction with Sodium.* Bring two small beakers—one containing about 2 ml of an alcohol, such as CH_3OH or C_2H_5OH, and the other containing about 2 ml of water—to your instructor, who will add a small piece of sodium metal (size of half a pea) to each. CAUTION: *do not point the beaker toward yourself or anyone else; wear goggles.* Compare the behavior of sodium with the two solvents. With C_2H_5OH the sodium compound formed is sodium ethoxide, C_2H_5ONa. With water the sodium compound formed is sodium hydroxide, NaOH. Write the equations for these reactions. Add 3 ml of water, and a drop of

phenolphthalein, to each mixture. Write the equation for the hydrolysis that occurs. Does the OH group of the alcohol itself impart basic characteristics? Try a 3-ml sample of C_2H_5OH with a drop of phenolphthalein.

3. Aldehydes, R—CHO. (a) *Preparation.* First write structural formulas for methyl alcohol, formaldehyde, and formic acid. Make a spiral of copper wire by inserting one end of a 10-inch length of medium-gauge wire into a cork (to serve as a handle) and winding the other end around a glass rod. Heat the spiral to red heat in the outer oxidizing Bunsen flame to form a black CuO deposit, and then plunge it into 1 ml of methyl alcohol in a small test tube. Repeat the procedure several times until a distinct characteristic odor can be detected in the test tube. Note also the color changes in the copper wire. Save the solution for (b).

(b) *Oxidation Steps.* Test this solution for the presence of an aldehyde, with Tollen's reagent, as follows. To 5 ml of 0.1 F $AgNO_3$ in a clean test tube add 6 F NH_3, drop by drop, with mixing, until the first precipitate of $Ag_2O(s)$ dissolves to form $Ag(NH_3)_2^+$. Divide this into two test tubes. To one add about 1 ml of the oxidized CH_3OH mixture. To the other, as a comparison, add 1–3 drops of 40% formaldehyde solution (formalin). Warm each slightly, if necessary, and observe the results. Write equations for the oxidation of the CH_3OH by the CuO and for the further oxidation of the aldehyde formed by $Ag(NH_3)_2^+$ (Tollen's reagent). Clean the test tubes with several drops of 6 F HNO_3.

4. Organic Acids, R—COOH. (a) *Preparation.* Organic acids may be prepared from their salts by adding a nonvolatile acid and distilling the volatile acid. Place about 1 g of $NaC_2H_3O_2$ in a test tube, and add 1 ml of concentrated H_2SO_4. Warm this gently, and cautiously note the odor of the vapors and their effect on moistened blue litmus paper. Is oxidation-reduction taking place here? Write the equation. (Save the acid prepared here for part 5.)

(b) *Oxidation Steps.* As a further example of the oxidation of an alcohol, first to an aldehyde and then to an acid, dissolve 1 g of $K_2Cr_2O_7(s)$ in 5 ml of 3 F H_2SO_4, and add 5 drops of ethyl alcohol. Warm the mixture very gently until a change in color occurs. Look up the several oxidation states, and the corresponding colors, of Cr in order to explain your observed results. Cautiously note the odor. (An odor somewhat like that of a pear is characteristic of acetaldehyde.) Warm the mixture again, and test the vapors with moistened blue litmus paper. Write equations for these reactions.

5. Esters, R—COOR′. *Preparation.* Esters are produced by the reaction of organic acids and alcohols in the presence of concentrated H_2SO_4 acting as a dehydrating agent.

To the $NaC_2H_3O_2$ and H_2SO_4 mixture from part 4, add 1 ml of ethyl alcohol. Warm this gently. After about three minutes pour the reaction mixture into a small beaker containing a little water. Note the characteristic odor.

Warm a mixture of 1 ml of 17 F (glacial) acetic acid, 1 ml of amyl alcohol, $C_5H_{11}OH$, and 1 ml of concentrated H_2SO_4. Pour this into a small beaker containing a little water. Note the odor.

Repeat, using 1 ml of methyl alcohol, 1 g of solid salicylic acid, ortho-HOC_6H_4COOH, and 1 ml of concentrated H_2SO_4. Do you recognize the odor? An ester is again produced by the reaction of the carboxylic acid group (COOH) on the salicylic acid and the hydroxyl group (OH) of the methyl alcohol as noted above. A slight variation of this reaction would produce one of the most widely used medicines. Heating salicylic acid with acetic anhydride ($CH_3COOCOCH_3$) yields acetylsalicylic acid (ortho-$CH_3COOC_6H_4COOH$), more commonly known as aspirin. This experiment is described in the procedure given in Experiment 37.

Write the equation for the preparation of ethyl acetate and the formulas for the other esters prepared. Why is sulfuric acid necessary? Contrast these reactions with that taking place between NaOH and HCl in aqueous solution by considering types of molecules involved and reaction rates.

REPORT

35

Introductory
Organic Chemistry

NAME

SECTION   LOCKER

INSTRUCTOR   DATE

1. Properties of Some Hydrocarbons

Fill in the data obtained from your experimental observations in parts (a) through (c).

Name	Formula	Type	Observation with Br_2 in $CHCl_3$ (*l*)	Type of flame
Methane				
Acetylene				
Benzene				
Naphthalene				

Describe, and write the equation for, the preparation of acetylene gas.

Contrast the characters of flames produced in combustion in the different types of hydrocarbons, comparing the ratio of C to H in the molecules of each hydrocarbon.

Describe the result of the addition of the $Ca(OH)_2$ solution to the products of the combustion of methane. Write the equation for this reaction.

Write equations for the complete combustion of methane and of acetylene.

Contrast the types of hydrocarbons by ease of reaction with Br_2 in $CHCl_3$ (*l*). Also write an equation for one type that did react.

2. Alcohols, R—OH

(a) Solubility. Enter the relative solubilities observed.

	In H_2O	In CCl_4
Methyl alcohol, CH_3OH	_____	_____
Amyl alcohol, $CH_3CH_2CH_2CH_2CH_2OH$	_____	_____

Explain the above differences in solubility, and also predict the relative solubilities of glycerol, $C_3H_5(OH)_3$, and of the alcohol $C_{16}H_{33}OH$, in water and in a hydrocarbon.

(b) Reaction with Sodium. Compare the reactions of sodium metal with an alcohol and with water. Also write the equations.

What evidence did you observe of any hydrolysis of the sodium ethoxide, C_2H_5ONa, and sodium hydroxide, NaOH, formed above when water was added? Write the hydrolysis equations.

What evidence did you observe that pure alcohol, C_2H_5OH, is, or is not, a base?

3. Aldehydes, R—CHO

(a) Preparation.
Structural formulas: _____   _____   _____

                      Methyl alcohol   Formaldehyde   Formic acid

Oxidation state of carbon: _____   _____   _____

What were the observed effects of plunging a hot Cu spiral, coated with CuO, into methyl alcohol? Write the equation for the reaction.

(b) Oxidation Steps. What were the observed effects of adding Tollen's reagent to the solution produced in (a)? Write the equation for the reaction.

4. Organic Acids, R—COOH

(a) Preparation. Write the observed results, and equation, for the reaction when sodium acetate and concentrated H_2SO_4 were heated together.

To what extent does oxidation-reduction occur in this reaction?_____

(b) Oxidation Steps. What were the observed results when a mixture of $K_2Cr_2O_7$, H_2SO_4, and C_2H_5OH was very slightly warmed (odor), then allowed to stand, and further warmed (litmus effect)?

Write balanced oxidation-reduction equations for the change of the following:

CH_3CH_2OH to the aldehyde _____

CH_3CH_2OH to the acid   _____

5. Esters, R—COOR'

Preparation. Write the observed results, and equation, for the reaction when ethyl alcohol was added to the $NaC_2H_3O_2$–H_2SO_4 mixture.

Note the odors, and write the formulas, for the esters formed for the following:

	Odor	Formula
Amyl alcohol, H_2SO_4 and acetic acid	_____	_____
Methyl alcohol, H_2SO_4, and salicylic acid	_____	_____

What is the role of sulfuric acid in these reactions?

Compare the conditions and rate of these reactions with those of an acid-base reaction.

SOME SIMPLE BIOCHEMICAL TESTS

PRE-STUDY

A major portion of the field of biochemistry is concerned with the important role of the essential food constituents—carbohydrates, proteins, and fats—in building body tissues and in supplying energy to the body. This experiment introduces you to simple tests for these organic compounds. Experiment 37, part E, presents an example of a biochemical synthesis in the human body—the synthesis of hippuric acid—which may be included with this experiment if desired.

Carbohydrates

Carbohydrates derive their name from their general formula, $C_n(H_2O)_n$. They formally seem to be hydrates of carbon. Many of them consist of macromolecules of very high molecular weight, and are synthesized by plants by the interaction of carbon dioxide from the air and water from the soil in the presence of chlorophyll and sunlight. *Cellulose,* the fibrous part of plants, is a polysaccharide consisting of possibly 1000 units of $C_6H_{10}O_5$. *Starch,* produced by many plants, contains possibly 30 simple saccharide units: $(C_6H_{10}O_5)_{30} \cdot H_2O$. Some important *sugars* are the monosaccharides glucose and fructose, having the common formula $C_6H_{12}O_6$, and the disaccharides sucrose (cane or beet sugar), maltose, and lactose, having the common formula $C_{12}H_{22}O_{11}$. Structural and stereochemical differences account for their individual behavior.

Monosaccharides and some disaccharides contain aldehyde or ketone groups. These are reducing groups, and, if they are not involved in the disaccharide linkage, they are characteristically tested for by *Fehling's solution,* an oxidizing agent consisting of the complex cupric tartrate ion in alkaline medium, which is thereby reduced to a red cuprous oxide precipitate. Carbohydrates that do not reduce Fehling's solution, such as sucrose, starch, and even cellulose, form simple reducing sugars when hydrolyzed by boiling with dilute acids or bases. With the aid of enzymes in the saliva, the digestive processes readily break down the large starch macromolecules by hydrolysis at bodily temperatures into maltose and finally into glucose.

A standard test in both quantitative and qualitative analysis for the presence of iodine, I_2, is the reaction of I_2 with starch in the presence of iodide ions to give a deep blue solution. The 30 or more glucose units that are connected together to form the starch molecule, coil up to give a long, hollow helix. The dimensions of this helix are such that the tri-iodide ions can enter and arrange themselves as shown in Figure 36-1. You will use this test both to detect the presence of starch and to measure the rate at which it is broken up into its constituent glucose molecules.

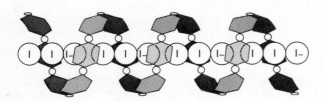

FIGURE 36-1
Tri-iodide ion-starch helical complex.

Proteins

Proteins are very complex macromolecular structures that contain about 52% carbon, 7% hydrogen, 16% nitrogen, 0–3% sulfur, 0–0.8% iodine, and sometimes phosphorus, iron, copper, and other transition metals. Molecular weights of proteins vary from about 10,000 to several million. The fundamental building units are amino acids, $CHR(NH_2)COOH$, which are combined by the union of the NH_2 group of one acid with the COOH group of another acid to form a peptide linkage or bond,

$$-\overset{\displaystyle O}{\overset{\displaystyle \|}{N}}-\underset{\displaystyle H}{C}-.$$

Owing to the presence of both acid groups (COOH) and basic groups (NH_2) in the side chain R groups, proteins are amphoteric and form a protein salt with either acids or bases. When heated, the polypeptide chains uncoil from their characteristic structure as native protein and thus lose many of their characteristic properties. They are then said to be denatured.

During digestive processes, enzymes catalyze the hydrolysis of proteins to simpler molecules and finally to amino acids. These are distributed throughout the body by the blood stream, and are resynthesized to new protein to replace the tissue that is constantly being broken down in life processes.

Fats

Fats are *fatty acid esters of glycerol,* produced in plants and animals by synthesis from carbohydrates. One of the fats present in olive oil is the oleic acid ($C_{17}H_{33}COOH$) ester of glycerol ($C_3H_5(OH)_3$), whose structure is

$$H_2C—O—CO—C_{17}H_{33}$$
$$HC—O—CO—C_{17}H_{33}$$
$$H_2C—O—CO—C_{17}H_{33}$$

When this molecule is hydrolyzed by boiling with NaOH it is split into glycerol and the sodium salt of the fatty acid, $C_{17}H_{33}COONa$, which is a soap.

When a fat is strongly heated in the presence of a dehydrating agent such as $KHSO_4$, the glycerol portion of the molecule is dehydrated to form the unsaturated aldehyde, acrolein, $CH_2{=}CH—CHO$, which has the peculiar odor of burnt grease. This procedure is a standard test for a fat.

EXPERIMENTAL PROCEDURE

Chemicals: Glucose, sucrose, starch, Fehling's solutions A and B, 0.05 F KI_3 egg albumin solution, 6 F CH_3COOH, 0.1 F $Pb(C_2H_3O_2)_2$, 0.1 F $HgCl_2$, 0.1 F $CuSO_4$, methyl orange and phenolphthalein indicators, olive oil, $KHSO_4(s)$, $C_2H_5OH(l)$, NaOH(s).

NOTE: Prepare your own experiment report form, including a neat summary of observations, your conclusions, and answers to questions. The optional section, step 3(b), may be performed as assigned, if time is available.

1. Carbohydrates. Prepare solutions of glucose and sucrose by dissolving 2 g of each in 20 ml of water. Prepare a colloidal suspension of starch: First make a paste of 1 g of starch with 5 ml of cold water; then add this to 50 ml of boiling water and stir the mixture for 1 minute.

(a) *The Fehling Test for Sugars with Reducing Groups.* Add 1 ml each of Fehling solutions A and B to 4 ml of water. Mix and divide this among three test tubes. To one of these, add 2 ml of glucose solution, to the second 2 ml of sucrose solution, and to the third 2 ml of starch suspension. Place the test tubes in a beaker containing water at about 60°C for several minutes. Which of the carbohydrates contained a reducing group?

Place a test tube containing 5 ml of the sucrose solution and 1 ml of 12 F HCl in a boiling water bath, and allow the sucrose to hydrolyze for 10 minutes. Neutralize the acid solution with 2 ml of 6 F NaOH, and make the Fehling test as outlined above. Account for the observed results.

(b) *The Hydrolysis of Starch.* First become familiar with the *iodine test for starch*: Dilute two drops of 0.05 F KI_3 with 1 ml of water, and add a drop of this diluted I_3^- solution to 1 ml of the prepared starch solution. Observe the color of the iodo-starch complex. Repeat the test with 1 ml of glucose solution.

Now mix 10 ml of the starch solution with 40 ml of water and 10 ml of 6 F HCl, and boil the mixture for 2 minutes. With a medicine dropper transfer about 1 ml of the solution to a small test tube, almost neutralize with a few drops of 6 F NaOH, and add a drop of iodine solution. Observe against a white background. Continue to boil the solution and withdraw samples at intervals until no blue color is obtained with the iodine solution. Now test 2 ml of this hydrolyzed starch solution, neutralized with NaOH, with 1 ml of freshly mixed Fehling's solution. What has happened to the starch molecule during hydrolysis?

(c) *Enzymatic Hydrolysis of Starch by Amylase from Saliva.* To another 10 ml of the starch solution add 40 ml of water. Suspend this in a slightly larger beaker containing water at body temperature (37°C), then add 2 ml of saliva and mix well. After 2 minutes, withdraw two 2-ml samples, and test these, respectively, for starch with iodine and for a reducing sugar with Fehling's solution. Repeat at intervals. Compare and report the rate of this enzymatic hydrolysis with that of the more strenuous conditions (acid solution at the boiling point) in part (b) above.

2. Proteins.

(a) *The Xanthoproteic Test.* Most proteins (those which have an aromatic ring) respond to this test. To 1 ml of egg albumin solution add 5 drops of 16 F HNO_3. Warm the mixture, and observe the color produced. Allow the tube to cool, and then *cautiously* add an excess of 15 F NH_3. Observe the intensification of color. Try this test on your fingernail by touching it with a drop of 16 F HNO_3 on a stirring rod. Wash off with water, and apply a drop of 16 F NH_3 solution. Try the test on a piece of wool yarn or cloth.

(b) *Coagulation of Proteins.* Add a drop of 6 F CH_3COOH to 3 ml of egg albumin solution, and heat this at the boiling point for several minutes. Account for the observed change in terms of the protein structure. Is the coagulated protein soluble in water? To 3-ml samples of egg albumin solution in each of three test tubes add, respectively, a few drops of 0.1 F $Pb(C_2H_3O_2)_2$, 0.1 F $HgCl_2$, and 0.1 F $CuSO_4$. What effect might these salts have if taken internally? Why are raw egg whites taken as an antidote for mercury poisoning?

(c) *Amphoteric Nature of Proteins.* Add a drop of 0.1 F HCl and a drop of methyl orange to a clean test tube, and a drop of 0.1 F NaOH and a drop of phenolphthalein to a second clean test tube. To each add about 2 ml of water and then 3 ml of egg albumin solution. In your own words, explain and interpret any changes in the indicator colors observed, and give typical equations. Can a protein act as a buffer?

3. Fats.

(a) *The Acrolein Test.* In a dry Pyrex test tube place about 1 g of $KHSO_4(s)$ and 5 drops of olive oil. Heat the tube rather strongly for a few minutes, then allow it to cool, and observe the characteristic acrolein odor. *Do not put your nose over the test tube while it is hot.* Write the equation for the dehydration of the glycerol portion of the fat molecule to the unsaturated aldehyde acrolein.

(b) *The Saponification of Fats. (Optional).* Mix 2 ml of olive oil, 1 g of NaOH(s), and 20 ml of C_2H_5OH in a 50-ml Erlenmeyer flask, and heat in a water bath to about 75°C for 15 minutes. Cool the solution in a beaker of cold water, then remove a small portion of the solid with a stirring rod, and dissolve it in some water in a test tube. Agitate this mixture, or blow through it with a glass tube. What are the results? Write an equation for the saponification, and give common names for the products.

Bibliography

Briggs, T. S., and Rauscher, W. C., "An Oscillating Iodine Clock," *J. Chem. Educ.* **50**, 496 (1973). The mechanism of the binding of I_3^- by starch is described, as well as an additional clever experiment, which you may find interesting to pursue.

Hearst, J. E., and Ifft, J. B., *Contemporary Chemistry,* Chapter 11, W. H. Freeman and Company, San Francisco (in press). The structures of the amino acids, monosaccharides, fatty acids, proteins, carbohydrates, and fats are discussed. In addition, the structures and properties of one of the most important biopolymers, the nucleic acids and their constituent monomers, the nucleotides, are also given.

SOME ORGANIC
AND BIOCHEMICAL SYNTHESES

The following experimental procedures illustrate the ways in which various functional groups interact to synthesize several interesting organic compounds. Carry out such preparations as your instructor directs. Prepare your own report, being sure to include the quantities of reagents, a summary of the method, with equations for the reactions, the percent yield obtained, and any pertinent observations.

A. UREA FROM AMMONIUM CYANATE

$$NH_4^+ + OCN^- \rightarrow O{=}C\begin{smallmatrix} NH_2 \\ \\ NH_2 \end{smallmatrix}$$

This preparation, made by the German chemist, Wohler, in 1828, represented the first synthesis of an organic compound produced in life processes from purely inorganic sources. It was of great historic significance in destroying the idea that some "vital life force" was essential for the production of such compounds. The preparation consists of a *molecular rearrangement*, which occurs readily on heating a mixture of ammonium and cyanate ions.

Directions

Chemicals: $(NH_4)_2SO_4(s)$, $KOCN(s)$, methanol, CH_3OH.

Dissolve 10.0 g of $KOCN(s)$ in 30 ml of water, and add, with mixing, a solution of 15.0 g of $(NH_4)_2SO_4$ in 25 ml of water. Let the mixture stand 1–2 minutes, and then filter off the precipitate of $K_2SO_4(s)$. (The solubility of potassium sulfate in water at 25°C is only 12 g $K_2SO_4/100$ ml solution.) Concentrate the filtrate by rapid boiling until it begins to bump, then transfer it to an evaporating dish, and complete the evaporation over a beaker of boiling water as a steam bath. Transfer the dry solid to a small flask. To this substance, add 20 ml of methanol (methyl alcohol). Immerse the flask partially in a beaker of hot water to heat the contents almost to boiling. Mix well to dissolve the urea ($K_2SO_4(s)$ is insoluble in methanol), and then filter quickly while hot. (A Büchner funnel, with suction, is preferable.) Return the solid material to the flask, and extract any remaining urea with a second 10-ml portion of hot

methanol. This can be poured through the same filter. Combine the two filtrates in a small beaker, and cool *thoroughly* in an ice-salt mixture. Filter the cold mixture rapidly over a fresh filter, and allow the crystals of urea to dry in the air. When they are dry, weigh the crystals to the nearest tenth gram on a triple-beam balance. Using as a basis the amount of original reagent that is not present in excess, calculate the maximum amount of urea that could be produced. By dividing the weight of the urea crystals by the maximum weight of urea that could be produced, compute the percentage of the theoretical yield that was actually obtained.

The *hydrolysis of urea* occurs readily either in acidic or basic solution. Place a few urea crystals in a small evaporating dish or beaker. Add 1–2 ml of 6 F NaOH, and warm gently. *Cautiously* note the odor (being careful to prevent the vapors from getting into the eyes), or test the vapors with moist red litmus. Write the equation for the hydrolysis of urea.

B. ASPIRIN (ACETYLSALICYLIC ACID)

Salicylic acid   +   Acetic anhydride   →   Aspirin   +   Acetic acid

Aspirin is an *ester* formed by the interaction of the OH group of salicylic acid with the COOH group of acetic acid. In the laboratory, acetic anhydride is used instead of acetic acid, because it is more efficient and gives a better yield.

Directions

Chemicals: Salicylic acid; acetic anhydride; phosphoric acid, 85% H_3PO_4.

Place 3 g of salicylic acid in a small flask. Add 6 ml of acetic anhydride, and then add 5–8 drops of 85% H_3PO_4 to act as a catalyst. Mix the reagents, and place the flask in a beaker of water, warmed to

70–80°C, for 10 minutes. While the mixture is still warm, cautiously add, drop by drop, about 1 ml of water. (The vigorous decomposition of excess acetic anhydride may cause spattering.) At once add 20 ml of water, and cool the flask in cold water. (An ice bath is advantageous, and scratching the walls of the flask with a stirring rod promotes crystallization.) Filter, and wash the crystals on the filter with a little cold water. Finally, let the crystals drain, and press them dry between several pieces of filter paper. Weigh the dry product,[1] and calculate the percent yield based on the amount of salicylic acid used.

---

[1]CAUTION: The product is not pure enough for internal use.

C. LUCITE

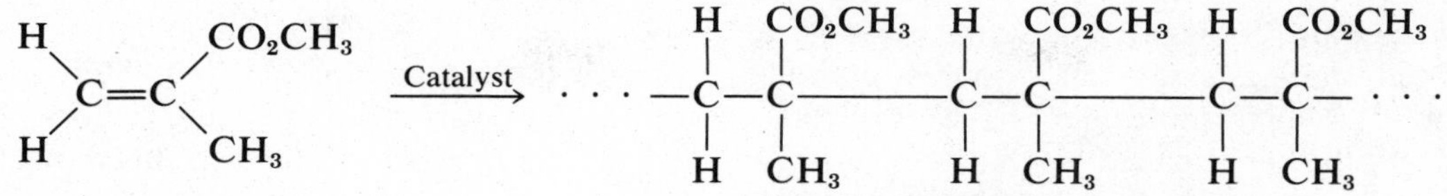

The monomer—
methyl methacrylate
(also called 2-methyl-
propenoic acid methyl ester)

Three units of the polymer—lucite

The synthesis of this *high polymer* results from the addition polymerization of the reactive carbon to carbon double bonds of the simple molecule or *monomer* to build up a long chain of very high molecular weight.

Directions

Chemicals: Methyl methacrylate, benzoyl peroxide, $(C_6H_5COO)_2$.

Place 200 ml of water in a 400-ml beaker, and heat the water to boiling. Meanwhile, place 5 ml of methyl methacrylate in a small test tube, and have your instructor add 25 mg of benzoyl peroxide as a catalyst. Place the test tube in the boiling water, stir the contents with a small stirring rod to effect solution, and then keep the solution boiling gently until the contents become quite viscous as polymerization takes place. Cool the water bath to 60–70°C by adding about 75 ml of cold water to the beaker and removing the flame. Let stand until the test tube contents harden. Finally, break the test tube and examine the rigid product. Does it shatter like glass? Is it soluble in water? Holding a piece of it with the tongs, warm it very gently over a flame and test its rigidity. Is lucite thermoplastic?

D. METHYL ORANGE, AN INDICATOR

Step 1. Diazotization

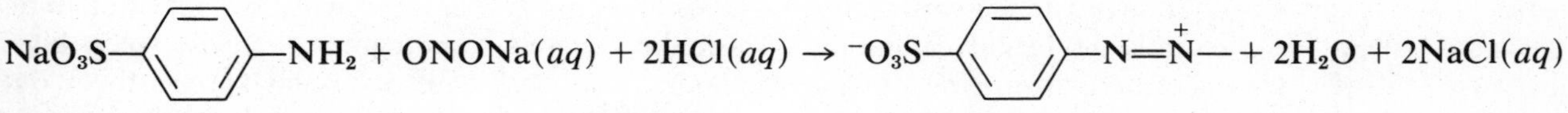

$$NaO_3S-\langle\rangle-NH_2 + ONONa(aq) + 2HCl(aq) \rightarrow {}^-O_3S-\langle\rangle-N{=}\overset{+}{N}- + 2H_2O + 2NaCl(aq)$$

Sodium salt of sulfanilic acid   Sodium nitrite   Diazonium compound

Step 2. Coupling

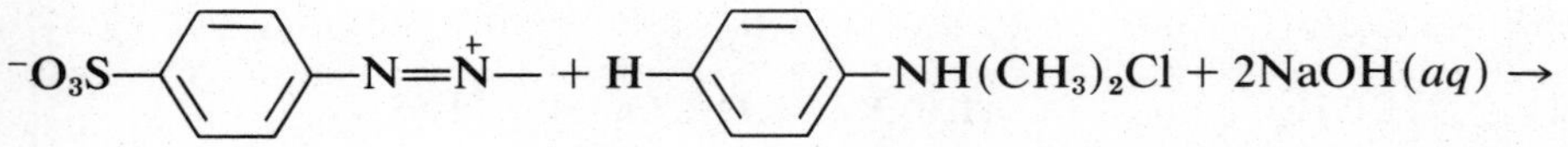

$${}^-O_3S-\langle\rangle-N{=}\overset{+}{N}- + H-\langle\rangle-NH(CH_3)_2Cl + 2NaOH(aq) \rightarrow$$

Dimethylaniline hydrochloride

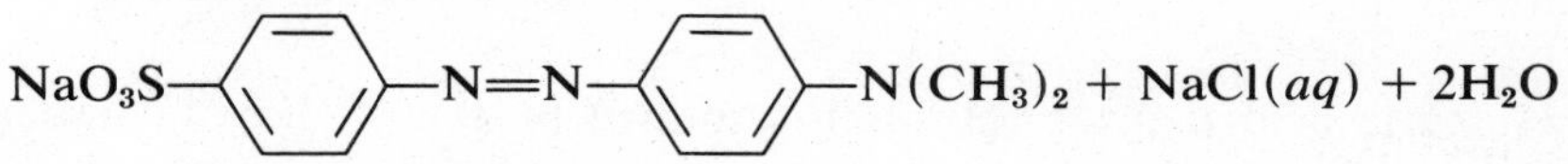

$$NaO_3S-\langle\rangle-N{=}N-\langle\rangle-N(CH_3)_2 + NaCl(aq) + 2H_2O$$

Methyl orange

Step 3. Indicator Action

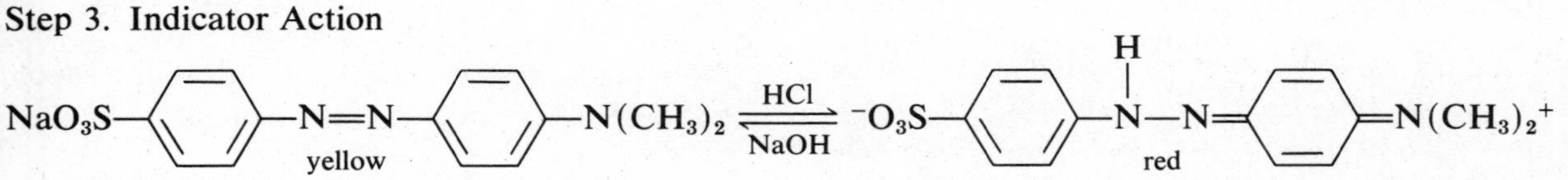

$$NaO_3S-\langle\rangle-N{=}N-\langle\rangle-N(CH_3)_2 \underset{NaOH}{\overset{HCl}{\rightleftharpoons}} {}^-O_3S-\langle\rangle-\overset{H}{N}-N{=}\langle\rangle{=}N(CH_3)_2{}^+$$

yellow   red

The structure of the indicator phenolphthalein was discussed in Experiment 20. As noted there, indicators contain conjugated systems, that is, a series of alternating single and double carbon-carbon bonds. Such molecules have electronic energy levels that are spaced so that their energy differences correspond to the energy contained in the photons of visible light. Indicators also possess the property that this conjugated system can undergo a slight change upon changes in pH. This alteration in the electronic structure of the molecule changes the absorption spectrum of the molecule enough to cause a change in the color of the indicator solution.

The diazo dyes contain the *diazo* group, $—N{=}N—$. The word is derived from the French word *azote,* which means nitrogen. These dyes, such as methyl red, methyl orange, and tropeolin 00 (orange IV), also serve as indicators by changing through various shades of red, orange, and yellow as the pH is increased. The exact shading of each form of the indicators depends on which groups are substituted on the benzene rings of the parent compound.

In this experiment, you are to make and study the properties of the indicator, methyl orange. The series of reactions you will carry out are shown on page 399. The diazo group is generated in step 1 from the water-soluble sodium salt of sulfanilic acid and nitrous acid, HNO_2. This acid is formed from the sodium nitrite, $NaNO_2$, and hydrochloric acid. (The presence of sodium nitrite in many of our packaged meats is currently under investigation because nitrous acid—formed from the nitrite and hydrochloric acid in the stomach—is a known mutagen for several organisms.) The diazonium

compound thus formed is coupled in step 2 with the soluble hydrochloride of dimethylaniline in the presence of NaOH. The sodium salt of methyl orange is then precipitated out. Step 3 shows the molecular rearrangements and colors that result when acid or base is added to this molecule.

Directions

Chemicals: Sulfanilic acid; dimethylaniline; sodium carbonate, $1\ F\ Na_2CO_3$; sodium nitrite, $NaNO_2(s)$; sodium chloride, NaCl (s); ethanol, $C_2H_5OH(l)$.

Dissolve 3 g of sulfanilic acid in 10 ml of $1\ F$ Na_2CO_3 mixed with 35 ml of water. Then add to this 100 g of cracked ice to cool the mixture to 0°C. Add 1.5 g of $NaNO_2(s)$ dissolved in 10 ml of water, and mix well. Dilute 4–5 ml of $6\ F$ HCl with 10 ml of water, and add this slowly to the above with stirring. Let this stand while you prepare a solution of 2 ml of dimethylaniline in 5 ml of $6\ F$ HCl. Add this, with stirring, to the above diazotized salt solution, and allow to stand for 10–15 minutes. Make the solution slightly basic (litmus paper test), by adding about 12 ml of $6\ F$ NaOH, in order to complete the coupling process. To crystallize out the methyl orange more completely, add 15–20 g of $NaCl(s)$, mix well, heat the mixture nearly to boiling, then cool it in an ice bath. The dye separates as orange crystals. Filter these (a Büchner funnel, with suction, is preferable) and rinse them once with 3–5 ml of ethanol. Let the crystals drain, press them between filter papers to absorb moisture, and spread them out to dry. Comment on the several colors you observed in this experiment.

E. THE BIOCHEMICAL SYNTHESIS OF HIPPURIC ACID

$$C_6H_5COOH + NH_2CH_2COOH \rightarrow C_6H_5CONHCH_2COOH + H_2O$$

Benzoic acid   Glycine   Hippuric acid

Hippuric acid (benzoyl glycine) is formed by the condensation of benzoic acid with glycine. It is synthesized in bodily processes in a similar manner. Certain fruits that contain benzoates (plums, cherries, cranberries) increase considerably the normal hippuric acid content of the urine (about 0.7 g daily). A temporary marked increase can also be obtained by ingesting benzoic acid or its salt

directly into the digestive system. This ingested benzoic acid reacts with the glycine that is produced by the normal biochemical hydrolysis of protein foods during digestive processes. The equation for the reaction is given above. (CAUTION: students with digestive disorders should not attempt to conduct this experiment.)

Directions

Chemicals: Sodium benzoate; ammonium sulfate, $(NH_4)_2SO_4$ (s); 18 F H_2SO_4; decolorizing charcoal, C(s).

Before you go to bed drink a solution of 5.0 g of sodium benzoate $C_6H_5COONa(s)$, in a half glass of water. Next morning preserve the urine, preferably in a stoppered 500-ml Erlenmeyer flask. For each 100 ml of urine add 25 g of $(NH_4)_2SO_4(s)$ and 1.5 ml of 18 F H_2SO_4. Stopper the flask, and mix to complete solution, and then store this in an ice bath or refrigerator, preferably overnight, until a maximum crystallization of hippuric acid is obtained.

Filter, using a fluted filter paper (or a Büchner funnel), and wash the crystals with a small amount of ice water. The product may be recrystallized as follows. Transfer it to a beaker, add 50 ml of water and 1–2 g of decolorizing charcoal, boil about 10 minutes, then filter again. Cool the filtered solution in an ice bath for at least an hour, and collect the purified hippuric acid crystals. Weigh your product on a triple-beam balance. Calculate the percent yield obtained, assuming complete conversion of the sodium benzoate to hippuric acid.

THE CHEMISTRY OF VITAMIN C

PRE-STUDY

Until late in the nineteenth century, it was common for ships on transoceanic voyages to lose more than half their crews: the deaths were the result of the dread disease, scurvy. Of those who survived, many were afflicted with massive hemorrhages, ulcerated gums, diarrhea and exhaustion. In 1795, the British navy ordered a daily ration of lime juice for every sailor. It was because of this measure that the nickname "Limey" was invented—but there were no more deaths from scurvy in the British navy. (The British merchant marine considered the practice quackery, and hundreds of its seamen perished in the course of the next 70 years before the Board of Trade passed a similar regulation, making such rations mandatory on all vessels.)

The vital constituent in lime juice, all other citrus juices, and, to a lesser degree, all fresh vegetables was not isolated until 1928. The brilliant Hungarian scientist, Albert Szent-Györgyi, made the discovery and named it hexuronic acid. Four years later, he renamed the substance ascorbic acid and determined its chemical formula to be $C_6H_8O_6$.

The English sugar chemist, W. M. Haworth, determined the structure of ascorbic acid in 1933 to be compound I shown in Figure 38-1. Haworth employed the standard convention for writing the structural formula of sugars with one ring. A somewhat more realistic representation for the same compound is shown in structure II of the figure.

For many years, public health organizations have recommended a minimum daily intake of ascorbic acid, also now known as vitamin C. The present daily dietary allowance recommended by the Food and Nutrition Board of the National Research Council is between 35 and 60 mg per day.

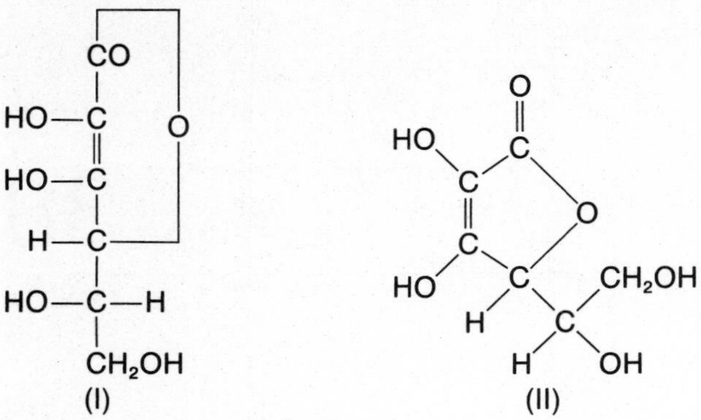

FIGURE 38-1
The structure of ascorbic acid.

TABLE 38-1
Ascorbic acid content in foodstuffs

Content of ascorbic acid, mg/100 g of food	Foodstuffs
100–350	Chili peppers (green and red), sweet peppers (green and red), parsley, turnip greens
25–100	Citrus juices, tomato juice, mustard greens, spinach, Brussels sprouts
10–25	Green beans and peas, sweet corn, asparagus, pineapple, cranberries, cucumbers, lettuce
less than 10	Eggs, milk, carrots, beets, cooked meat

Professor Linus Pauling, twice a winner of the Nobel Prize, has studied and compiled most of the original research reports on vitamin C. His recently published small book, *Vitamin C and the Common Cold*,[1] has stimulated further work in the subject (and the idea for this experiment). Pauling has surveyed a number of earlier studies on the effectiveness of this vitamin in reducing the incidence of common colds. He has concluded that if taken in sufficiently large quantities, Vitamin C can dramatically curb the frequency of colds. He recommends a daily intake of between 250 mg and 10 g, depending on the size and physiology of the person. To help put the magnitudes of these numbers in better perspective, we can say that the minimum daily requirements of the essential amino acids vary from 2 to 3 g per day.

The controversy over the merits of the proposal has been evident in the debates that have appeared in the press between Pauling and various members of the scientific and medical communities. It is very much to be hoped that one of our national research organizations will soon fund a major effort to investigate Pauling's theory, so that it can be either verified or disproved. Such verification is important because, if the proposal is wrong, people may be doing themselves serious harm by ingesting such massive doses and, at the very least, they are wasting money. If the proposal is indeed correct, major changes should be made in the prevailing dietary patterns, so that larger numbers of people will benefit.

This experiment is not intended as such a test. We will look at one of the unique chemical proper-

ties of this now famous molecule and determine the amounts of it present in a variety of foodstuffs. Table 38-1 gives you some average values for the vitamin C contents of a variety of foods. This will help you in your selection of foods to study in part 2 of the experimental procedure, and it will also tell you how much you would have to eat of a particular food to ingest Pauling's recommended amounts. He recommends doses of powder or tablets of ascorbic acid (though not the organically grown substance, which is no better but is much more expensive). It would be difficult to obtain such large amounts of the substance from foodstuffs — unless you wanted to consume large quantities of green peppers daily!

Experimental Method

In this method,[2,3] we will use the standard procedure for the detection of vitamin C. The method was developed some years ago by King of the University of Pittsburgh (Bessey and King, 1933; King, 1941). It utilizes one of the most distinctive features of this vitamin, its powerful reducing properties. The most common oxidizing agent used for this analysis is an interesting dye molecule called 2,6-dichlorophenol-indophenol, or 2,6-dichloro-indophenol. The reaction with ascorbic acid is shown in Figure 38-2.

The oxidizing agent is an especially interesting compound because it acts both as an acid-base

---

[1]Linus Pauling, *Vitamin C and the Common Cold*, W. H. Freeman and Company, San Francisco, 1970.

[2]Based in part on an experiment originally written by Professor William Jolly, University of California, Berkeley.

[3]A simple alternate procedure, using potassium hexacyanoferrate(III) in acid medium to titrate ascorbic acid, has been described by G. S. Sastry and G. G. Rao, *Talanta*, **19**, 212(1972).

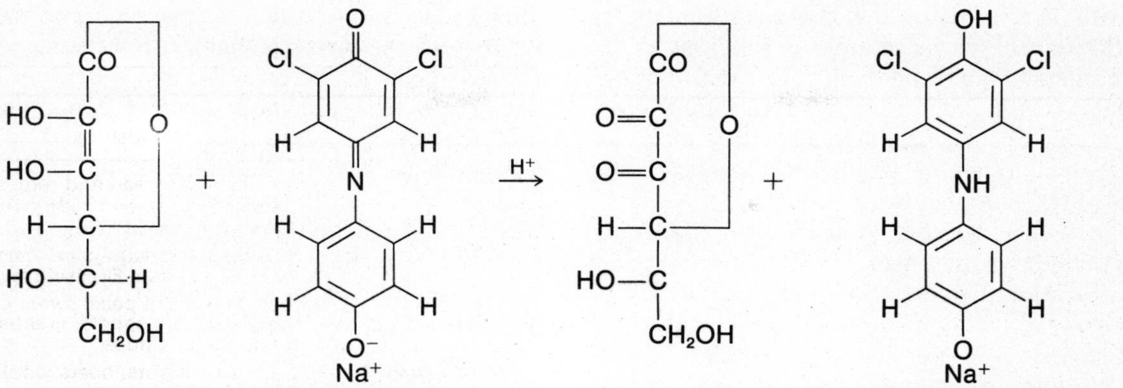

FIGURE 38-2
The reaction of the dye molecule 2,6-dichloroindophenol with ascorbic acid.

Ascorbic Acid

2,6-dichloroindophenol
(sodium salt)
(red in acid solution)

Dehydro-ascorbic
acid

Reduced 2,6-dichloroindophenol
(sodium salt)
(colorless in acid solution)

indicator and as a redox indicator. The solution you will receive is slightly basic and is a deep blue. HCl or metaphosphoric acid is added to the solution to be titrated so that, as the indicator enters this solution, it will turn from blue to red. If vitamin C is present in the solution, it will rapidly be decolorized to a colorless solution.

The acids serve a second vital purpose. Vitamin C is oxidized fairly rapidly by atmospheric oxygen in neutral or alkaline solutions. The HCl or metaphosphoric acid ensures that the O_2 won't get to the vitamin C before the indicator does.

Since dichloroindophenol itself is not particularly stable, it is necessary to measure the concentration of the indicator solution. We will use commercial tablets of vitamin C for this standardization.

Metaphosphoric acid is a phosphorus compound having the nominal formula

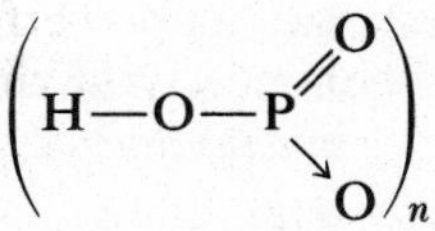

$$\left(H-O-P \underset{O}{\overset{O}{\diagup}} \right)_n$$

where n is a very large number. Na_3PO_4 is added to the compound to permit compression into rods. The result is a colorless, transparent, glass-like solid, which dissolves slowly in water. It spontaneously converts to H_3PO_4 in the course of a 3–4 day period. For some reason, biochemists have preferred it as a source of acid when titrating biological liquids for vitamin C content.

EXPERIMENTAL PROCEDURE

Special Supplies: 25- or 50-ml buret; mortar and pestle or Waring blender; citrus juices; dried breakfast drink, such as Tang; natural foodstuffs containing vitamin C.

Chemicals: 2,6-dichloroindophenol indicator solution, 0.25 g/liter; ascorbic acid tablets, 250 mg; metaphosphoric acid solution, 3% by weight.

1. Standardization of the Indophenol Solution. Check out a 100-ml graduated cylinder from the stockroom. Obtain a 250-mg tablet of vitamin C from your instructor and 100 ml of 2,6-dichloroindophenol solution. This solution nominally contains 0.250 g of indicator per liter of solution. The molecular weight of the indicator is 326.11. Calculate the molarity of this solution.

Place the vitamin C tablet on a sheet of paper and weigh on an analytical balance. Think about your result and comment in your notebook about your finding. If you don't find anything peculiar about this weight, think about it again after the next step.

Dissolve the vitamin C tablet in about 30–40 ml of water. Vitamin C is very soluble in water. *Don't wait for everything in the tablet to dissolve.* Quantitatively transfer this solution to your graduated cylinder, and dilute this solution carefully to 100 ml. Use your 1-ml pipet to transfer 1.00 ml of this solution to a clean beaker. Add exactly 99.0 ml of water and thoroughly mix. Transfer two 50.0-ml portions to each of two Erlenmeyer flasks. Add about ½ ml of 3% metaphosphoric acid to each flask. Rinse your buret first with water and then with *a few ml* of the indophenol solution. Then titrate each of the vitamin C solutions to the first permanent pink end point. The end point is reasonably sharp and your two values should agree within about 2%. Use the average value to calculate the milligrams of vitamin C corresponding to 1 ml of the indophenol solution.

Use molecular weights, your data, and the equation given above to check on the stoichiometry of your titration. Comment on your findings.

2. Determination of Vitamin C Content of a Foodstuff. You are to determine the vitamin C content of a foodstuff you have some interest in. *You are responsible for bringing this food to the laboratory for analysis.* Try to obtain a sample of a fresh vegetable, fruit, juice, or meat for this part of the experiment. Cereals or breads will also be all right.

Remember that you will be titrating to a faint pink end point. Therefore the technique is not too successful with beets and cranberry juice, but it has been successfully used with green beans and parsley.

Since vitamin C is quite water soluble and is stabilized by the presence of metaphosphoric acid, you are to extract the vitamin in an aqueous solution of $(HPO_3)_n$. Two different procedures are given for the extraction process followed by titration directions for the 50-ml solutions that result from either process.

(a)(1) *Extraction from Juices.* Amounts of 2–5 ml of citrus juices will provide enough vitamin for an accurate titration. Use the information in Table 38-1 as a basis for guessing what amounts of other juices might be needed to react with an accurately measurable quantity of indophenol solution.

Centrifuge the sample to remove most of the particulate matter. Then accurately pipet your predicted volume of the juice required for an accurate titration into a graduated cylinder containing 10 ml of 3% aqueous metaphosphoric acid. Dilute the sample to 50.0 ml with distilled water.

(a)(2) *Extraction from Solids.* Weigh out 2–4 grams of the solid. Place this and 15 ml of 3% aqueous metaphosphoric acid in a mortar. Grind it up with the mortar and pestle or use a Waring blender if available. Centrifuge the mixture and decant the supernatant (the solution on top) into a graduated cylinder. Put the wet solid residue back in the mortar, and continue the grinding with

another 15 ml of the 3% metaphosphoric acid. Centrifuge the mixture again, using the same centrifuge tubes as before. Decant the supernatant again into the graduated cylinder containing the first extract. Be careful not to lose any solution during these several transfers. If an appreciable amount of solids is suspended in the nominal 30 ml in the graduated cylinder, centrifuge the solution one more time and discard the solids. Finally dilute the solution in the cylinder to 50.0 ml with distilled water.

(b) *Titration.* Stir the solution in the graduated cylinder to insure that it is homogeneous. Transfer two 25.0-ml portions of the vitamin C solution to separate Erlenmeyer flasks. Titrate with the standardized 2,6-dichloroindophenol solution until a pink color that lasts several seconds appears. Proceed very slowly during the first titration, because only a mililiter or so may be required to reach the endpoint if your sample contained very little vitamin C. If less than 1 ml is required for the titration, repeat, using a larger food sample or a more dilute indicator solution. The two titration volumes should agree within about 4%. Average the two values and calculate the vitamin C content of your foodstuff. Report your results in milligrams per 100 ml if your sample was a juice, or in milligrams per 100 g if your sample was a solid.

3. Independent Experiment. Think up an interesting experiment for study of some additional aspect of the chemistry of vitamin C. Your independent project should not require more than 30 minutes to an hour of laboratory time. Students who have chosen to analyze a solid for part 2, or who must repeat part 2 because the initial concentration of vitamin was too low will not be expected to produce as much in this part of the experiment as will those who have managed to analyze a juice sample with little or no difficulty. Some suggestions for projects follow, but you are encouraged to try your own ideas.

(a) Analyze two peppers, green or red—a fresh one and one that's been sitting at room temperature for a week or so.

(b) Determine the pK_a of vitamin C. In one interesting article (Hurd, 1970) it is suggested that the acidity of ascorbic acid is due to the lack of a carboxylic group (see structures I and II in Figure 38-1). Nevertheless, as anyone who has ever drunk pure vitamin C solution can attest, the acid is a weak one. To determine its pK_a measure the pH of a solution that you have obtained by titrating to the end point with 1 M NaOH and then by adding a quantity of 1 M HCl that was equivalent to exactly one-half of the NaOH solution used. The pK_a will be equivalent to the pH you have measured. The 1 M HCl and 1 M NaOH will be provided.

(c) Determine and compare the vitamin C content of two or more juices—such as fresh orange juice and the frozen variety; fresh juice and canned juice; orange juice and boiled juice; orange juice and orange soda; or other pairs that occur to you.

(d) Quantitatively examine the vitamin C content of a dried breakfast juice, such as Tang.

Bibliography

Bessey, O. A., "A Method for the Determination of Small Quantities of Ascorbic Acid and Dehydroascorbic Acid in Turbid and Colored Solutions in the Presence of Other Reducing Substances," *J. Biol. Chem.* **126**, 771 (1938).

Bessey, O. A., and C. G. King, "The Distribution of Vitamin C in Plant and Animal Tissues, and Its Determination," *J. Biol. Chem.* **103**, 687 (1933).

Hurd, C. D., "The Acidities of Ascorbic and Sialic Acids," *J. Chem. Ed.* **47**, 481 (1970).

King, C. G., "Chemical Methods for Determination of Vitamin C." *Ind. and Eng. Chem.* **13**, 225 (1941).

Pauling, L., *Vitamin C and the Common Cold*, W. H. Freeman and Company, San Francisco (1970).

REPORT

38

The Chemistry of
Vitamin C

NAME

SECTION   LOCKER

INSTRUCTOR   DATE

DATA

1. Standardization of the Indophenol Solution

Calculate the molarity of the indophenol solution.

Molarity _____ M

Give the following in grams: the weight of the tablet and the paper _____ g; the weight of the paper alone _____ g; the weight of the tablet alone _____ g.

Comment on the weight of the tablet.

Give the following data.   1   2

Buret reading, 2nd   _____ ml   _____ ml

Buret reading, 1st   _____ ml   _____ ml

Volume of indophenol solution   _____ ml   _____ ml

Average volume of indophenol solution   _____ ml

Calculate the milligrams of vitamin C per milliliter of indophenol solution.

_____ mg

Check on the stoichiometry of the titration.

2. Determination of Vitamin C Content of a Foodstuff

Give the name of the foodstuff you analyzed. _____

Give the mass or volume of foodstuff. _____

Give the following data.

	1	2
Buret reading, 2nd	_____ml	_____ml
Buret reading, 1st	_____ml	_____ml
Volume of indophenol solution	_____ml	_____ml
Average volume of indophenol solution		_____ml

Calculate the vitamin C content of the foodstuff (in milligrams of vitamin C per 100 ml or 100 g of sample).

_____mg

3. Independent Experiment

Carefully describe all aspects of this portion of your experiment.

PROTEINS AND POLYSACCHARIDES

PRE-STUDY

There are three important classes of molecules in biological systems—proteins, polysaccharides, and nucleic acids. This experiment is designed to acquaint you with certain molecules from the first two classes.

Proteins

Proteins have numerous functions in all living systems. For example, hemoglobin transports oxygen from the lungs to myoglobin in the muscle, a variety of cytochromes cause biological oxidations, and the structural element of hair and nail is a protein. Perhaps the most important (and most amazing) of the proteins are the *enzymes*.

Enzymes are catalysts—molecules that increase the rate of a reaction. They are crucial for living organisms because, without their acceleration of chemical reactions, the metabolic processes would

be slowed so much that the organism would virtually cease to function. Catalysts perform such functions by lowering the activation energy, E_a, which is the minimum energy required to convert reactants to products (see Figure 39-1).

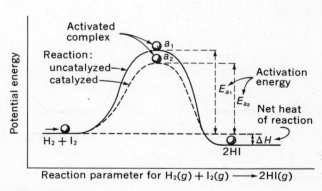

FIGURE 39-1
A potential energy diagram, illustrating the meaning of the terms "activation energy" and "activated complex," as well as the effect of a catalyst and the heat of reaction.

Some of the structural features of proteins are discussed in Experiment 36. Like all proteins, enzymes are large molecules composed of amino acids that are joined by peptide bonds, —NH—CO—. They have very specific three-dimensional geometries, which are responsible for their remarkable specificities. Because of their sizes and geometries, they are quite susceptible to extremes of temperature or *p*H, which destroy their effectiveness as catalysts.

The names of most enzymes end with the suffix "-ase". The two enzymes you will study in this experiment are *amylase* from your own saliva and *catalase* from a potato.

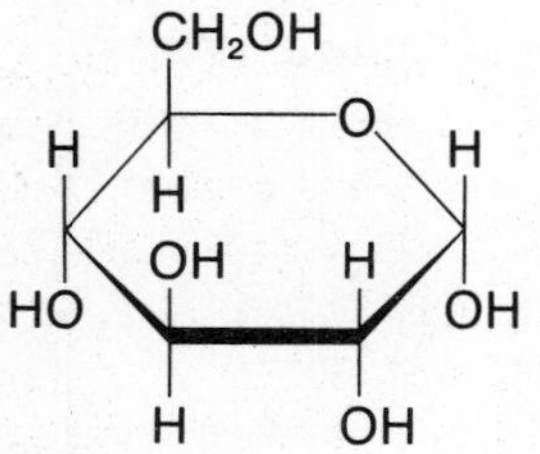

FIGURE 39-2
The structure of α-D-glucose.

Polysaccharides

The second class of molecules you will observe in this experiment are the polysaccharides. These molecules are polymers (or linked units) of the monosaccharides (organic compounds containing carbon, hydrogen, and oxygen in the ratio $(CH_2O)_x$, so that they belong to the class of compounds known as carbohydrates). The most common monosaccharide is glucose, $C_6H_{12}O_6$. A number of monosaccharides have this formula, but only glucose has the specific configuration shown in Figure 39-2.

In Figure 39-3, glucose molecules are joined by —C—O—C— bonds, called *glycosidic linkages,* to form α-maltose. In addition to the α-maltose structure, two glucose molecules can combine to yield α-isomaltose (Figure 39-4).

Thus the possibility exists that when additional molecules of glucose are added to these dimers (*two* linked molecules) to form starch, two different polymers may be produced. The two types of structures, linear and branched, are represented schematically in Figure 39-5.

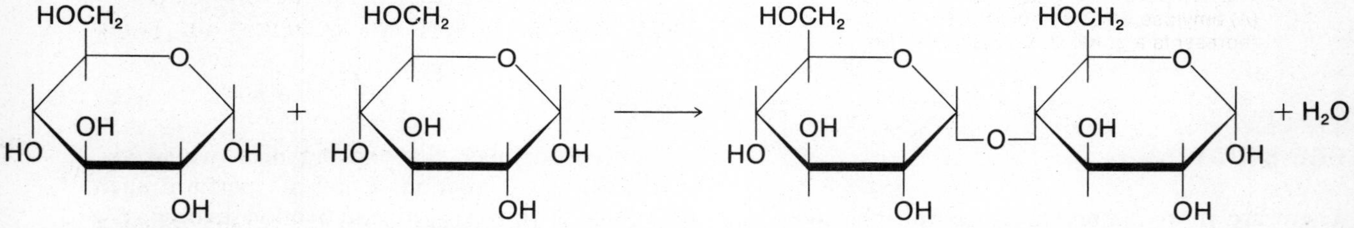

α-glucose α-glucose α-maltose

FIGURE 39-3
The formation of α-maltose from α-glucose.

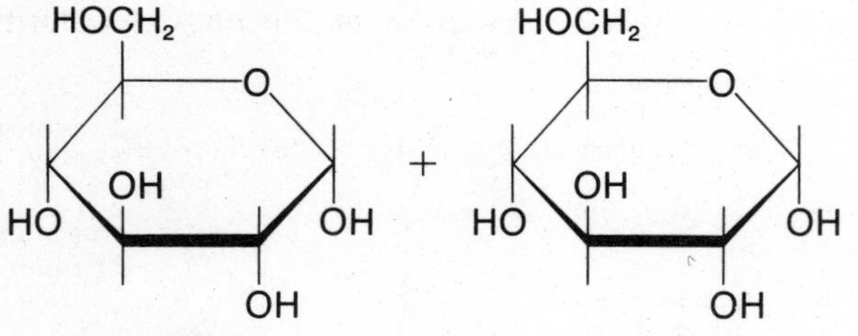

α-glucose α-glucose

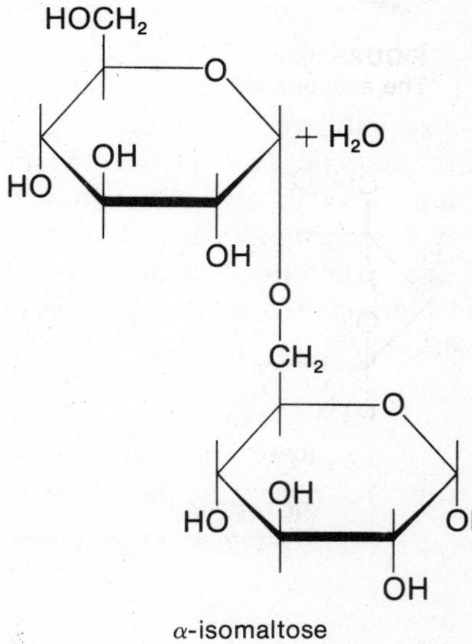

α-isomaltose

FIGURE 39-4
The formation of α-isomaltose from α-glucose.

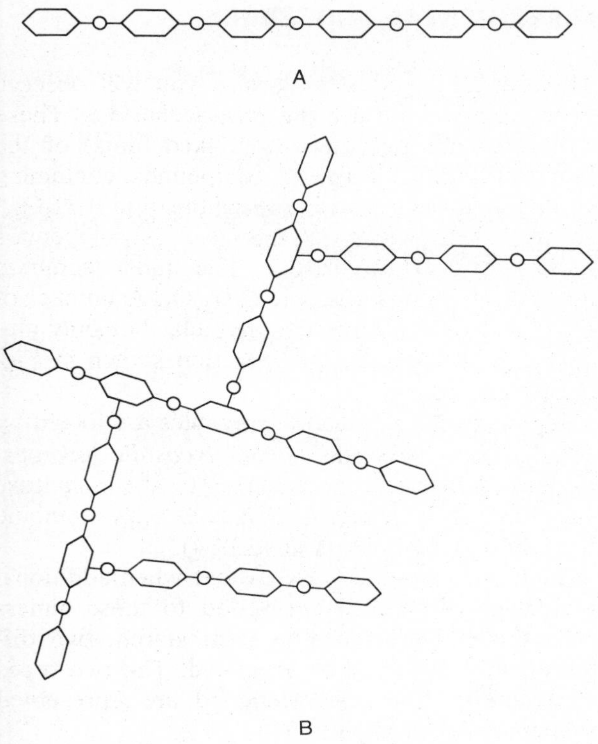

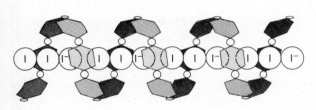

A

FIGURE 39-5
The two polysaccharides found in starch:
(A) amylose; (B) amylopectin. Each unit
represents a glucose molecule.

B

FIGURE 39-6
The amylose-iodine complex.

Experimental Method

In your study of the enzyme amylase, you will be concerned with its action on the polysaccharides amylose and amylopectin. This action can be observed in two ways. First, we can determine the rate of disappearance of starch by studying the rate at which the blue color of a colored starch complex fades. Amylose reacts with the I_3^- ion (produced by adding I^- to a saturated solution of I_2) by forming a helix around the linear I_3^- ions, as shown in Figure 39-6.[1] This helical compound is an intense blue color. Amylopectin forms similar complexes, which are a lighter, reddish color. Starch contains both of these polymers. Therefore, as the amylase in saliva hydrolyzes the polysaccharides back to glucose molecules (the reverse of the reaction shown in Figure 39-3), the starch–I_3^- solution turns from a very dark blue to violet to reddish brown and finally to a pale yellow, when there is no polymer left.

The second way to measure the action of amylase is to follow the rate of appearance of glucose molecules. A common biochemical test for the presence of most monosaccharides is Benedict's test. This test indicates the presence of reducing sugars—those monosaccharides which contain the group

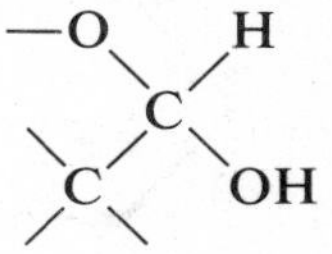

Reaction of this group with the citrate complex of Cu(II) in a hot alkaline solution yields gluconic acid and Cu_2O, which is rust colored and is insoluble in this solution (Figure 39-7). Thus if you observe the appearance of increasing amounts of this precipitate as the starch is converted to glucose, you will be able to follow the amylase activity.

[1]This structure is described also in Experiment 36.

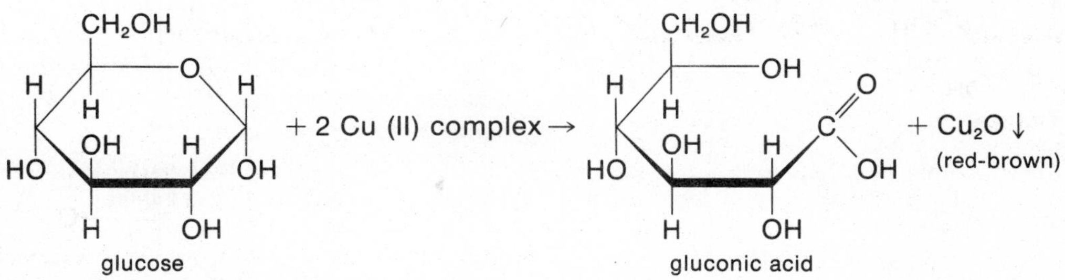

CH$_2$OH
glucose
+ 2 Cu (II) complex →
CH$_2$OH
gluconic acid
+ Cu$_2$O ↓
(red-brown)

FIGURE 39-7
The formula for the reaction that takes place when Benedict's test is applied.

The second enzyme you will study is catalase. Catalase is found in the tissues of most living organisms. Its function is to convert hydrogen peroxide to oxygen and water:

$$H_2O_2 \rightarrow H_2O + \tfrac{1}{2}O_2$$

It does this job exceedingly well. Enzymologists have tabulated turnover numbers for a variety of enzymes. This number gives the number of molecules of substrate (in this case, H_2O_2) that are converted to produce (water and oxygen) by each enzyme molecule per unit time. Catalase has a value of 50,000 sec^{-1} at 0°C. Can you visualize 50,000 molecules of hydrogen peroxide diffusing to the surface of one catalase molecule and being converted to products in one second?

Catalase belongs to a large class of proteins called the heme proteins. These proteins contain heme groups—a large organic molecule called a porphyrin ring with an iron atom in the center (Figure 39-8). The iron atom has an important function in catalysis. When anions, such as the cyanide or sulfide ions, are added, they form a complex with the iron and totally inhibit enzymatic activity.

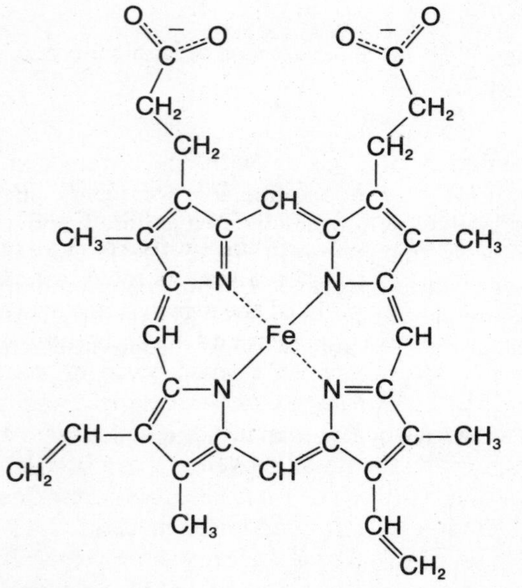

FIGURE 39-8
Structural formula for the heme group. This group is an essential component in hemoglobins, cytochromes, and enzymes such as catalase and peroxidase.

EXPERIMENTAL PROCEDURE[2]

Special Supplies: A potato; a knife; a flashlight; U-tube manometer; felt-tip pen; ruler.

Chemicals: 1% starch solution, 0.01 F KI_3, Benedict's reagent, 0.1 F $Hg(NO_3)_2$, 3% H_2O_2, 6 F $(NH_4)_2S$

1. The Hydrolysis of Starch by Salivary Amylase.

(a) *Preparation of Solution of Saliva.* Spend a few minutes thinking about a T-bone steak, or other delicacy of your choosing. Then transfer 2 ml of saliva to a 25-cc graduated cylinder. Dilute this fluid to 20 ml with distilled water, and thoroughly mix the solution.

(b) *Measurement of Rate of Hydrolysis by Disappearance of Starch–I_3^- Complex.* You will qualitatively measure the rate at which the blue color of this complex disappears. Four samples as described below, will be employed:

(1) An unhydrolyzed starch solution;
(2) The same solution, hydrolyzed with saliva for half a minute;
(3) The same solution, hydrolyzed with saliva for 2 minutes;
(4) The same solution, hydrolyzed with saliva for 10 minutes.

Label four test tubes 1, 2, 3, and 4 to correspond to the four samples. Add one drop of the 0.01 F I_3^- solution to four labeled test tubes. Add 15 ml of the starch solution to a 50-ml Erlenmeyer flask.

Add 1 ml of the 1% starch solution to tube 1 and note the color.

While noting the second hand of a clock, add 1.0 ml of the saliva solution to the starch solution in the flask, and immediately swirl. At 30 seconds, 2 minutes, and 10 minutes after mixing, pipet 1-ml samples into the appropriately numbered tubes. Mix and note the color of each. Compare the four colors, and describe qualitatively in your report.

(c) *Measurement of Rate of Hydrolysis by Appearance of $Cu_2O(s)$.* The same concept as that in (b) will be employed in this step, except that now you are to observe the appearance of glucose by its reaction with Benedict's solution to produce Cu_2O. Label the test tubes 1 through 4, as in (b). Each of the numbered labels will correspond to the same time intervals. Add 5 ml of Benedict's reagent to each. Again add 15 ml of the starch solution to a 50-ml Erlenmeyer flask.

---

[2]This procedure is based in part on an experiment originally written by Professor William Jolly, University of California, Berkeley.

Add 1 ml of the 1% starch solution to tube 1, shake, and set it aside.

Repeat the addition of 1.0 ml of the saliva solution to the 15 ml of starch, and remove 1-ml samples after 30 seconds, 2 minutes, and 10 minutes. Immediately add these samples to the test tubes containing the Benedict's solution and mix. The high pH of the Benedict's reagent stops the enzymatic hydrolysis.

The Benedict's test for glucose is now completed for all four solutions. Place the four test tubes in a boiling water bath for 5 minutes. This "develops" the color. Note and record the relative amounts of Cu_2O precipitate. You may find a flashlight helpful to detect a small amount of precipitate. Why?

(d) *Effects of Heat and of Heavy Metals on Amylase Activity.* Add 1.0 ml of saliva solution to each of three test tubes, numbered 1 through 3, and treat them as follows:

(1) Keep as a control test tube.
(2) Heat for 5 minutes at 100°C in a boiling water bath. Cool to room temperature.
(3) Add 5 drops of 0.1 F $Hg(NO_3)_2$, mix, and allow to stand 10 minutes.

Then add 5 ml of the starch solution to each test tube and thoroughly mix. After 10 minutes, perform the I_3^- and Benedict's tests on 1-ml samples from each test tube. (In one of the tests, the iodine color may be bleached out. Add more I_3^- solution until a color results.)

Tabulate your results on the report form, and try to explain them.

2. The Catalysis of H_2O_2 Decomposition by Catalase from Potato.

(a) *Preparation of Catalase Solution.* Peel a potato and weigh out a piece that is approximately 5 g. Cut it up into small 0.5-cm cubes. Add 10 ml of distilled water to the potato pieces in a small beaker. Allow the mixture to stand for 10 minutes, swirling occasionally. Decant the supernatant liquid into a centrifuge tube, and centrifuge the solid matter to the bottom of the tube.

(b) *Measurement of Rate of Decomposition of H_2O_2 upon Catalysis.* Obtain the U-tube shown in Figure 39-9, and add water to it as shown. Make sure that water does not block any other portion of the U-tube.

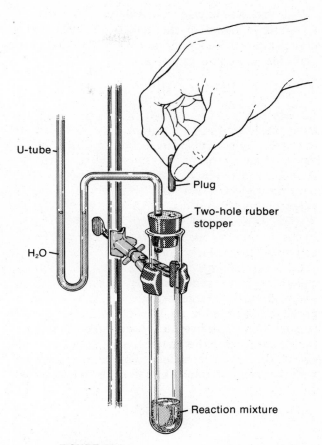

FIGURE 39-9
Apparatus for measuring catalase activity.

Transfer 2 ml of the clear supernatant potato extract from the upper part of the centrifuge tube to the test tube shown in the figure, add 3 ml of 3% H_2O_2, and swirl. After a few minutes, insert the plug in the open hole of the stopper, and mark the level of water in the open end of the U-tube, using either a felt-tip pen or a small piece of masking tape. Mark the level at one-minute intervals until the water level has reached the top. Record the distance the water level moved as a function of time.

(c) *Effect of Sulfide on Catalase Activity.* Repeat part (b), but this time add a drop of 6 F $(NH_4)_2S$ and wait a few minutes before adding the peroxide. Again record the position of the water level as a function of time.

OBSERVATIONS AND DATA

1. The Hydrolysis of Starch by Salivary Amylase

Measurement of rate of hydrolysis by disappearance of starch–I_3^- complex:

Tube	Time of hydrolysis (minutes)	Qualitative description of color
1	0	
2	0.5	
3	2	
4	10	

Comment on your results:

Measurement of rate of hydrolysis by appearance of $Cu_2O(s)$:

Tube	Time of hydrolysis (minutes)	Qualitative description of amount of precipitate
1	0	
2	0.5	
3	2	
4	10	

Comment on your results.

Effects of heat and of heavy metals on amylase activity:

Tube	Treatment	Results	
		I_3^- test	Benedict's test
1	Control	_____	_____
2	Heated	_____	_____
3	Hg^{2+} added	_____	_____

Comment on your results.

2. The Catalysis of H_2O_2 Decomposition by Catalase from Potato

Measurement of rate of decomposition of H_2O_2 upon catalase catalysis.

Time (minutes)	Water level	Time (minutes)	Water level
1	_____	6	_____
2	_____	7	_____
3	_____	8	_____
4	_____	9	_____
5	_____	10	_____

Make a graph plotting these data, and determine an approximate value of the rate of this reaction in the arbitrary units of cm/minute.

Rate_____cm/minute.

Effect of sulfide on catalase activity:

Time (minutes)	Water level	Time (minutes)	Water level
1	_____	6	_____
2	_____	7	_____
3	_____	8	_____
4	_____	9	_____
5	_____	10	_____

Again plot your data and determine an approximate value for the rate of this reaction in cm/minute.

Rate _____ cm/minute

APPENDIXES

A REVIEW OF FUNDAMENTAL MATHEMATICAL OPERATIONS

HOW TO APPROACH A CHEMICAL PROBLEM

Problems in first year chemistry are no more difficult than other simple mathematical relationships that you have encountered. If they seem to be, it is only because much of the terminology, ideas, and units of measurement is new to you. Start this way:

1. *Think about the problem until you really understand the underlying chemical principles,* and can pick out the data pertinent to its solution. *Then state these essential facts and relationships as a mathematical equation.* Do not substitute rote memory for logical reasoning. In this manual, as each new type of problem is encountered, you will find a discussion of the method, together with solved examples. Study each of these again and again until you understand thoroughly how the particular mathematical equation expresses the fundamental idea most concisely. It should not surprise you that the natural laws of chemistry and physics are most easily stated in terms of simple mathematical relationships.

2. *Develop the fundamental equation, together with its proper units or dimensions. Then apply arithmetic operations (multiplication, division, addition, and subtraction) to calculate the desired quantity.* Do not proceed in a random trial-and-error fashion. Be logical. Consider these two illustrations of an important physical property of a substance, the density.

Example 1. A rectangular bar of iron has the dimensions 1.10 cm by 2.17 cm by 6.41 cm. It weighs 121.7 g. What is the density of iron?

Recall that density is an intrinsic physical property, independent of the amount of the sample. It is defined as the ratio of the mass to volume, that is, mass per unit volume, as expressed by the equation

$$\text{density} = \frac{\text{mass}}{\text{volume}}$$

The mass is given, and the volume is simply the product of the three dimensions given. Note that *units* of measurement, as well as the numerical values, must be included. We therefore write

$$\text{density} = \frac{121.7 \text{ g}}{1.10 \text{ cm} \times 2.17 \text{ cm} \times 6.41 \text{ cm}}$$

Performing these operations gives

$$\text{density} = \frac{121.7 \text{ g}}{15.3 \text{ cm}^3} = 7.89 \text{ g/cm}^3$$

Note that the units obtained, g/cm³, are the correct units for density, confirming that the correct mathematical operation has been performed. Compare this example with the following.

Example 2. An experiment calls for 204 g of mercury. What volume of liquid mercury should be measured out in a graduated cylinder to obtain the desired weight? (The density of mercury at room temperature is 13.6 g/cm³.)

The student confronted with this problem, Mr. M. I. Lucky (Figure AA-1) is hopelessly confused—he has gambled by the "Monte Carlo" or "Las Vegas" method, hoping for the best. But note that since there are four possible ways of setting up the data, he has only one chance in four of being right.

The alternative to this method of chance is, of course, to reason out a solution based on an understanding of the terms and concepts involved, followed by a *factor-the-units* check. Ms. I. M. Shoor (Figure AA-2) follows this logical reasoning. Since density is defined as mass per unit volume

$$\text{density} = \frac{\text{mass}}{\text{volume}}$$

the volume will be the mass divided by the density, or

$$\text{volume} = \frac{\text{mass}}{\text{density}} = \frac{\text{g}}{\text{g/cm}^3} = \text{cm}^3$$

Inserting the numerical data of this problem gives

$$\text{volume} = \frac{204 \text{ g}}{13.6 \text{ g/cm}^3} = 15.0 \text{ cm}^3$$

Since the *factor-the-units* process gives the correct units, Ms. Shoor is "sure" her method is correct. (See also Appendix B, The Use of Dimensional Units.)

If the simple mathematics required troubles you, review the following sections until you understand the reason for each mathematical operation.

FIGURE AA-1
M. I. Lucky has one chance in four. [From J. W. Hagen, *Empirical Chemistry*, W. H. Freeman and Company. Copyright © 1972.]

FIGURE AA-2
I. M. Shoor uses unit analysis as a check on calculations.

HOW TO SOLVE
AN ALGEBRAIC EQUATION

An equation represents two quantities that are equal, or equivalent. Most student difficulties in handling simple equations arise from a failure to realize that if the two sides of the equation are to remain equal, *the same operations must be performed on each side of the equation.* The following rules apply:

1. *Each side of an equation may be multiplied or divided by the same quantity.*

Example 3. If we wish to know the weight of a liter of carbon tetrachloride, after we have found by experiment that 50 ml of it weighs 80 grams, we may proceed as follows.

State the facts as a unit conversion factor equation, which may be written as

$$\frac{50 \text{ ml } CCl_4}{80 \text{ g } CCl_4} = \frac{0.62 \text{ ml}}{g}$$

This is entirely equivalent to writing

$$\frac{80 \text{ g } CCl_4}{50 \text{ ml } CCl_4} = \frac{1.6 \text{ g}}{ml}$$

The problem can be solved by the use of these unit conversion factors by writing

$$1 \text{ liter} \times \frac{1000 \text{ ml}}{1 \text{ liter}} \times \frac{1.6 \text{ g}}{ml} = 1600 \text{ g}$$

Notice that we have in effect multiplied both sides of our conversion factor equations by 1000.

Example 4. Transpose the gas law equation

$$\frac{P_1 V_1}{T_1} = \frac{P_2 V_2}{T_2}$$

in order to obtain values for P_1 and for T_2.

Dividing both sides by V_1 and multiplying both sides by T_1, and cancelling the common factors, we have

$$\frac{P_1 V_1 \times T_1}{T_1 \times V_1} = \frac{P_2 V_2 \times T_1}{T_2 \times V_1} \quad \text{or} \quad P_1 = \frac{P_2 V_2 T_1}{T_2 V_1}$$

Similarly, to solve for T_2, multiply both sides by T_1 and T_2, divide both sides by P_1 and V_1, and cancel the common factors:

$$\frac{P_1 V_1 \times T_1 \times T_2}{T_1 \times P_1 \times V_1} = \frac{P_2 V_2 \times T_1 \times T_2}{T_2 \times P_1 \times V_1}$$

or

$$T_2 = \frac{P_2 V_2 T_1}{P_1 V_1}$$

Note again how this operation results in transposing the various quantities from numerator to denominator, or vice versa, on opposite sides of the equation.

---

2. *Any given quantity may be added to, or subtracted from, each side of the equation.* This is frequently used to transpose a sum or difference to the opposite side of the equation, with the sign changed.

Example 5. Given

$$12x + 4 = 10x$$

Subtracting 4 from each side of the equation, we have

$$12x + 4 - 4 = 10x - 4$$
or
$$12x = 10x - 4$$

We may similarly transpose the $10x$ to the other side, with its sign changed:

$$12x - 10x = -4$$

If we combine by terms, we have

$$2x = -4$$

and if we then divide by 2, we have

$$x = -2$$

---

3. *First-degree equations involve only the first power of the unknown.* Note the preceding example.

4. *Second-degree equations involve the square of the unknown quantity,* and are called *quadratic equations.* The general form of such an equation may be written

$$ax^2 + bx + c = 0$$

The solution of this results in the following formula for the value of x:

$$x = \frac{-b \pm \sqrt{b^2 - 4ac}}{2a}$$

For example,

$$5x^2 + 3x - 10 = 0$$

$$x = \frac{-3 \pm \sqrt{3^2 - (4)(5)(-10)}}{(2)(5)}$$

$$= \frac{-3 \pm 14.4}{10} = 1.14 \quad \text{or} \quad -1.74$$

(In scientific work a negative root is generally without physical significance, and is, therefore, discarded.)

5. *One may take the square, or square root, of both sides of an equation.* For example,

$$25x^2 = 100$$

$$5x = \pm 10$$

$$x = \pm 2$$

PROPORTION

A fraction expresses a division—that is, a ratio between the numerator and the denominator. Two such ratios (or fractions), which are equal to one another, constitute *a proportion*.

Example 6. Direct Proportion: The corresponding volumes and absolute temperatures of a given quantity of gas have a constant ratio, or are directly proportional. This is Charles' law. It may be expressed by the equations

$$V = kT \qquad \text{or} \qquad \frac{V}{T} = k$$

where T represents the absolute temperature, and k is a constant. The absolute or Kelvin scale of temperature is related to the celsius (centigrade) scale by the equation $K = °C + 273$, or $T = t + 273$. Since any two corresponding values of V/T are equal to k, we may write the proportion

$$\frac{V_1}{T_1} = \frac{V_2}{T_2}$$

or, by transposing, as

$$\frac{V_1}{V_2} = \frac{T_1}{T_2}$$

If the temperature of 45 ml of hydrogen gas is decreased from 20°C (293°K) to −100°C (173°K), the decreased volume, V_1, is given by

$$\frac{V_1}{45 \text{ ml}} = \frac{173°}{293°}$$

Multiplying both sides of this equation by 45 ml, we have

$$V_1 = \frac{173°}{293°} \times 45 \text{ ml} = 27 \text{ ml } H_2$$

Calculate the volume at 100°C. (*Answer:* 57 ml.)

Example 7. Inverse Proportion: The corresponding volumes and pressures of a given quantity of gas bear an inverse proportional relationship to one another. This is known as Boyle's law, and may be expressed by the equations

$$V = \frac{k}{P} \qquad \text{or} \qquad PV = k$$

Since any two corresponding values of PV are equal to k, we may write the proportion

$$P_1V_1 = P_2V_2$$

or, by transposing, as

$$\frac{V_1}{V_2} = \frac{P_2}{P_1}$$

If the pressure of 45 ml of oxygen gas is increased from 1.0 atmosphere to 3.0 atmospheres, the decreased volume, V_1, is given by

$$\frac{V_1}{45 \text{ ml}} = \frac{1.0 \text{ atm}}{3.0 \text{ atm}}$$

or

$$V_1 = \frac{1.0 \text{ atm}}{3.0 \text{ atm}} \times 45 \text{ ml } O_2 = 15 \text{ ml } O_2$$

You may verify other corresponding values as follows. When $P_1 = 5.0$ atm, $V_1 = 9.0$ ml O_2, and when $P_1 = 0.50$ atm, $V_1 = 90$ ml O_2.

GRAPHICAL RELATIONS AND EXPERIMENTAL DATA

Scientists very frequently make a *graph* of their experimental data as means of discovering fundamental relationships, or of comparing their data with some known law. Also, the extent to which their individual experimental values lie on a smooth curve is an indication of the precision of their measurements.

The data of Example 6, in the preceding discussion on proportion, are plotted in the graph in Figure AA-3. The graph shows that a *direct proportion plots as a straight line*. Only two points are needed to fix the line, from which any other values then may be read. If the line were extrapolated to a volume of zero, it would cross the temperature axis at −273°C. Of course, hydrogen would behave less like a perfect gas as the temperature is decreased and would change to the liquid (and finally to the solid) state. By plotting precise experimental

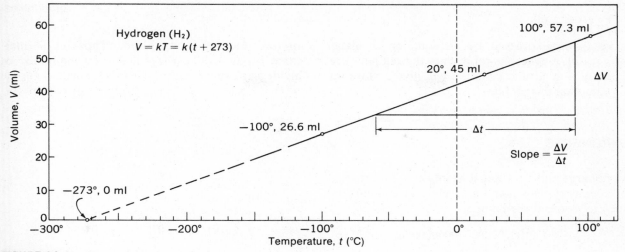

FIGURE AA-3
The relation between the temperature and the corresponding volume of a sample of gas, such as hydrogen. A direct proportion, as represented by Charles' law, $V = kT = k(t + 273)$, plots as a straight line.

data for the relative volumes and temperatures, or pressures and temperatures, of a sample of hydrogen, one can determine the value of absolute zero.

The change in volume for a given change in temperature is called the *slope* of the graph. In the equation, $V = k(t + 273)$ (see Figure AA-3), the slope is equal to k since we may write

$$V_1 = k(t_1 + 273)$$

and

$$V_2 = k(t_2 + 273)$$

so that subtraction of the first equation from the second gives

$$V_2 - V_1 = k(t_2 - t_1)$$

or

$$k = \frac{V_2 - V_1}{t_2 - t_1} = \frac{\Delta V}{\Delta t}$$

where the greek letter delta, Δ, is used to symbolize the change in a variable.

The point at which the straight line plot crosses the temperature axis is called the *intercept* on this axis. In Figure AA-3 the intercept of the plot is at $-273°C$. This point is taken as zero in the Kelvin or absolute scale of temperature so that $K = °C + 273$.

In Figure AA-4, the graph for the data of Example 7 (inverse proportion) shows that *an inverse proportion plots as a hyperbola*. A number of points are necessary to fix the curve.

Figure AA-5 shows a characteristic curve for the change in solubility of a salt with temperature. This is not a first-order or linear curve, and would

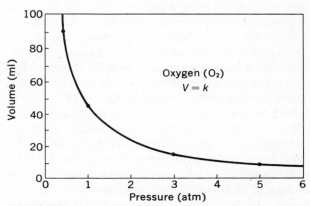

FIGURE AA-4
The relation between the pressure and the corresponding volume of a sample of a gas, such as oxygen. The graph of an inverse proportion, as represented by Boyle's law, $PV = k$, is a hyperbola.

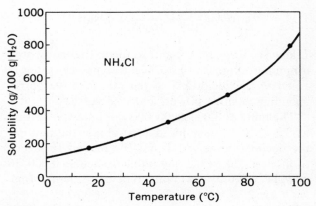

FIGURE AA-5
The change in solubility of ammonium chloride with temperature. Such a solubility graph is seldom linear, but often can be expressed by an equation of the type $y = a + bx + cx^2 + \cdots$.

have to be expressed by an equation in higher powers of x, where x represents the temperature. Usually a quadratic equation will give a close approximation to the data.

EXPONENTS

Very Large and Very Small Numbers. The student of chemistry uses many extremely large numbers, such as Avogadro's number:

$$602210000000000000000000$$

and many extremely small numbers, such as a sulfide ion concentration of 0.00000000000000031 M. The reason for such numbers obviously is due to the extremely minute sizes of atoms and molecules, and to the correspondingly enormous numbers of them that can exist even in a small space.

Since such numbers are awkward to handle, they are often expressed more conveniently by writing them as some simple number times ten raised to a power (the power is indicated by an *exponent*). Thus, Avogadro's number may be written 0.60221×10^{24}, or 6.0221×10^{23}. The exponent 23 means that 10 is raised to the 23rd power. The number 6.0221×10^{23} means that when the decimal point in 6.0221 is moved 23 places to the right, the indicated number is obtained. Any number such as 129,000 may be written in a number of equivalent forms, as, for example: 129×10^3, 1.29×10^5, or 0.129×10^6.

The Meaning of a Negative Exponent. In handling negative exponents, the student should note that

$$10^{-4} = \frac{1}{10^4} = \frac{1}{10000} = 0.0001$$

Likewise, $10^3 = 1/10^{-3}$, and so forth. In other words a number expressed as a negative power of ten is equal to the reciprocal of ten raised to the corresponding positive power, and vice versa.

The number 0.00000000000000031 may be written as 3.1×10^{-16}, meaning that to obtain the number the decimal point in 3.1 must be moved 16 places to the left. Likewise, any number such as 0.000736 may be written in a number of equivalent forms, such as: 7.36×10^{-4}, 736×10^{-6}, or 0.0736×10^{-2}.

Multiplication and Division of Exponential Numbers. *To multiply* numbers written in exponential notation, *add* the exponents; and *to divide* the numbers,

subtract the exponents. Other factors in the indicated product or quotient are to be multiplied or divided as usual.

EXAMPLES OF MULTIPLICATION

$2^2 \times 2^3 \times 2^7 = 2^{2+3+7} = 2^{12}$,
$10^5 \times 10^{-2} = 10^3$,
$(2 \times 10^3)(4 \times 10^2) = 8 \times 10^5$.

EXAMPLES OF DIVISION

$10^7/10^3 = 10^{7-3} = 10^4$,
$10^2/10^{-5} = 10^7$,
$(6 \times 10^2)/(2 \times 10^3) = 3 \times 10^{-1} = 0.3$.

Powers of Exponential Numbers. The exponent is multiplied by the power desired, as in the following examples.

$$(10^2)^3 = 10^2 \times 10^2 \times 10^2 = 10^6,$$

$$(2 \times 10^3)^4 = 16 \times 10^{12}.$$

Roots of Exponential Numbers. The exponent is divided by the root desired. In case it cannot be divided evenly, the exponent number must be changed to a form so that it *can* be divided, as in the following examples.

The square root of $5^6 = (5^6)^{1/2} = 5^{6/2} = 5^3$,
   The cube root of $8 \times 10^{12} = (8 \times 10^{12})^{1/3}$
      $= 2 \times 10^4$,
   The square root of $5 \times 10^{-7} = (5 \times 10^{-7})^{1/2}$
      $= (50 \times 10^{-8})^{1/2} = 7.1 \times 10^{-4}$.

Addition and Subtraction of Exponential Numbers. The numbers first must be changed to a nonexponential form, or to the same power of ten, as in the following examples.

$$10^2 + 10^3 = 100 + 1000 = 1100 = 1.1 \times 10^3,$$

$$(6 \times 10^{-2}) - (2 \times 10^{-3})$$

$$= (6 \times 10^{-2}) - (0.2 \times 10^{-2})$$

$$= 5.8 \times 10^{-2} = 0.058$$

LOGARITHMS

The common logarithm of a number is the power to which 10 must be raised to obtain the number; that is, *a logarithm is an exponent and the preceding rules for exponents are applicable.*

$$\log_{10} x = y$$

$$x = 10^y$$

The logarithm of all numbers that are integral powers of 10 are whole numbers:

Log of: 100, or 10^2 = 2,

10, or 10^1 = 1,

1, or 10^0 = 0,

0.1, or 10^{-1} = -1,

0.01, or 10^{-2} = -2, and so on.

The logarithms of all other intervening numbers, which are not integral powers of 10, will be made up of two parts: a whole number (the *characteristic*) and a decimal fraction (the *mantissa*). A table of logarithms gives only the mantissas.

The use of logarithms in the solution of problems shortens the labor of calculation, particularly where a series of multiplications and divisions is to be performed.

Example 8. Using the four-place table of logarithms (Appendix C, Table 15), find the logarithm of 174.5.

The mantissa depends on the series of digits, without regard to the decimal point. Mantissas are always given as positive numbers in logarithm tables.

Find the number in the table opposite 17, in column 4 2405
Find the number in "Proportional parts" 5, row 17 12
The sum, with a decimal point prefixed, is the mantissa .... .2417

The characteristic depends only on the decimal point in the number. For numbers greater than 1, it is a *positive integer* that is one less than the number of digits to the left of the decimal point. For numbers less than 1, it is a *negative integer* that is one more than the number of zeros immediately following the decimal point. In this example, the characteristic is 2.

The log of 174.5 is therefore 2.2417.

The log of 0.001745 is $\bar{3}.2417$. This means $-3 + 0.2417$ or, written as one negative number, it means -2.7583.

Example 9. Find the number corresponding to the logarithm 1.5280.

In the table, the mantissa next lower than 5280 is 5276, which is 4 units too small, and which corresponds to the digits 337. In the proportional parts columns of the same row 33, find 4, which is in proportional parts column 3. This is the fourth digit in the number. The sequence of

digits is 3373. Since the characteristic is 1, there are two digits to the left of the decimal point, and the number is 33.73.[1]

To multiply numbers, add their logarithms and find the antilogarithm (number corresponding to a logarithm) of this sum. *To divide numbers*, subtract their logarithms and find the antilogarithm of this difference. In a series of consecutive multiplications and divisions, it is convenient to add the cologarithms of the divisors, instead of subtracting their logarithms. The *cologarithm* of a number is found by subtracting its logarithm from 10, and then appending -10. For example, the logarithm of 174.5 is 2.2417. The cologarithm is 7.7583 $-$ 10.

Example 10. Find the volume at standard conditions of 253 ml of oxygen measured at 25°C (298°K) and 742 Torr.

Volume at 0°C and 760 Torr

$$= 253 \text{ ml} \times \frac{273°}{298°} \times \frac{742 \text{ Torr}}{760 \text{ Torr}} = ?$$

log 253 = 2.4031
log 273 = 2.4362
log 742 = 2.8704
log 298 = 2.4742, colog 298 = 7.5258 $-$ 10
log 760 = 2.8808, colog 760 = 7.1192 $-$ 10
22.3547 $-$ 20
= 2.3547

Antilog of 2.3547 = 226.3, or 226 ml volume at standard conditions.

To find a given power of a number, multiply its logarithm by the power desired, and find the antilogarithm of this product. For example, find the fifth power of 15, or 15^5.

The log of 15 is 1.1761. Multiplying by 5, we have 5.8805.

Antilog of 5.8805 = 759500, or $7.6 \times 10^5 = 15^5$

To find a given root of a number, divide its logarithm by the root desired, and find the antilogarithm of this quotient. Note that if the characteristic is negative, the entire logarithm must be changed to

[1]It is possible, by interpolation in the proportional parts columns, to determine the logarithm of five-digit numbers, and vice versa. However, most experimental work in this course is significant to three figures only, so that it is not even necessary to use the proportional parts columns at all.

a negative number before dividing by the root. For example, find the cube root of 0.00248.

$\log 0.00248 = \bar{3}.3945 = -3 + 0.3945 = -2.6055$
$\frac{1}{3}$ of $-2.6055 = -0.8685 = \bar{1}.1315$
antilog $\bar{1}.1315 = 0.1354$, which is the cube root of 0.00248.

(An alternate and perhaps simpler procedure is to transform the number 0.00248 to 2.48×10^{-3}, and then take the cube root of each of these factors separately.)

THE SLIDE RULE

The slide rule scales that are used for multiplication and division are graduated in lengths proportional to the logarithms of the digits from 1 to 10, but the digits are placed on the scale, rather than the logarithms. Hence, to multiply numbers, add the lengths; and to divide numbers, subtract the lengths of the scales. This is accomplished by sliding the movable scale into positions so that the lengths can be added or subtracted mechanically. The details of operation, which are quite easily mastered, are given in the manual accompanying the instrument.

Since a course in chemistry necessitates a considerable amount of calculation, you are encouraged to obtain and use a slide rule. An ordinary inexpensive 10-inch rule is satisfactory. It can be read to about three significant figures, a precision sufficient for most calculations in a basic course.

APPROXIMATE CALCULATIONS

A quick, mental, and quite accurate estimate of the answer for a fairly complicated calculation can easily be done. A quick check on the proper position of the decimal point (a common error) saves much time. *This technique is worth learning.* Just round off the numbers and express them in convenient exponential form, usually with one digit to the left of the decimal. Then estimate mentally the required multiplication and division.

Example 11. Suppose a large rectangular golf course area is 754 ft by 2528 ft. How many acres does it contain? (43,560 ft² equals 1 acre.)

The area will be $\dfrac{754 \text{ ft} \times 2528 \text{ ft}}{43,560 \text{ ft}^2/\text{acre}}$

First round off the numbers and express exponentially:

$$\frac{8 \times 10^2 \times 2 \times 10^3}{4 \times 10^4} \frac{\text{ft}^2}{\text{ft}^2/\text{acre}}$$

Now combining the terms and estimating mentally the indicated multiplication and division, we have

$$\frac{8 \times 2 \times 10^5}{4 \times 10^4} = \frac{16}{4} \times 10^1 = 4 \times 10^1 \text{ acres}$$

Compare this result with the actual calculation, 43.8 acres.

Example 12. Solve the following:

$$\frac{6148 \times 0.0285 \times 3.246}{0.000489 \times 8.75}$$

As before, round off the numbers and express exponentially:

$$\frac{6 \times 10^3 \times 3 \times 10^{-2} \times 3}{5 \times 10^{-4} \times 9}$$

Then collect these together and estimate the multiplication and division:

$$\frac{6 \times 3 \times 3 \times 10^1}{5 \times 9 \times 10^{-4}} = \frac{6}{5} \times 10^5 = 1.2 \times 10^5$$

The exact result is 1.33×10^5.

EXERCISES

NOTE: You may refer to the answers to these problems in Appendix D, *after* you have solved them.

1. $d = \dfrac{m}{v}$. What does m equal?

2. $x - 2 = 4 + 4y$. What does x equal?

3. $\dfrac{2}{x} - \dfrac{1}{y}$. Reduce to a common denominator.

4. $y - (y - b + a)$. Simplify.

5. $x^2 - 9 = 16$. What does x equal?

6. $y = \dfrac{4 \times 10^4 \times 6 \times 10^{-1}}{3 \times 10^2}$. What does y equal?

7. What is the square root of 16,900?
8. What is 0.0096 divided by 0.06?
9. What is 8% of 12?
10. 7 is 5% of what number?
11. Express these numbers in the exponential form:
    - (a) 1000
    - (b) 200
    - (c) 5500
    - (d) 0.1
    - (e) 0.001
    - (f) 0.004
    - (g) 0.000000000003
    - (h) 750,000,000,000
    - (i) 2,006,000
    - (j) 0.00204

12. Change the decimal point in these expressions without changing the value of the number, so that there is one digit to the left of the decimal point.
    - (a) 42.6×10^3
    - (b) $41,075 \times 10^{-5}$
    - (c) 0.375×10^{-4}
    - (d) 0.07287×10^2
    - (e) 0.000465×10^3
    - (f) 6022×10^{20}
    - (g) $30,103$
    - (h) 0.625×10^4

13. Simplify these expressions as indicated, by multiplying, dividing, taking a power, extracting a root, adding, or subtracting.
    - (a) $10^2 \times 10^4$
    - (b) $10^3 \times 10^3 \times 10^{-2}$
    - (c) $(4.6 \times 10^4) - (3 \times 10^3)$
    - (d) $(5.42 \times 10^{-1}) + (1.3 \times 10^2)$
    - (e) $(10^3)^2 (10^2) (10^{-5})$
    - (f) $10^6 / 10^4$
    - (g) $(10^2)(10^{-3})(10^{-2})$
    - (h) $(2 \times 10^5)(4 \times 10^6)$
    - (i) $(2.5 \times 10^3)^2$
    - (j) $(25 \times 10^6)^{1/2}$
    - (k) $\dfrac{(10^3)(16 \times 10^6)^{1/2}}{10^4}$
    - (l) $\sqrt{3.6 \times 10^5}$
    - (m) $\dfrac{(4.6 \times 10^4)(2.1 \times 10^{-2})}{(3 \times 10^2)(2.3 \times 10^{-4})}$
    - (n) $\dfrac{(x^2)(y^3)(2x^3)}{(x^4)^{1/2}(y^{-2})(z^{-1})}$

14. Express the following scientific laws as mathematical equations.
    - (a) The pressure (P) of a gas, at constant volume (V), is directly porportional to its absolute temperature (T). (Note: Where a proportion is stated, one may replace the proportionality sign by an equal sign and a proportionality constant; that is, if x is directly proportional to y, we may write it $x = ky$.)
    - (b) The kinetic energy of a particle in motion is equal to one-half the product of its mass times the square of its velocity.
    - (c) The atomic number (Z) of an element is inversely proportional to the square root of the wave length (λ) of its characteristic X-ray spectra.
    - (d) Two electrically charged particles (ions) of opposite charge will attract one another with a force which is directly proportional to the product of their charges (e_1 and e_2), and inversely proportional to the square of the distance (d) between them.
    - (e) At constant pressure, the volume of a gas changes by 1/273 of its volume at 0°C, for each degree of temperature change.

15. (a) At constant temperature, the masses (m_1 and m_2) and the squares of the average velocities (v_1 and v_2) of the molecules of two gases are inversely proportional to one another. State this relation as an equation, and then use it to solve the following problem.
    - (b) If the average velocity of methane molecules (CH_4) is 4×10^5 cm/sec at a given temperature, what will be the average velocity of sulfur dioxide molecules at this same temperature?

16. Look up the logarithm of each of the following.
    - (a) 146.8
    - (b) 7.408
    - (c) 0.003682
    - (d) 50
    - (e) 0.6023×10^{24}
    - (f) 1.8×10^{-5}

17. Find the antilogarithms of which the following are the logarithms.
   (a) 2.4829
   (b) 1.0654
   (c) 4.5542
   (d) $9.8451 - 10$
   (e) $6.7410 - 10$
   (f) $18.6275 - 20$

18. Solve the following by logarithms, and then check your answers by the slide rule.
   (a) The volume of a sample of gas at standard conditions:

   $$267 \text{ ml} \times \frac{273°K}{305°K} \times \frac{734 \text{ Torr}}{760 \text{ Torr}} =$$

   (b) The number of molecules in a drop of water:

   $$\frac{0.050 \text{ g}}{18 \text{ g/mole}} \times 0.6023 \times 10^{24} \text{ molecules/mole} =$$

   (c) The volume of a drop of water as steam at 100°C and 760 Torr pressure:

   $$\frac{0.050 \text{ g}}{18 \text{ g/mole}} \times 22{,}400 \frac{\text{ml}}{\text{mole}} \times \frac{373°K}{273°K} =$$

19. Solve by the simplest procedure practical.
   (a) $(6.5 \times 10^{-1})^2$
   (b) $(3.75 \times 10^2)^3$
   (c) $(4.025)^2(1.234)^3$
   (d) $(3125)^{1/5}$
   (e) $(3.1 \times 10^2)^3(1.4 \times 10^{-3})^2$
   (f) $(0.02478)^{1/2}$
   (g) $(8.1 \times 10^{11})^{1/2}$
   (h) $(4.9 \times 10^{-3})^{1/2}(3.0 \times 10^4)$

20. Find the value of x in the following binomial expressions.
   (a) $x^2 + 6x - 12 = 0$
   (b) $2x^2 + 7x - 14 = 0$
   (c) $x^2 + (1.7 \times 10^{-5})x - (1.7 \times 10^{-5}) = 0$
   (d) $x(x + 0.01) = 3 \times 10^{-3}$

THE USE OF DIMENSIONAL UNITS.
THE TREATMENT OF EXPERIMENTAL ERRORS

THE USE OF DIMENSIONAL UNITS

The measurement of any physical quantity involves two factors: the numerical quantity itself, and the unit or units in which the measurement is made. The units are the *dimensions* of the measurement. For example, if a piece of silver is measured, and the dimensions are reported as 3.00 wide and 5.00 long, the information is of little use—even the shape is uncertain—for we do not know if both quantities are expressed in the same units. But if the dimensions are reported as 3.00 cm wide and 5.00 cm long, the information adequately describes both the size and the shape.

In the solution of all problems on physical quantities the units must be given. As with any other quantity, units may be subjected to the mathematical operations of multiplication and division. Thus, the surface area (A) of one side of the above piece of silver in square centimeters (cm^2) is obtained by multiplying the units as well as the numbers:

$$A = 3.00 \text{ cm} \times 5.00 \text{ cm} = 15.0 \text{ cm}^2$$

If the thickness is given as 2.00 mm and we wish to compute the volume (V), we must first convert all dimensions to the same units; that is, 2.00 mm = 0.200 cm. We then calculate the volume in cubic centimeters (cm^3) by multiplying both the numbers and the units:

$$V = 15.0 \text{ cm}^2 \times 0.200 \text{ cm} = 3.00 \text{ cm}^3$$

If the mass (M) of this piece of silver is found to be 31.5 g, we may calculate the density (D) of silver in accordance with the defining formula for density:

$$D = \frac{M}{V} = \frac{31.5 \text{ g}}{3.00 \text{ cm}^3} = 10.5 \frac{\text{g}}{\text{cm}^3}$$

The units, or dimensions, of density, are g/cm^3; they are read as "grams per cubic centimeter." Since these units are unlike the dimensions of weight and volume, they cannot be reduced further and are left simply with the indicated division.

Units frequently may cancel out in the processes of division, as shown in the following examples.

Example 1. Compare the relative weights of equal volumes of the heaviest metal, osmium (density 22.48 g/cm³), and of the lightest metal, lithium (density 0.53 g/cm³).

$$\frac{\text{Density of osmium}}{\text{Density of lithium}} = \frac{22.48 \text{ g/cm}^3}{0.53 \text{ g/cm}^3} = 42$$

The units cancel out, and the result, 42, a ratio of two densities, is a pure number without dimensions.

Example 2. If the density of silver is 10.5 g/cm³, what will a block of silver weigh if its measurements are 4.00 cm, 5.00 cm, and 10.0 cm?

The volume is 4.00 cm × 5.00 cm × 10.0 cm = 200 cm³. If we transpose the formula for density,

$$M = D \times V = 10.5 \frac{\text{g}}{\text{cm}^3} \times 200 \text{ cm}^3 = 2100 \text{ g}$$

Canceling the cm³ in both numerator and denominator, we can give the dimension correctly in grams.

Students who have trouble in deciding whether to multiply or to divide in the solution of a given problem will be helped by a consideration of dimensions, as the next example illustrates.

Example 3. The density of gasoline is 7.0 lb/gal. How many gallons would it take to weigh 100 lb?

If we multiply, the dimensions will be

$$\text{lb} \times \frac{\text{lb}}{\text{gal}} = \frac{\text{lb}^2}{\text{gal}}$$

which is obviously incorrect. If we divide, the dimensions will be

$$\text{lb} \div \frac{\text{lb}}{\text{gal}} = \text{lb} \times \frac{\text{gal}}{\text{lb}} = \text{gal}$$

Since volume is required, this is obviously the correct procedure. Therefore,

$$100 \text{ lb} \div 7.0 \frac{\text{lb}}{\text{gal}} = 100 \text{ lb} \times \frac{\text{gal}}{7.0 \text{ lb}} = 14 \text{ gal}$$

THE TREATMENT OF EXPERIMENTAL ERRORS

All generalizations or laws of science are based on regularities derived from experimental observations. Consequently, it is important for a scientist to take into account any limitations in the reliability of the data from which he makes his deductions.

Precision and Accuracy

The limitations of both precision and accuracy will contribute to uncertainty in the measurement. The *error* in a measurement or result is the difference between the true value of the quantity measured and the measured value. The smaller the error, the closer the measured value is to the true value and the more accurate is the result. *Accuracy is a measure of the correctness of a measurement.*

Unless we have precise standards against which we can test our measurement, we often do not know the "true value" of a measured quantity. If we do not, we can obtain only the mean or average value of a number of measurements, and measure the spread or dispersion of the measurements. A measure of the spread of individual values from the mean value is the *deviation*, δ — defined as the difference between the measured value, x_i, and the arithmetic mean, $\bar{x}$, of a number, n, of measurements.

$$\delta_i = x_i - \bar{x} \qquad (1)$$

The mean value is given by adding up all of the individual measurements and dividing by the total number of measurements

$$\bar{x} = \frac{\Sigma \, x_i}{n} \qquad (2)$$

where Σ represents the operation of summation.[1] The smaller the deviations in a series of measurements, the more precise is the measurement. *Precision is a measure of the reproducibility of a measurement.*

The difference between accuracy and precision is shown in Figures AB-1 and AB-2. Figure AB-1 illustrates the distribution of a number of weighings of a sample on a less precise balance (curve A) and on a more precise balance (curve B). In this particular series of measurements, both balances give the same mean value so that their accuracy is the same. However, the precision of the measurement is much better for balance B, and therefore we can have more confidence in the result of that measurement.

Systematic Errors. Errors are of two general types, systematic (determinate) and random (indeterminate). A *systematic error* causes an error in the same direction in each measurement and diminishes

---

[1]Therefore $\Sigma \, xi$ means to form the sum $x_1 + x_2 + x_3 + x_4 \, \ldots \,$, and so on.

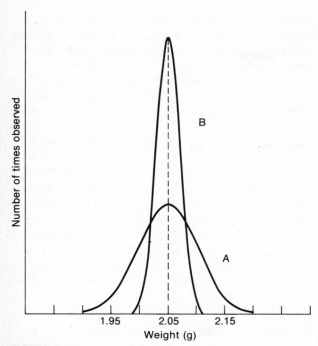

FIGURE AB-1
The distribution of a number of weighings of a sample on a less precise balance (Curve A), and on a more precise balance (Curve B). The standard deviation for Curve A is ±0.05 g and for Curve B, ±0.02 g.

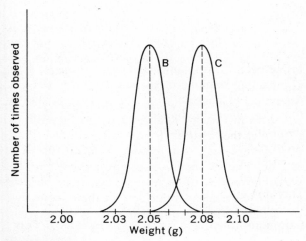

FIGURE AB-2
The distribution of a number of weighings of a sample on two balances of equal precision (B and C). However, there is a systematic error in at least one of the balances.

accuracy, although the precision of the measurement may remain good. A miscalibrated scale on a ruler, for example, would cause a systematic error in the measurement of length.

Similarly, in Figure AB-2 the result of weighings of a sample on two balances of equal precision is shown, but in at least one of the balances there is a systematic error, perhaps due to a miscalibrated weight that was used in each measurement. Ordinarily we can detect an inaccurate (but precise) measurement only by use of precision standards. Such standards are produced and tested by the National Bureau of Standards. Even if all the systematic errors in a measurement are found by careful work with precise standards, the possibility remains that the measurement will be in error because it is impossible to make any measurement with infinite accuracy.

Random Errors and Standard Deviation. If the measurement is made a large number of times, you will obtain a range of values (like those shown in Figure AB-1) that is due to the *random error* inherent in any measurement. Of the random errors, small errors are more probable than large errors and negative deviations are as likely as positive ones. The resulting distribution of the measurements (like that shown in Figure AB-1) is called a normal error distribution. The mean value of the set of measurements is the most probable value, corresponding to the center of the distribution curve. The spread or dispersion of the results is expressed by the *standard deviation*, s, as follows:

$$s = \left[\frac{\Sigma \ (x_i - \bar{x})^2}{n - 1} \right]^{1/2} \qquad (3)$$

This formula actually gives only an estimate of the standard deviation unless the number of measurements is very large (in principle, an infinite number). We must recognize that when we repeat a measurement only two or three times,[2] we are not obtaining a very large sample of measurements and the confidence that we can place in the mean value of a small number of measurements is correspondingly reduced.

Although the formula may look forbiddingly complex, the steps are very simple. First calculate the arithmetic mean, or average, value, $\bar{x}$, of the measurements. Then subtract the mean value, $\bar{x}$, from each one of the individual values, x_i, to obtain the deviation. Square each deviation, and add up all of the squares. Then divide the total by $n - 1$,

---

[2] If only two or three measurements are made, the standard deviation may be approximated by the average deviation which is the mean value of the absolute values of the deviations, δ_i;
$$\bar{\delta} = \frac{\Sigma |x_i - \bar{x}|}{n}$$

Measured value (x_i)	Deviation ($x_i - \overline{x}$)	Square of deviation ($x_i - \overline{x}$)^2
4.28	−0.01	0.0001
4.21	−0.08	0.0064
4.30	0.01	0.0001
4.36	0.07	0.0049
4.26	−0.03	0.0009
4.33	0.04	0.0016
		$\Sigma = 0.0140$

mean, $\overline{x} = 4.29$

standard deviation, $s = \left[\dfrac{0.0140}{6-1}\right]^{1/2} = 0.053$

The best value of the measurements is written as 4.29 ± 0.05 g.

FIGURE AB-3
The procedure for calculating the standard deviation. The measured values are those obtained from a series of six replicate measurements of the weight of a sample on a triple-beam balance.

where n is the total number of measurements. Finally take the square root of the result to obtain the estimate of the standard deviation. The procedure is illustrated for the calculation of the standard deviation in Figure AB-3.

The curves like those shown in Figure AB-1 correspond to the distributions that would be obtained from a very large number of measurements. Curve A corresponds to a mean value of 2.05 g with a standard deviation of 0.05 g. Curve B corresponds to a mean value of 2.05 g with a standard deviation of 0.02 g. The standard deviation can be related to the *confidence interval*, or the range about the mean value in which one of a group of measurements may be expected to fall. If we recognize that the normal error distribution is a distribution of the probabilities of obtaining a particular measurement, we see that the probability of measurements occurring close to the mean will be greater than the probability of a measurement occurring far away from the mean. In fact there is a 0.68 or 68% probability that a given measurement will fall within plus or minus *one* standard deviation of the mean value. There is a 0.95 (95%) probability that a measurement will fall within plus or minus *two* standard deviations of the mean value. This means that for curve A of Figure B-1, we can expect that 95% of the measurements will fall between 1.95 and 2.15 g. If we were to measure the total area under the curve, we would also see that this interval of ±2s

corresponds to 95% of the area under the normal error curve. Note that we cannot make a definite prediction about any single measurement. We can only say that, if we make the measurement a large number of times, we can expect that 95% of the values obtained will fall within ± two standard deviations of the mean.

Personal Errors. In addition to the types of errors already described— systematic and random—we might add a third category, the personal error or *blunder*. Such errors are all too common in student work. Thus if the numbers on a scale are misread, or recorded incorrectly, or if part of a solution is spilled in a titration, the result will contain an error. Careful work will not contain any blunders, and any work suspected of containing one should be repeated.

Precision of Laboratory Operations

The precision associated with various pieces of equipment you may use in the laboratory is summarized in the following table. These uncertainties express limitations in the reading of the instruments and do not reflect *systematic* errors.

INSTRUMENT	TYPICAL UNCERTAINTY
Platform balance	±0.5 g
Triple-beam (centigram) balance	±0.01 g
Top loading semimicro balance	±0.001 g
Analytical balance	±0.0001 g
100-ml graduated cylinder	±0.2 ml
10-ml graduated cylinder	±0.1 ml
50-ml buret	±0.02 ml
25-ml pipet	±0.02 ml
10-ml pipet	±0.01 ml
Thermometer (10°C to 110°C, graduated to 1°C)	±0.2°C
Barometer (mercury)	±0.5 Torr

Significant Figures

Another way of indicating the uncertainty of a measurement is by the method known as *significant figures*. In this method, all digits that are certain and one additional uncertain digit are used to express the measurement. For example, in the weighings plotted in Figure AB-1 the average value, 2.05 g, contains two digits—2.0—which are certain, but the third digit is not; thus 2.05 is said to have three significant figures. Note that the amount by which the last digit is uncertain is not specified when only the significant figure notation—2.05 g—is used. The numbers 2.05 ± 0.01 g, 2.05 ± 0.02 g, 2.05 ± 0.03 g, 2.05 ± 0.04 g, in which the standard deviation varies from 0.5% to 2.0%, are all written 2.05 with three significant figures. However, when strong emphasis on precision and on quantitative results is not desired, the approximate method of significant figures is usually adequate.

The Use of Significant Figures in Recording Measurements. (a) Record the integers that are certain and one more in which there is some uncertainty. For example, 2.05 ± 0.02 g would be recorded simply as 2.05 g.

(b) The number of significant figures has nothing to do with the magnitude of the number. For example, in the numbers 0.2056, 2.056, 20.56, 205.6 and 2056, each has four significant figures.

(c) Zeros that merely indicate the magnitude of the measurement are not significant. Thus, the statement that a Faraday of electricity is 96,500 coulombs does not necessarily mean that there are five significant figures in this value. The last two zeros may indicate only the magnitude of the term. Actually, the digit 5 is not certain and there are only three significant figures. This can be made entirely clear by expressing the value as 965×10^2 or 9.65×10^4, when the exponential notation is used to denote the magnitude. If the last zero in a number is certain—for example, in 4650—this certainty can be indicated by the expression 4.650×10^3, in which there are four significant figures.

The zeros in the number 0.00342 are not significant since they merely denote the magnitude. If this value is written 3.42×10^{-3} it becomes clear that there are only three significant figures. Obviously, too, the zeros inserted between integers—for example, 300.4 g—are significant because they are measured values. Any zeros purposely added at the end of a decimal expression, such as 3.50 g, are also measured and significant or they would not have been recorded.

Rounding Off Numbers. Calculated results usually contain more digits than can be justified from the standard deviation in the data. For example, consider the following problem. A rectangle was measured with a desk ruler and found to be 15.2 ± 0.1 cm $\times$ 13.2 ± 0.1 cm. The value for the area of the rectangle—(15.2 cm $\times$ 13.2 cm)$= 200.64$ cm²—has five significant figures, whereas the original

linear measurements had only three significant figures. Suppose one student used the data giving the largest product—(15.3 cm $\times$ 13.3 cm)$= 203.49$ cm²—and another used the data giving the smallest product—(15.1 cm $\times$ 13.1 cm) $= 197.81$ cm²; it becomes obvious that there is a range in the area of about 6 cm² and that only three significant figures, 201 cm², should be used. The value 200.64 cm² was therefore *rounded off* to 201 cm². The conventions used in rounding off significant figures can be best understood by studying the following examples.

> 56.24 rounded off to three digits is 56.2, since 0.24 is nearer 0.2 than 0.3.
> 234.57 rounded off to four digits is 234.6.
> 234.12 rounded off to four digits is 234.1.
> 234.546 rounded off to five digits is 234.55.
> 234.546 rounded off to four digits is 234.5.

When a 5 is dropped, the last digit remaining is raised to the next higher even number if the digit is odd—for example, 234.75 rounds off to 234.8; but the last digit remaining is left unchanged if it is even—thus 234.45 becomes 234.4.

The Propagation of Errors in Calculated Results

Addition or Subtraction. When measured quantities are added or subtracted, the standard deviation in the result is the *sum of the standard deviation in each measurement.* For example, in data such as these,

Weight of crucible plus sample	24.265 ± 0.001 g
Weight of crucible empty	23.144 ± 0.001 g
Weight of sample	1.121 ± 0.002 g

the uncertainty in the derived result can be simply verified by using the data which will give the greatest and the smallest values for the weight of the sample:

$$
\begin{array}{ccc}
24.266 \text{ g} & & 24.264 \text{ g} \\
\underline{23.143 \text{ g}} & \text{and} & \underline{23.145 \text{ g}} \\
1.123 \text{ g} & & 1.119 \text{ g}
\end{array}
$$

Therefore the average value 1.121 has a standard deviation of ± 0.002 g.

Using the method of significant figures, one would calculate the uncertainty as follows:

$$
\begin{array}{c}
24.26\overline{5} \text{ g} \\
\underline{23.14\overline{4} \text{ g}} \\
1.12\overline{1} \text{ g}
\end{array}
$$

Here the last digit, 1, is uncertain, but the magnitude of the uncertainty, ±0.002 g, is not shown.

Multiplication or Division. When measured quantities undergo the operations of multiplication or division, the derived result has an uncertainty that is the sum of the *percentage standard deviation* in each measurement. For example, what is the weight of 20.2 ± 0.1 ml of mercury that has a density, at 20°C, of 13.4562 ± 0.0001 g/ml. The volume measurement has a percentage standard deviation of $(0.1/20.2) \times 100 = 0.5\%$, whereas the density has a percentage uncertainty of only $(0.0001/13.5462) \times 100 = 0.0008\%$. The sum of the percentage standard deviations is 0.5008%, but obviously the main error is in the volume measurement, so that this factor alone can be used to determine the uncertainty in the result.

If you paid no attention to the percentage standard deviations $(13.5462 \text{ g/ml})(20.2 \text{ ml}) = 273.63324$ g would be obtained. Applying the percentage standard deviation in volume, one would express the result as 274 ± 0.5% g or 274 ± 1 g. When the error is large in one term, as in the volume measurement above, one can use the following rule: use the same number of significant figures in the product or dividend as in the least precise measurement. Thus in the example above, one should carry out the operation (20.2 ml)(13.5462 g/ml) using (20.2 ml)(13.55 g/ml) = 273.7 g or 274 g, rounded off to three significant figures. This rule is not satisfactory if the percentage standard deviation in both terms of a product is about the same. For example, what is the weight of a sample of mercury having a volume of 20.2 ± 0.1 ml and a density of 13.5 ± 0.1 g/ml?

> The percentage standard deviation in volume = ±0.5%
> The percentage standard deviation in density = ±0.8%
> The sum of the % standard deviations = ±1.3%
> Therefore the weight of mercury = (20.2 ml)(13.5 g/ml) = 273 ± 4 g

Using the significant figure rule, one would obtain (20.2 ml)(13.5 g/ml) = 272.70 g or 273 rounded off to three significant figures, but there is no indication that the uncertainty is ±4 in the last digit.

In calculations requiring several steps in multiplication and division, the same procedure applies as was used in the above example; that is, the percentage error in the derived result is the sum of the percentage errors in each of the factors.

EXERCISES

NOTE: After you have solved these problems, refer to Appendix D for the correct answers.

1. Carry out the operations on the data given in each of the following problems to calculate the quantity called for. Show your method, including the dimensions of measurement. (These units will tell you which mathematical operation to perform.)
   (a) Velocity = 50 mi/hr, time = 0.5 hr. Distance = ?
   (b) Velocity = 186,000 mi/sec, distance = 93,000,000 mi. Time = ?
   (c) Time = 9.3 sec, distance = 100 yd. Velocity = ?
   (d) Density of Al = 2.70 g/cm³, weight = 2700 g. Volume = ? If this were shaped as a cube, length of one edge = ?
   (e) Weight of apples = 240 lb, amount in each box = 40 lb/box. Number of boxes = ?
   (f) Weight of H_2O = 180 g, molecular weight = 18 g/mole. Number of moles = ?

2. Perform the operations, and give the *standard deviation of the result* for each:

   (a) $51.2 \pm 0.2°C$
   $\underline{-23.4 \pm 0.2°C}$

   (b) 142.24 ± 0.01 g
   $+27.32 \pm 0.01$ g
   $\underline{+\ 6.23 \pm 0.01\ \text{g}}$

   (c) 24.6732 ± 0.0001 g
   $\underline{+19.2435 \pm 0.0001\ \text{g}}$

   (d) 500.0 ± 0.2 ml
   $\underline{+10.0 \pm 0.1\ \text{ml}}$

3. Consider the following data to be used to calculate the density of mercury at 20°C.
   (a) Where both the volume and weight have a large percentage standard deviation:

   volume = 20.2 ml ± 0.1 ml (±0.5%)
   weight = 274 g ± 2 g (±0.7%)

   (b) Where there is a large percentage standard deviation in one term only:

   volume = 20.2 ml ± 0.1 ml (±0.5%)
   weight = 273.633 g ± 0.001 g (±0.0004%)

   Calculate the density in each example, using the rules for rounding off significant figures. Note that the sum of the percentage errors in the data for (a) is nearly three times that for the sum in the data for (b). Is this reflected in your result for the density you have calculated with the use of significant figure rules?

4. Draw a rectangle on your notebook paper. Measure it with a metric ruler and indicate the uncertainty of each dimension. Calculate the area of the rectangle and express your result to the proper number of significant figures.

5. A quantity of water was weighed in a beaker on a platform balance. The temperature of the water was 15.6 ± 0.2°C. It was then heated to 24.7 ± 0.2°C. The weight of the beaker plus water = 142.4 ± 0.5 g, and the weight of the empty beaker = 61.8 ± 0.5 g. How many calories of heat were absorbed by this quantity of water?

6. What is the density of a metal if a cube of the metal measuring 5.8 cm ± 0.1 cm on each side weighs 3939.0 g ± 0.2 g at 20°C?

7. What is the mean value you are justified in recording as the weight of an object, given the following four weighings?

Weight
40.225 g
40.198
40.176
40.245

Give the mean value and the standard deviation.

TABLES OF DATA

TABLE 1
The International System (SI) of units and conversion factors

Basic SI units			Common conversion factors

Basic SI units

Physical quantity	Unit	Symbol
Length	meter	m
Mass	kilogram	kg
Time	second	s (sec)
Electric current	ampere	A (amp)
Temperature	kelvin	K (°K)

Common derived units

Physical quantity	Unit	Symbol	Definition
Frequency	hertz	Hz	s^{-1}
Energy	joule	J	$kg\ m^2\ s^{-2}$
Force	newton	N	$kg\ m\ s^{-2} = J\ m^{-1}$
Pressure	pascal	Pa	$kg\ m^{-1}\ s^{-2} = N\ m^{-2}$
Power	watt	W	$kg\ m^2\ s^{-3} = J\ s^{-1}$
Electric charge	coulomb	C	As
Electric potential difference	volt	V	$kg\ m^2\ s^{-3}\ A^{-1} =$ $JA^{-1}\ s^{-1}$

Decimal fractions and multiples

Factor	Prefix	Symbol	Factor	Prefix	Symbol
10^{-18}	atto	a	10^{-1}	deci	d
10^{-15}	femto	f	10	deca	da
10^{-12}	pico	p	10^2	hecto	h
10^{-9}	nano	n	10^3	kilo	k
10^{-6}	micro	μ	10^6	mega	M
10^{-3}	milli	m	10^9	giga	G
10^{-2}	centi	c	10^{12}	tera	T

Common conversion factors

LENGTH
1 Ångstrom unit (Å) = 10^{-8} cm
2.54 cm = 1 inch
1 meter = 39.4 inches

MASS
453.5 grams = 1 pound
1 kilogram = 2.20 lb
28.3 g = 1 ounce (avoirdupois)

VOLUME
1 milliliter (ml) = 1 cubic centimeter (cm^3)
   (note that the ml is now defined precisely equal to 1 cm^3)
1 liter = 1.06 quarts
28.6 ml = 1 fluid ounce

PRESSURE
1 atm = 1.013×10^5 pascal (newton/m^2)
   = 760 Torr (mm Hg); pressure of a mercury column 760 mm or 29.92 inches high at 0°C.
   = 1.0133 bar (dyne/cm^2)
   = 14.70 lb/in^2.

TEMPERATURE
Absolute zero (°K) = −273.16 °C
   K = °C + 273.16
   °F = $\frac{9}{5}$ °C + 32
   °C = $\frac{5}{9}$ (°F − 32)

ENERGY
1 joule = 1 watt-sec = 10^7 erg
1 erg = 1 dyne-cm = 1 g cm^2 sec^{-2}
1 calorie = 4.184 joule
1 electron volt/molecule = 23.06 kcalorie/mole

TABLE 2
Fundamental physical and mathematical constants

Physical constants		
Symbol	Name	Numerical value
N_A	Avogadro's Number	6.0221×10^{23} mole^{-1}
F	Faraday constant	96,487 coulombs per mole of electrons transferred
h	Planck's constant	6.63×10^{-27} erg sec per particle
c	Speed of light (in vacuo)	3.00×10^{10} cm/sec
R	The gas constant	0.08206 liter-atm/mole-K
		82.06 ml-atm/mole-K
		8.314 joule/mole-K
e	Charge of the electron	1.602×10^{-19} coulomb
	Volume of 1 mole of an ideal gas	
	at 1 atm	0°C = 22.41 liters
	at 1 atm	25°C = 24.46 liters

Mathematical constants
$\pi = 3.1416$
$\ln x = 2.303 \log_{10} x$

TABLE 3
Apparent radii of selected elements and ions in angstroms[1]

Ag^0	1.34	C^0	0.77	F^0	0.64	K^0	2.03	NH_4^+	1.43	Sb^{5+}	0.62
Ag^+	1.26	C^{4+}	0.16	$F^\times$	1.33	K^+	1.33	Ni^0	1.15	Se^0	1.17
Ag^{2+}	0.89	Ca^0	1.74	Fe^0	1.17			Ni^{2+}	0.69	Se^{2-}	1.98
Al^0	1.25	Ca^{2+}	0.99	Fe^{2+}	0.74					Se^{4+}	0.50
Al^{3+}	0.51	Cd^0	1.41	Fe^{3+}	0.64			O^0	0.66	Se^{6+}	0.42
As^0	1.21	Cd^{2+}	0.97			Li^0	1.23	O^{2-}	1.40	Si^0	1.17
As^{3+}	0.58	Cl^0	0.99	Ga^0	1.25	Li^+	0.68			Si^{4+}	0.42
As^{5+}	0.46	Cl^{2-}	1.81	Ga^{3+}	0.62			P^0	1.10	Sn^0	1.40
Au^0	1.34	Cl^{5+}	0.34	Ge^0	1.22			P^{3+}	0.44	Sn^{2+}	0.93
Au^+	1.37	Cl^{7+}	0.27	Ge^{2+}	0.73			P^{5+}	0.35	Sn^{4+}	0.71
Au^{3+}	ca. 0.9	Co^0	1.16	Ge^{4+}	0.53	Mg^0	1.36	Pb^0	1.54	Sr^0	1.91
		Co^{2+}	0.72			Mg^{2+}	0.66	Pb^{2+}	1.20	Sr^{2+}	1.12
B^0	0.81	Co^{3+}	0.63	Hg^0	1.44	Mn^0	1.17	Pb^{4+}	0.84		
B^{3+}	0.23	Cr^0	1.18	Hg^{2+}	1.10	Mn^{2+}	0.80			Te^0	1.37
Ba^0	1.98	Cr^{3+}	0.63			Mn^{3+}	0.66	Ra^{2+}	1.43	Te^{2-}	2.21
Ba^{2+}	1.34	Cr^{6+}	0.52	I^0	1.33	Mn^{4+}	0.60	Rb^0	2.16	Te^{4+}	ca. 0.7
Be^0	0.89	Cs^0	2.35	I^-	2.20	Mn^{7+}	0.46	Rb^+	1.47	Te^{6+}	0.56
Be^{2+}	0.35	Cs^+	1.67	I^{5+}	0.62					Tl^0	1.55
Bi^0	ca. 1.5	Cu^0	1.17	I^{7+}	0.50	N^0	0.70	S^0	1.04	Tl^+	1.47
Bi^{3+}	0.96	Cu^+	0.96	In^0	1.50	N^{3+}	0.16	S^{2-}	1.84	Tl^{3+}	0.95
Bi^{5+}	0.74	Cu^{2+}	0.72	In^{3+}	0.81	N^{5+}	0.13	S^{4+}	0.37		
Br^0	1.14					Na^0	1.57	S^{6+}	0.30	Zn^0	1.25
Br^-	1.96					Na^+	0.97	Sb^0	1.41	Zn^{2+}	0.74
Br^{5+}	0.47							Sb^{3+}	0.76		

[1]Selected from the compilation in M. J. Sienko, R. Plane and Ronald E. Hester, *Inorganic Chemistry: Principles and Problems,* Benjamin, Menlo Park, Calif., 1965.

TABLE 4
Emission lines of mercury in the visible spectrum

Shown below is a recording, made with a photomultiplier detector, of the visible emission from a fluorescent lamp. Note the sharp and intense emission lines of mercury superimposed on the much lower continuum emission from the phosphor coating on the inside of the lamp. Through a spectroscope like that described in Experiment 8, most observers will see four lines: violet, blue, green and yellow. The yellow "line" is actually an unresolved multiplet of three lines.

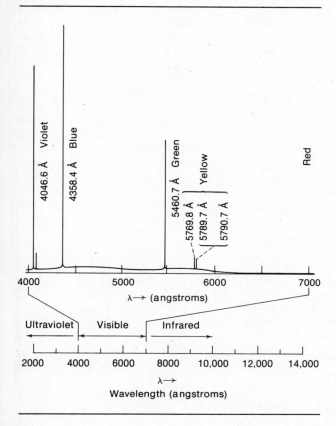

TABLE 5
Vapor pressure of water at different temperatures

Temperature (°C)	Vapor pressure (torr)
−10 (ice)	1.0
− 5 (ice)	3.0
0	4.6
5	6.5
10	9.2
15	12.8
16	13.6
17	14.5
18	15.5
19	16.5
20	17.5
21	18.6
22	19.8
23	21.1
24	22.4
25	23.8
26	25.2
27	26.7
28	28.3
29	30.0
30	31.8
35	42.2
40	55.3
45	71.9
50	92.5
60	149.4
70	233.7
80	355.1
90	525.8
100	760.0
110	1074.6
150	3570.5
200	11659.2
300	64432.8

TABLE 6
Formula weights of compounds[1]

Ag	107.87	H	1.008	N	14.01
AgCl	143.32	H_2	2.016	N_2	28.02
AgI	234.77	$HC_2H_3O_2$ (acetic)	60.05	NH_3	17.03
$AgNO_3$	169.88	$HC_7H_5O_2$ (benzoic)	122.12	NH_4Cl	53.49
		HCl	36.46	$(NH_4)_2C_2O_4 \cdot H_2O$	142.12
Al	26.98	$H_2C_2O_4 \cdot 2H_2O$ (oxalic)	126.07	$(NH_4)_2HPO_4$	132.06
Al_2O_3	101.96	HNO_3	63.02	$(NH_4)_3PO_4 \cdot 12MoO_3$	1876.50
$Al(OH)_3$	78.00	H_2O	18.02	$(NH_4)_2SO_4$	132.14
		H_2O_2	34.02		
As	74.92	H_3PO_4	98.00	Na	22.99
As_2O_3	197.84	H_2SO_3	82.08	NaBr	102.90
As_2O_5	229.84	H_2SO_4	98.08	NaCl	58.44
				NaCN	49.01
Ba	137.34	Hg	200.59	Na_2CO_3	105.99
$BaBr_2$	297.22	HgO	216.59	$Na_2C_2O_4$	134.00
$BaCl_2$	208.25			$NaHCO_3$	84.01
$BaCl_2 \cdot 2H_2O$	244.28	I	126.90	$Na_2HPO_4 \cdot 12H_2O$	358.15
$BaCO_3$	197.35	I_2	253.81	$NaNO_2$	69.00
BaF_2	175.34			Na_2O	61.98
BaO	153.34	K	39.10	Na_2O_2	77.98
$Ba(OH)_2$	171.36	$KAl(SO_4)_2 \cdot 12H_2O$	474.40	NaOH	40.00
$Ba(OH)_2 \cdot 8H_2O$	315.48	KBr	119.01	$Na_2S_2O_3$	158.11
$BaSO_4$	233.40	$KBrO_3$	167.01	$Na_2S_2O_3 \cdot 5H_2O$	248.19
		KCl	74.55		
Br	79.91	$KClO_4$	138.55	O	16.00
Br_2	159.82	KCN	65.12	O_2	32.00
		KCNS	97.18		
C	12.01	K_2CO_3	138.21	P	30.97
CO_2	44.01	K_2CrO_4	194.20	P_2O_5	141.95
		$K_2Cr_2O_7$	294.20		
Ca	40.08	$K_4Fe(CN)_6 \cdot 3H_2O$	422.41	Pb	207.19
$CaCO_3$	100.09	$KFe(SO_4)_2 \cdot 12H_2O$	503.27	$PbCl_2$	278.10
CaF_2	78.08	$KHC_4H_4O_6$ (tartrate)	188.18	$PbCrO_4$	323.19
CaO	56.08	$KHC_8H_4O_4$ (phthalate)	204.23	PbO_2	239.19
$Ca(OH)_2$	74.10	$KHCO_3$	100.12	Pb_2O_3	462.38
$Ca_3(PO_4)_2$	310.18	$KHC_2O_4 \cdot H_2O$	146.14	Pb_3O_4	685.57
$CaSO_4$	136.14	$KHSO_4$	136.17	$Pb_3(PO_4)_2$	811.52
		KI	166.01	$PbSO_4$	303.25
Cl	35.45	KIO_3	214.01		
Cl_2	70.91	$KMnO_4$	158.04	S	32.06
		KNO_2	85.11	SO_2	64.06
Cr	52.00	K_2O	94.20	SO_3	80.06
Cr_2O_3	152.00	KOH	56.11		
				Sb	121.75
Cu	63.54	Li	6.94	Sb_2O_4	307.50
CuO	79.54	LiCl	42.39		
Cu_2O	143.08	Li_2CO_3	73.89	Si	28.09
CuS	95.60	LiOH	23.95	SiF_4	104.09
$CuSO_4 \cdot 5H_2O$	249.68			SiO_2	60.09
		Mg	24.31		
Fe	55.85	$MgCl_2$	95.22	Sn	118.69
$Fe(NO_3)_3 \cdot 9H_2O$	404.02	$MgCO_3$	84.32	$SnCl_2$	189.61
FeO	71.85	MgO	40.31	SnO_2	150.69
Fe_2O_3	159.70	$Mg(OH)_2$	58.33		
Fe_3O_4	231.55			Ti	47.90
$Fe(OH)_3$	106.87	Mn	54.94	TiO_2	79.90
FeS_2	119.97	MnO	70.94		
Fe_2Si	139.79	MnO_2	86.94	Zn	65.37
$FeSO_4 \cdot 7H_2O$	278.02	Mn_3O_4	228.82	ZnO	81.37
$FeSO_4 \cdot (NH_4)_2SO_4 \cdot 6H_2O$	392.15				

[1]This table includes all principal elements and compounds that you have studied in the experiments and study assignments of this manual.

TABLE 7
Concentration of desk acid and base solutions

Reagent	Formula	Formality	Density	Percent solute
Acetic acid, glacial	$HC_2H_3O_2$	17 F	1.05 g/ml	99.5%
Acetic acid, dil.		6	1.04	34
Hydrochloric acid, conc.	HCl	12	1.18	36
Hydrochloric acid, dil.		6	1.10	20
Nitric acid, conc.	HNO_3	16	1.42	72
Nitric acid, dil.		6	1.19	32
Sulfuric acid, conc.	H_2SO_4	18	1.84	96
Sulfuric acid, dil.		3	1.18	25
Ammonia solution, conc. (ammonium hydroxide)	NH_3	15	0.90	58
Ammonia solution, dil. (ammonium hydroxide)		6	0.96	23
Sodium hydroxide, dil.	NaOH	6	1.22	20

TABLE 8
The color changes and pH intervals of some important indicators

Name of indicator	pH interval	Color change	Solvent
Methyl violet	0.2– 3.0	Yellow, blue, violet	Water
Thymol blue	1.2– 2.8	Red to yellow	Water (+ NaOH)
Orange IV (tropeolin OO)	1.3– 3.0	Red to yellow	Water
Benzopurpurin 4B	1.2– 4.0	Violet to red	20% alcohol
Methyl orange	3.1– 4.4	Red to orange to yellow	Water
Bromphenol blue	3.0– 4.6	Yellow to blue violet	Water (+ NaOH)
Congo red	3.0– 5.0	Blue to red	70% alcohol
Bromcresol green	3.8– 5.4	Yellow to blue	Water (+ NaOH)
Methyl red	4.4– 6.2	Red to yellow	Water (+ NaOH)
Chlorphenol red	4.8– 6.8	Yellow to red	Water (+ NaOH)
Bromcresol purple	5.2– 6.8	Yellow to purple	Water (+ NaOH)
Litmus	4.5– 8.3	Red to blue	Water
Bromthymol blue	6.0– 7.6	Yellow to blue	Water (+ NaOH)
Phenol red	6.8– 8.2	Yellow to red	Water (+ NaOH)
Thymol blue	8.0– 9.6	Yellow to blue	Water (+ NaOH)
Phenolphthalein	8.3–10.0	Colorless to red	70% alcohol
Thymolphthalein	9.3–10.5	Yellow to blue	70% alcohol
Alizarin yellow R	10.0–12.0	Yellow to red	20% alcohol
Indigo carmine	11.4–13.0	Blue to yellow	50% alcohol
Trinitrobenzene	12.0–14.0	Colorless to orange	70% alcohol

TABLE 9
Equilibrium constants for the dissociation of acids and bases (25°C)

Compound	Dissociation reaction	K_a	pK_a
WATER			
	$H_2O = H^+ + OH^-$ (25°C)	1.00×10^{-14}	14.00
	(0°C)	0.11×10^{-14}	14.94
	(60°C)	9.61×10^{-14}	13.02
WEAK ACIDS			
Acetic	$HC_2H_3O_2 = H^+ + C_2H_3O_2^-$	1.76×10^{-5}	4.75
Boric	$H_3BO_3 = H^+ + H_2BO_3^-$	6.0×10^{-10}	9.22
Carbonic ($CO_2 + H_2O$)	$CO_2 + H_2O = H^- + HCO_3^-$	K_1: 4.4×10^{-7}	6.35
	$HCO_3^- = H^+ + CO_3^{2-}$	K_2: 4.7×10^{-11}	10.33
Chromic	$H_2CrO_4 = H^+ + HCrO_4^-$	K_1: 2×10^{-1}	0.7
	$HCrO_4^- = H^+ + CrO_4^{2-}$	K_2: 3.2×10^{-7}	6.50
Formic	$HCHO_2 = H^+ + CHO_2^-$	2.1×10^{-4}	3.68
Hydrogen cyanide	$HCN = H^+ + CN^-$	4×10^{-10}	9.4
Hydrofluoric	$HF = H^+ + F^-$	6.9×10^{-4}	3.16
Hydrogen peroxide	$H_2O_2 = H^+ + HO_2^-$	2.4×10^{-12}	11.62
Hydrogen sulfate ion	$HSO_4^- = H^+ + SO_4^{2-}$	K_2: 1.2×10^{-2}	1.92
Hydrogen sulfide	$H_2S = H^+ + HS^-$	K_1: 1.0×10^{-7}	7.00
	$HS^- = H^+ + S^{2-}$	K_2: 1.3×10^{-13}	12.89
Nitrous	$HNO_2 = H^+ + NO_2^-$	4.5×10^{-4}	3.50
Oxalic	$H_2C_2O_4 = H^+ + HC_2O_4^-$	K_1: 3.8×10^{-2}	1.42
	$HC_2O_4^- = H^+ + C_2O_4^{2-}$	K_2: 5.0×10^{-5}	4.30
Phosphoric	$H_3PO_4 = H^+ + H_2PO_4^-$	K_1: 7.1×10^{-3}	2.15
	$H_2PO_4^- = H^+ + HPO_4^{2-}$	K_2: 6.3×10^{-8}	7.20
	$HPO_4^{2-} = H^+ + PO_4^{3-}$	K_3: 4.4×10^{-13}	12.36
Phosphorous	$H_2HPO_3 = H^+ + HHPO_3^-$	K_1: 1.6×10^{-2}	1.80
Sulfurous ($SO_2 + H_2O$)	$H_2SO_3 = H^+ + HSO_3^-$	K_1: 1.2×10^{-2}	1.92
	$HSO_3^- = H^+ + SO_3^{2-}$	K_2: 5.6×10^{-8}	7.25
CATION ACIDS — HYDRATED METAL IONS			
Aluminum ion	$Al(H_2O)_6^{3+} = H^+ + Al(H_2O)_5OH^{2+}$	1.1×10^{-5}	4.96
Chromium(III) ion	$Cr(H_2O)_6^{3+} = H^+ + Cr(H_2O)_5OH^{2+}$	1.6×10^{-4}	3.80
Iron(III) ion	$Fe(H_2O)_6^{3+} = H^+ + Fe(H_2O)_5OH^{2+}$	6.7×10^{-3}	2.17
Zinc ion	$Zn(H_2O)_4^{2+} = H^+ + Zn(H_2O)_3OH^+$	2.5×10^{-10}	9.60
WEAK BASES			
Ammonia	$NH_3 + H_2O = NH_4^+ + OH^-$	1.8×10^{-5}	4.75
Methylamine	$CH_3NH_2 + HOH = CH_3NH_3^+ + OH^-$	5×10^{-4}	3.3
Barium hydroxide	$Ba(OH)_2 = BaOH^+ + OH^-$	strong	
	$BaOH^+ = Ba^{2+} + OH^-$	K_2: 1.4×10^{-1}	0.85
Calcium hydroxide	$Ca(OH)_2 = CaOH^+ + OH^-$	strong	
	$CaOH^+ = Ca^{2+} + OH^-$	K_2: 3.5×10^{-2}	1.5

TABLE 10
Equilibrium constants for the dissociation of complex ions, amphoteric hydroxides, and weak salts ($\approx 25°C$)

The formation of complex ions undoubtedly occurs in steps. It is only in the presence of a large excess of the coordinating ion or molecule that the complete, cumulative ionization constant can be used with any measure of quantitative reliability. Furthermore, the total ionic strength of the solution exerts a major influence in modifying equilibrium values. Thus, with the high concentration of ions usually present in the formation of the iron(III) thiocyanate complex ion, the calculated value of $K_{FeSCN^{2+}}$ will be increased three- to fivefold.

Compound	Dissociation reaction	K	pK
AMMINE (AMMONIA) COMPLEX IONS			
Tetraamminecadmium(II)	$Cd(NH_3)_4^{2+} = Cd^{2+} + 4NH_3$	2×10^{-7}	6.7
Tetraamminecopper(II)	$Cu(NH_3)_4^{2+} = Cu^{2+} + 4NH_3$	8×10^{-13}	12.1
Diamminesilver(I)	$Ag(NH_3)_2^{+} = Ag^{+} + 2NH_3$	6×10^{-8}	7.2
Tetraamminezinc(II)	$Zn(NH_3)_4^{2+} = Zn^{2+} + 4NH_3$	1×10^{-9}	9.0
HYDROXIDE COMPLEX IONS—AMPHOTERIC HYDROXIDES			
Tetrahydroxoaluminate	$Al(OH)_4^{-} = Al(OH)_3(s) + OH^-$	3×10^{-2}	1.5
Tetrahydroxochromate(III)	$Cr(OH)_4^{-} = Cr(OH)_3(s) + OH^-$	2.5	−0.40
Trihydroxoplumbate(II) ion	$Pb(OH)_3^{-} = Pb(OH)_2(s) + OH^-$	2×10^{1}	−1.3
Trihydroxostannate(II)	$Sn(OH)_3^{-} = Sn(OH)_2(s) + OH^-$	2.6	−0.41
Tetrahydroxozincate	$Zn(OH)_4^{2-} = Zn(OH)_2(s) + 2OH^-$	4×10^{1}	−1.6
CHLORIDE COMPLEX IONS AND WEAK SALTS			
Dichlorocadmium	$CdCl_2(aq) = Cd^{2+} + 2Cl^-$	2.5×10^{-3}	2.60
Tetrachloroaurate(III) ion	$AuCl_4^{-} = Au^{3+} + 4Cl^-$	5×10^{-22}	21.3
Trichloroiron(III)	$FeCl_3(aq) = Fe^{3+} + 3Cl^-$	8×10^{-2}	1.9
Dichloroiron(III) ion	$FeCl_2^{+}(aq) = Fe^{3+} + 2Cl^-$	8×10^{-3}	2.9
Chloroiron(III) ion	$FeCl^{2+} = Fe^{3+} + Cl^-$	3.5×10^{-2}	1.46
Mercury(II) chloride	$HgCl_2(aq) = HgCl^+ + Cl^-$	K_1: 3.3×10^{-7}	6.48
Chloromercury(II) ion	$HgCl^+ = Hg^{2+} + Cl^-$	K_2: 1.8×10^{-7}	6.74
Tetrachloromercurate(II)	$HgCl_4^{2-} = Hg^{2+} + 4Cl^-$		15.07
Tin(II) chloride	$SnCl_2(aq) = Sn^{2+} + 2Cl^-$	5.7×10^{-3}	2.24
Tetrachlorostannate(II) ion	$SnCl_4^{2-} = Sn^{2+} + 4Cl^-$	3.3×10^{-2}	1.48
Hexachlorostannate(IV) ion	$SnCl_6^{2-} = Sn^{4+} + 6Cl^-$	$? \; 10^{-4}$	4
Dichloroargentate(I) ion	$AgCl_2^{-} = Ag^{+} + 2Cl^-$	5×10^{-6}	5.3
OTHER COMPLEX IONS AND WEAK SALTS			
Tetracyanocadmate(II) ion	$Cd(CN)_4^{2-} = Cd^{2+} + 4CN^-$	8×10^{-18}	17.1
Thiocyanatoiron(III) ion	$FeSCN^{2+} = Fe^{3+} + SCN^-$	1×10^{-3}	3.0
Lead(II) acetate	$Pb(C_2H_3O_2)_2(aq) = Pb^{2+} + 2C_2H_3O_2^-$	1×10^{-4}	4.0
Triacetatoplumbate(II) ion	$Pb(C_2H_3O_2)_3^{-} = Pb^{2+} + 3C_2H_3O_2^-$	2.5×10^{-7}	6.60
Dicyanoargentate(I) ion	$Ag(CN)_2^{-} = Ag^{+} + 2CN^-$	1×10^{-20}	20.0
Dithiosulfatoargentate(I) ion	$Ag(S_2O_3)_2^{3-} = Ag^{+} + 2S_2O_3^{2-}$	4×10^{-14}	13.4

TABLE 11
General solubility rules for common salts and bases

1. NO_3^-		All *nitrates* are soluble.
2. $C_2H_3O_2^-$		All *acetates* are soluble, ($AgC_2H_3O_2$ only moderately).
3. Cl^-		All *chlorides* are soluble, except AgCl, Hg_2Cl_2, and $PbCl_2$. ($PbCl_2$ is slightly soluble in cold water, moderately soluble in hot water.)
4. SO_4^{2-}		All *sulfates* are soluble, except $BaSO_4$ and $PbSO_4$. ($CaSO_4$, Hg_2SO_4, and Ag_2SO_4 are slightly soluble; the corresponding bisulfates are more soluble.)
5. CO_3^{2-}, and PO_4^{3-}		All *carbonates* and *phosphates* are insoluble, except those of Na^+, K^+, and NH_4^+. (Many acid phosphates are soluble, as $Mg(H_2PO_4)_2$, and $Ca(H_2PO_4)_2$.)
6. OH^-		All *hydroxides* are insoluble, except NaOH, KOH, and $Ba(OH)_2$. ($Ca(OH)_2$ is slightly soluble.)
7. S^{2-}		All *sulfides* are insoluble, except those of Na^+, K^+, and NH_4^+, and those of the alkaline earths: Mg^{2+}, Ca^{2+}, Sr^{2+}, and Ba^{2+}. (Sulfides of Al^{3+} and Cr^{3+} hydrolyze and precipitate the corresponding hydroxides.)
8. Na^+, K^+, NH_4^+		All salts of *sodium, potassium,* and *ammonium* are soluble, except several uncommon ones, as $Na_4Sb_2O_7$, $K_2NaCo(NO_2)_6$, $(NH_4)_2NaCo(NO_2)_6$, K_2PtCl_6, $(NH_4)_2PtCl_6$.
9. Ag^+		All *silver* salts are insoluble, except $AgNO_3$ and $AgClO_4$. ($AgC_2H_3O_2$ and Ag_2SO_4 are only moderately soluble.)

TABLE 12
Solubility product constants (18–25°C)[1]

Compound	K_{sp}	Compound	K_{sp}	Compound	K_{sp}
ACETATES		CHROMATES		SULFATES	
$AgC_2H_3O_2$	4×10^{-3}	Ag_2CrO_4	2×10^{-12}	Ag_2SO_4	1.7×10^{-5}
		$BaCrO_4$	1.2×10^{-10}	$BaSO_4$	1.5×10^{-9}
HALIDES AND CYANIDES		$PbCrO_4$	2×10^{-16}	$CaSO_4$	2.4×10^{-5}
$AgCN$	10^{-16}	$SrCrO_4$	3.6×10^{-5}	$PbSO_4$	1.3×10^{-8}
$AgCl$	1.8×10^{-10}			$SrSO_4$	7.6×10^{-7}
$AgBr$	5×10^{-13}	HYDROXIDES			
AgI	8.5×10^{-17}	$Al(OH)_3$	10^{-33}	SULFIDES	
$CuCl$	3.2×10^{-7}	$Ca(OH)_2$	1.3×10^{-6}	Ag_2S	10^{-50}
Hg_2Cl_2	1.1×10^{-18}	$Cr(OH)_3$	10^{-30}	CdS	10^{-26}
$PbCl_2$	1.6×10^{-5}	$Cu(OH)_2$	2×10^{-19}	CoS	10^{-21}
PbI_2	8.3×10^{-9}	$Fe(OH)_2$	2×10^{-15}	CuS	10^{-36}
MgF_2	8×10^{-8}	$Fe(OH)_3$	10^{-37}	FeS	10^{-17}
CaF_2	1.7×10^{-10}	$Mg(OH)_2$	9×10^{-12}	HgS	10^{-50}
		$Mn(OH)_2$	2×10^{-13}	MnS	10^{-13}
CARBONATES		$Pb(OH)_2$	4×10^{-15}	NiS	10^{-22}
Ag_2CO_3	8×10^{-12}	$Sn(OH)_2$	10^{-27}	PbS	10^{-26}
$BaCO_3$	1.6×10^{-9}	$Zn(OH)_2$	5×10^{-17}	SnS	10^{-27}
$CaCO_3$	4.8×10^{-9}			ZnS	10^{-20}
$CuCO_3$	2.5×10^{-10}	OXALATES			
$FeCO_3$	2×10^{-11}	BaC_2O_4	1.5×10^{-8}		
$MgCO_3$	4×10^{-5}	CaC_2O_4	1.3×10^{-9}		
$MnCO_3$	9×10^{-11}	MgC_2O_4	8.6×10^{-5}		
$PbCO_3$	1.5×10^{-13}	SrC_2O_4	5.6×10^{-8}		
$SrCO_3$	7×10^{-10}				

[1]These values are approximate. The solubility is affected by the concentration of the metal ion (about 10^{-3} to 10^{-2} M), by the temperature, and by the presence of substances that cause complex ion formation or that result in rather stable colloidal suspensions. The rate of hydrolysis and precipitation is often quite slow.

TABLE 13
Oxidation-reduction potentials (25°C)[1]

Strongest oxidizing agents	Half-reaction	$E°$ (volts)	Weakest reducing agents
Increasing oxidizing strength	$\frac{1}{2}F_2(g) + H^+ + e^- \rightleftharpoons HF(aq)$	+3.06	Increasing reducing strength
	$\frac{1}{2}F_2(g) + e^- \rightleftharpoons F^-$	+2.87	
	$H_2O_2(aq) + 2H^+ + 2e^- \rightleftharpoons 2H_2O$	+1.77	
	$PbO_2(s) + 4H^+ + SO_4^{2-} + 2e^- \rightleftharpoons 2H_2O + PbSO_4(s)$	+1.685	
	$Ce^{4+} + e^- \rightleftharpoons Ce^{3+}$	+1.61	
	$Bi_2O_4(s) + 4H^+ + 2e^- \rightleftharpoons 2H_2O + 2BiO^+$	+1.6	
	$MnO_4^- + 8H^+ + 5e^- \rightleftharpoons 4H_2O + Mn^{2+}$	+1.51	
	$Mn^{3+} + e^- \rightleftharpoons Mn^{2+}$	+1.51	
	$Au^{3+} + 3e^- \rightleftharpoons Au$	+1.50	
	$HClO(aq) + H^+ + 2e^- \rightleftharpoons H_2O + Cl^-$	+1.49	
	$ClO_3^- + 6H^+ + 5e^- \rightleftharpoons 3H_2O + \frac{1}{2}Cl_2(g)$	+1.47	
	$Co^{3+} + e^- \rightleftharpoons Co^{2+}$	+1.45[2]	
	$\frac{1}{2}Cl_2 + e^- \rightleftharpoons Cl^-$	+1.3595	
	$Cr_2O_7^{2-} + 14H^+ + 6e^- \rightleftharpoons 7H_2O + 2Cr^{3+}$	+1.33	
	$MnO_2(s) + 4H^+ + 2e^- \rightleftharpoons 2H_2O + Mn^{2+}$	+1.23	
	$O_2(g) + 4H^+ + 4e^- \rightleftharpoons 2H_2O$	+1.229	
	$\frac{1}{2}Br_2(g) + e^- \rightleftharpoons Br^-$	+1.0652	
	$AuCl_4^- + 3e^- \rightleftharpoons 4Cl^- + Au$	+1.00	

(continued on next page.)

TABLE 13 *(continued)*

Strongest oxidizing agents	Half-reaction	$E°$ (volts)	Weakest reducing agents
	$NO_3^- + 4H^+ + 3e^- \rightleftharpoons 2H_2O + NO\,(g)$	+0.96	
	$2Hg^{2+} + 2e^- \rightleftharpoons Hg_2^{2+}$	+0.92	
	$ClO^- + H_2O + 2e^- \rightleftharpoons 2OH^- + Cl^-$	+0.89	
	$HO_2^- + H_2O + 2e^- \rightleftharpoons 3OH^-$	+0.88	
	$Hg^{2+} + 2e^- \rightleftharpoons Hg\,(l)$	+0.854	
	$O_2 + 4H^+ (10^{-7}\,M) + 4e^- \rightleftharpoons 2H_2O$	+0.815	
	$Ag^+ + e^- \rightleftharpoons Ag$	+0.79991	
	$Hg_2^{2+} + 2e^- \rightleftharpoons 2Hg\,(l)$	+0.789	
	$Fe^{3+} + e^- \rightleftharpoons Fe^{2+}$	+0.771	
	$O_2\,(g) + 2H^+ + 2e^- \rightleftharpoons H_2O_2\,(aq)$	+0.682	
	$MnO_4^- + 2H_2O + 3e^- \rightleftharpoons 4OH^- + MnO_2\,(s)$	+0.60	
	$I_3^- + 2e^- \rightleftharpoons 3I^-$	+0.54	
	$\frac{1}{2}I_2\,(aq) + e^- \rightleftharpoons I^-$	+0.536	
	$MnO_2\,(s) + H_2O + NH_4^+ + e^- \rightleftharpoons NH_3 + Mn(OH)_3\,(s)$	+0.50	
	$O_2\,(g) + 2H_2O + 4e^- \rightleftharpoons 4OH^-$	+0.401	
	$Cu^{2+} + 2e^- \rightleftharpoons Cu$	+0.337	
	$BiO^+ + 2H^+ + 3e^- \rightleftharpoons H_2O + Bi$	+0.32	
	$HAsO_2\,(aq) + 3H^+ + 3e^- \rightleftharpoons 2H_2O + As$	+0.2475	
	$AgCl\,(s) + e^- \rightleftharpoons Cl^- + Ag$	+0.2222	
	$SbO^+ + 2H^+ + 3e^- \rightleftharpoons H_2O + Sb$	+0.212	
	$SO_4^{2-} + 4H^+ + 2e^- \rightleftharpoons H_2O + H_2SO_3\,(aq)$	+0.17	
	$Sn^{4+} + 2e^- \rightleftharpoons Sn^{2+}$	+0.15	
	$S + 2H^+ + 2e^- \rightleftharpoons H_2S\,(g)$	+0.141	
	$2H^+ + 2e^- \rightleftharpoons H_2\,(g)$	0.000	
	$O_2\,(g) + H_2O + 2e^- \rightleftharpoons OH^- + HO_2^-$	−0.076	
	$Pb^{2+} + 2e^- \rightleftharpoons Pb$	−0.126	
	$CrO_4^{2-} + 4H_2O + 3e^- \rightleftharpoons 5OH^- + Cr(OH)_3\,(s)$	−0.13	
	$Sn^{2+} + 2e^- \rightleftharpoons Sn$	−0.136	
	$Ni^{2+} + 2e^- \rightleftharpoons Ni$	−0.250	
	$Co^{2+} + 2e^- \rightleftharpoons Co$	−0.277	
	$PbSO_4\,(s) + 2e^- \rightleftharpoons SO_4^{2-} + Pb$	−0.356	
	$Cd^{2+} + 2e^- \rightleftharpoons Cd$	−0.403	
	$Cr^{3+} + e^- \rightleftharpoons Cr^{2+}$	−0.41	
	$2H^+ (10^{-7}\,M) + 2e^- \rightleftharpoons H_2\,(g)$	−0.414	
	$Fe^{2+} + 2e^- \rightleftharpoons Fe$	−0.44	
	$S + 2e^- (1\,M\,OH^-) \rightleftharpoons S^{2-}$	−0.48	
	$2CO_2\,(g) + 2H^+ + 2e^- \rightleftharpoons H_2C_2O_4\,(aq)$	−0.49	
	$Cr^{3+} + 3e^- \rightleftharpoons Cr$	−0.74	
	$Zn^{2+} + 2e^- \rightleftharpoons Zn$	−0.763	
	$2H_2O + 2e^- \rightleftharpoons 2OH^- + H_2\,(g)$	−0.828	
	$SO_4^{2-} + H_2O + 2e^- \rightleftharpoons 2OH^- + SO_3^{2-}$	−0.93	
	$Mn^{2+} + 2e^- \rightleftharpoons Mn$	−1.18	
	$Al^{3+} + 3e^- \rightleftharpoons Al$	−1.66	
	$Mg^{2+} + 2e^- \rightleftharpoons Mg$	−2.37	
	$Na^+ + e^- \rightleftharpoons Na$	−2.714	
	$Ca^{2+} + 2e^- \rightleftharpoons Ca$	−2.87	
	$Sr^{2+} + 2e^- \rightleftharpoons Sr$	−2.89	
	$Ba^{2+} + 2e^- \rightleftharpoons Ba$	−2.90	
	$Cs^+ + e^- \rightleftharpoons Cs$	−2.92	
	$K^+ + e^- \rightleftharpoons K$	−2.925	
	$Li^+ + e^- \rightleftharpoons Li$	−3.045	

Increasing oxidizing strength (left) / *Increasing reducing strength* (right)

[1]The reduction potential convention recommended by the IUPAC is used. The potentials listed are those which would be obtained for the given "half-cell" when measured with respect to the standard hydrogen (H^+–H_2) half-cell. All species are at unit activity unless otherwise specified. Values in the table are from W. M. Latimer, Oxidation Potentials, 2nd ed., Prentice-Hall, Englewood Cliffs, New Jersey, 1952.
[2]Newly revised value based on the work of A. L. Rotinjan et al., *Electrochimica Acta*, **19**, 43(1974).

TABLE 14
Electronegativities of the elements[1]

H																
2.1																

Li	Be											B	C	N	O	F
1.0	1.5											2.0	2.5	3.0	3.5	4.0

Na	Mg											Al	Si	P	S	Cl
0.9	1.2											1.5	1.8	2.1	2.5	3.0

K	Ca	Sc	Ti	V	Cr	Mn	Fe	Co	Ni	Cu	Zn	Ga	Ge	As	Se	Br
0.8	1.0	1.3	1.5	1.6	1.6	1.5	1.8	1.8	1.8	1.9	1.6	1.6	1.8	2.0	2.4	2.8

Rb	Sr	Y	Zr	Nb	Mo	Tc	Ru	Rh	Pd	Ag	Cd	In	Sn	Sb	Te	I
0.8	1.0	1.2	1.4	1.6	1.8	1.9	2.2	2.2	2.2	1.9	1.7	1.7	1.8	1.9	2.1	2.5

Cs	Ba	La–Lu	Hf	Ta	W	Re	Os	Ir	Pt	Au	Hg	Tl	Pb	Bi	Po	At
0.7	0.9	1.1–1.2	1.3	1.5	1.7	1.9	2.2	2.2	2.2	2.4	1.9	1.8	1.8	1.9	2.0	2.2

Fr	Ra	Ac	Th	Pa	U	Np–Lw										
0.7	0.9	1.1	1.3	1.5	1.7	1.3										

[1]These values are based on those given in L. Pauling, *The Nature of the Chemical Bond*, 3rd ed., Cornell University Press, Ithaca, New York, 1960, p. 93.

TABLE 15
Logarithms

N	0	1	2	3	4	5	6	7	8	9	1	2	3	4	5	6	7	8	9
														Proportional parts					
10	0000	0043	0086	0128	0170	0212	0253	0294	0334	0374	4	8	12	17	21	25	29	33	37
11	0414	0453	0492	0531	0569	0607	0645	0682	0719	0755	4	8	11	15	19	23	26	30	34
12	0792	0828	0864	0899	0934	0969	1004	1038	1072	1106	3	7	10	14	17	21	24	28	31
13	1139	1173	1206	1239	1271	1303	1335	1367	1399	1430	3	6	10	13	16	19	23	26	29
14	1461	1492	1523	1553	1584	1614	1644	1673	1703	1732	3	6	9	12	15	18	21	24	27
15	1761	1790	1818	1847	1875	1903	1931	1959	1987	2014	3	6	8	11	14	17	20	22	25
16	2041	2068	2095	2122	2148	2175	2201	2227	2253	2279	3	5	8	11	13	16	18	21	24
17	2304	2330	2355	2380	2405	2430	2455	2480	2504	2529	2	5	7	10	12	15	17	20	22
18	2553	2577	2601	2625	2648	2672	2695	2718	2742	2765	2	5	7	9	12	14	16	19	21
19	2788	2810	2833	2856	2878	2900	2923	2945	2967	2989	2	4	7	9	11	13	16	18	20
20	3010	3032	3054	3075	3096	3118	3139	3160	3181	3201	2	4	6	8	11	13	15	17	19
21	3222	3243	3263	3284	3304	3324	3345	3365	3385	3404	2	4	6	8	10	12	14	16	18
22	3424	3444	3464	3483	3502	3522	3541	3560	3579	3598	2	4	6	8	10	12	14	15	17
23	3617	3636	3655	3674	3692	3711	3729	3747	3766	3784	2	4	6	7	9	11	13	15	17
24	3802	3820	3838	3856	3874	3892	3909	3927	3945	3962	2	4	5	7	9	11	12	14	16
25	3979	3997	4014	4031	4048	4065	4082	4099	4116	4133	2	3	5	7	9	10	12	14	15
26	4150	4166	4183	4200	4216	4232	4249	4265	4281	4298	2	3	5	7	8	10	11	13	15
27	4314	4330	4346	4362	4378	4393	4409	4425	4440	4456	2	3	5	6	8	9	11	13	14
28	4472	4487	4502	4518	4533	4548	4564	4579	4594	4609	2	3	5	6	8	9	11	12	14
29	4624	4639	4654	4669	4683	4698	4713	4728	4742	4757	1	3	4	6	7	9	10	12	13
30	4771	4786	4800	4814	4829	4843	4857	4871	4886	4900	1	3	4	6	7	9	10	11	13
31	4914	4928	4942	4955	4969	4983	4997	5011	5024	5038	1	3	4	6	7	8	10	11	12
32	5051	5065	5079	5092	5105	5119	5132	5145	5159	5172	1	3	4	5	7	8	9	11	12
33	5185	5198	5211	5224	5237	5250	5263	5276	5289	5302	1	3	4	5	6	8	9	10	12
34	5315	5328	5340	5353	5366	5378	5391	5403	5416	5428	1	3	4	5	6	8	9	10	11
35	5441	5453	5465	5478	5490	5502	5514	5527	5539	5551	1	2	4	5	6	7	9	10	11
36	5563	5575	5587	5599	5611	5623	5635	5647	5658	5670	1	2	4	5	6	7	8	10	11
37	5682	5694	5705	5717	5729	5740	5752	5763	5775	5786	1	2	3	5	6	7	8	9	10
38	5798	5809	5821	5832	5843	5855	5866	5877	5888	5899	1	2	3	5	6	7	8	9	10
39	5911	5922	5933	5944	5955	5966	5977	5988	5999	6010	1	2	3	4	5	7	8	9	10
40	6021	6031	6042	6053	6064	6075	6085	6096	6107	6117	1	2	3	4	5	6	8	9	10
41	6128	6138	6149	6160	6170	6180	6191	6201	6212	6222	1	2	3	4	5	6	7	8	9
42	6232	6243	6253	6263	6274	6284	6294	6304	6314	6325	1	2	3	4	5	6	7	8	9
43	6335	6345	6355	6365	6375	6385	6395	6405	6415	6425	1	2	3	4	5	6	7	8	9
44	6435	6444	6454	6464	6474	6484	6493	6503	6513	6522	1	2	3	4	5	6	7	8	9
45	6532	6542	6551	6561	6571	6580	6590	6599	6609	6618	1	2	3	4	5	6	7	8	9
46	6628	6637	6646	6656	6665	6675	6684	6693	6702	6712	1	2	3	4	5	6	7	7	8
47	6721	6730	6739	6749	6758	6767	6776	6785	6794	6803	1	2	3	4	5	5	6	7	8
48	6812	6821	6830	6839	6848	6857	6866	6875	6884	6893	1	2	3	4	4	5	6	7	8
49	6902	6911	6920	6928	6937	6946	6955	6964	6972	6981	1	2	3	4	4	5	6	7	8
50	6990	6998	7007	7016	7024	7033	7042	7050	7059	7067	1	2	3	3	4	5	6	7	8
51	7076	7084	7093	7101	7110	7118	7126	7135	7143	7152	1	2	3	3	4	5	6	7	8
52	7160	7168	7177	7185	7193	7202	7210	7218	7226	7235	1	2	2	3	4	5	6	7	7
53	7243	7251	7259	7267	7275	7284	7292	7300	7308	7316	1	2	2	3	4	5	6	6	7
54	7324	7332	7340	7348	7356	7364	7372	7380	7388	7396	1	2	2	3	4	5	6	6	7

N	0	1	2	3	4	5	6	7	8	9	1	2	3	4	5	6	7	8	9
														Proportional parts					
55	7404	7412	7419	7427	7435	7443	7451	7459	7466	7474	1	2	2	3	4	5	5	6	7
56	7482	7490	7497	7505	7513	7520	7528	7536	7543	7551	1	2	2	3	4	5	5	6	7
57	7559	7566	7574	7582	7589	7597	7604	7612	7619	7627	1	2	2	3	4	5	5	6	7
58	7634	7642	7649	7657	7664	7672	7679	7686	7694	7701	1	1	2	3	4	4	5	6	7
59	7709	7716	7723	7731	7738	7745	7752	7760	7767	7774	1	1	2	3	4	4	5	6	7
60	7782	7789	7796	7803	7810	7818	7825	7832	7839	7846	1	1	2	3	4	4	5	6	6
61	7853	7860	7868	7875	7882	7889	7896	7903	7910	7917	1	1	2	3	4	4	5	6	6
62	7924	7931	7938	7945	7952	7959	7966	7973	7980	7987	1	1	2	3	4	4	5	6	6
63	7993	8000	8007	8014	8021	8028	8035	8041	8048	8055	1	1	2	3	3	4	5	5	6
64	8062	8069	8075	8082	8089	8096	8102	8109	8116	8122	1	1	2	3	3	4	5	5	6
65	8129	8136	8142	8149	8156	8162	8169	8176	8182	8189	1	1	2	3	3	4	5	5	6
66	8195	8202	8209	8215	8222	8228	8235	8241	8248	8254	1	1	2	3	3	4	5	5	6
67	8261	8267	8274	8280	8287	8293	8299	8306	8312	8319	1	1	2	3	3	4	5	5	6
68	8325	8331	8338	8344	8351	8357	8363	8370	8376	8382	1	1	2	3	3	4	4	5	6
69	8388	8395	8401	8407	8414	8420	8426	8432	8439	8445	1	1	2	2	3	4	4	5	6
70	8451	8457	8463	8470	8476	8482	8488	8494	8500	8506	1	1	2	2	3	4	4	5	5
71	8513	8519	8525	8531	8537	8543	8549	8555	8561	8567	1	1	2	2	3	4	4	5	5
72	8573	8579	8585	8591	8597	8603	8609	8615	8621	8627	1	1	2	2	3	4	4	5	5
73	8633	8639	8645	8651	8657	8663	8669	8675	8681	8686	1	1	2	2	3	4	4	5	5
74	8692	8698	8704	8710	8716	8722	8727	8733	8739	8745	1	1	2	2	3	4	4	5	5
75	8751	8756	8762	8768	8774	8779	8785	8791	8797	8802	1	1	2	2	3	3	4	5	5
76	8808	8814	8820	8825	8831	8837	8842	8848	8854	8859	1	1	2	2	3	3	4	5	5
77	8865	8871	8876	8882	8887	8893	8899	8904	8910	8915	1	1	2	2	3	3	4	4	5
78	8921	8927	8932	8938	8943	8949	8954	8960	8965	8971	1	1	2	2	3	3	4	4	5
79	8976	8982	8987	8993	8998	9004	9009	9015	9020	9025	1	1	2	2	3	3	4	4	5
80	9031	9036	9042	9047	9053	9058	9063	9069	9074	9079	1	1	2	2	3	3	4	4	5
81	9085	9090	9096	9101	9106	9112	9117	9122	9128	9133	1	1	2	2	3	3	4	4	5
82	9138	9143	9149	9154	9159	9165	9170	9175	9180	9186	1	1	2	2	3	3	4	4	5
83	9191	9196	9201	9206	9212	9217	9222	9227	9232	9238	1	1	2	2	3	3	4	4	5
84	9243	9248	9253	9258	9263	9269	9274	9279	9284	9289	1	1	2	2	3	3	4	4	5
85	9294	9299	9304	9309	9315	9320	9325	9330	9335	9340	1	1	2	2	3	3	4	4	5
86	9345	9350	9355	9360	9365	9370	9375	9380	9385	9390	1	1	2	2	3	3	4	4	5
87	9395	9400	9405	9410	9415	9420	9425	9430	9435	9440	0	1	1	2	2	3	3	4	4
88	9445	9450	9455	9460	9465	9469	9474	9479	9484	9489	0	1	1	2	2	3	3	4	4
89	9494	9499	9504	9509	9513	9518	9523	9528	9533	9538	0	1	1	2	2	3	3	4	4
90	9542	9547	9552	9557	9562	9566	9571	9576	9581	9586	0	1	1	2	2	3	3	4	4
91	9590	9595	9600	9605	9609	9614	9619	9624	9628	9633	0	1	1	2	2	3	3	4	4
92	9638	9643	9647	9652	9657	9661	9666	9671	9675	9680	0	1	1	2	2	3	3	4	4
93	9685	9689	9694	9699	9703	9708	9713	9717	9722	9727	0	1	1	2	2	3	3	4	4
94	9731	9736	9741	9745	9750	9754	9759	9763	9768	9773	0	1	1	2	2	3	3	4	4
95	9777	9782	9786	9791	9795	9800	9805	9809	9814	9818	0	1	1	2	2	3	3	4	4
96	9823	9827	9832	9836	9841	9845	9850	9854	9859	9863	0	1	1	2	2	3	3	4	4
97	9868	9872	9877	9881	9886	9890	9894	9899	9903	9908	0	1	1	2	2	3	3	4	4
98	9912	9917	9921	9926	9930	9934	9939	9943	9948	9952	0	1	1	2	2	3	3	4	4
99	9956	9961	9965	9969	9974	9978	9983	9987	9991	9996	0	1	1	2	2	3	3	3	4

TABLE 16
International atomic weights

Based on $C^{12} = 12$ exactly

Element	Symbol	Atomic number	Atomic weight	Element	Symbol	Atomic number	Atomic weight
Actinium	Ac	89	[227]*	Mercury	Hg	80	200.59
Aluminum	Al	13	26.9815	Molybdenum	Mo	42	95.94
Americium	Am	95	[243]*	Neodymium	Nd	60	144.24
Antimony	Sb	51	121.75	Neon	Ne	10	20.183
Argon	Ar	18	39.948	Neptunium	Np	93	[237]*
Arsenic	As	33	74.9216	Nickel	Ni	28	58.71
Astatine	At	85	[210]*	Niobium	Nb	41	92.906
Barium	Ba	56	137.34	Nitrogen	N	7	14.0067
Berkelium	Bk	97	[247]*	Nobelium	No	102	[254]*
Beryllium	Be	4	9.0122	Osmium	Os	76	190.2
Bismuth	Bi	83	208.980	Oxygen	O	8	15.9994[1]
Boron	B	5	10.811[1]	Palladium	Pd	46	106.4
Bromine	Br	35	79.909[2]	Phosphorus	P	15	30.9738
Cadmium	Cd	48	112.40	Platinum	Pt	78	195.09
Calcium	Ca	20	40.08	Plutonium	Pu	94	[242]*
Californium	Cf	98	[247]*	Polonium	Po	84	[210]
Carbon	C	6	12.01115[1]	Potassium	K	19	39.102
Cerium	Ce	58	140.12	Praseodymium	Pr	59	140.907
Cesium	Cs	55	132.905	Promethium	Pm	61	[147]*
Chlorine	Cl	17	35.453[2]	Protactinium	Pa	91	[231]*
Chromium	Cr	24	51.996[2]	Radium	Ra	88	[226]*
Cobalt	Co	27	58.9332	Radon	Rn	86	[222]*
Copper	Cu	29	63.54	Rhenium	Re	75	186.2
Curium	Cm	96	[247]*	Rhodium	Rh	45	102.905
Dysprosium	Dy	66	162.50	Rubidium	Rb	37	85.47
Einsteinium	Es	99	[254]*	Ruthenium	Ru	44	101.07
Erbium	Er	68	167.26	Samarium	Sm	62	150.35
Europium	Eu	63	151.96	Scandium	Sc	21	44.956
Fermium	Fm	100	[253]*	Selenium	Se	34	78.96
Fluorine	F	9	18.9984	Silicon	Si	14	28.086[1]
Francium	Fr	87	[223]*	Silver	Ag	47	107.870[2]
Gadolinium	Gd	64	157.25	Sodium	Na	11	22.9898
Gallium	Ga	31	69.72	Strontium	Sr	38	87.62
Germanium	Ge	32	72.59	Sulfur	S	16	32.064[1]
Gold	Au	79	196.967	Tantalum	Ta	73	180.948
Hafnium	Hf	72	178.49	Technetium	Tc	43	[97]*
Helium	He	2	4.0026	Tellurium	Te	52	127.60
Holmium	Ho	67	164.930	Terbium	Tb	65	158.924
Hydrogen	H	1	1.00797[1]	Thallium	Tl	81	204.37
Indium	In	49	114.82	Thorium	Th	90	232.038
Iodine	I	53	126.9044	Thulium	Tm	69	168.934
Iridium	Ir	77	192.2	Tin	Sn	50	118.69
Iron	Fe	26	55.847[2]	Titanium	Ti	22	47.90
Krypton	Kr	36	83.80	Tungsten	W	74	183.85
Lanthanum	La	57	138.91	Uranium	U	92	238.03
Lawrencium	Lw	103	[257]*	Vanadium	V	23	50.942
Lead	Pb	82	207.19	Xenon	Xe	54	131.30
Lithium	Li	3	6.939	Ytterbium	Yb	70	173.04
Lutetium	Lu	71	174.97	Yttrium	Y	39	88.905
Magnesium	Mg	12	24.312	Zinc	Zn	30	65.37
Manganese	Mn	25	54.9380	Zirconium	Zr	40	91.22
Mendelevium	Md	101	[256]*				

[1]The atomic weight varies because of natural variations in the isotopic composition of the element. The observed ranges are: boron, ±0.003; carbon, ±0.00005; hydrogen, ±0.00001; oxygen, ±0.0001; silicon, ±0.001, sulfur, ±0.003.

[2]The atomic weight is believed to have an experimental uncertainty of the following magnitude: bromine, ±0.002; chlorine, ±0.001; chromium, ±0.001; iron, ±0.003; silver, ±0.003. For other elements the last digit given is believed to be reliable to ±0.5.

*A number in brackets designates the mass number of a selected isotope of the element, usually the one of longest known half-life.
Atomic weights: Courtesy the International Union of Pure and Applied Chemistry.
Bracketed numbers: Courtesy the National Bureau of Standards.

ANSWERS TO PROBLEMS

ANSWERS TO EXERCISES FROM APPENDIX A

1. $m = dv$
2. $x = 6 + 4y$
3. $\dfrac{2y - x}{xy}$
4. $b - a$
5. $x = \pm 5$
6. $y = 80$
7. ± 130
8. 0.16
9. 0.96
10. 140
11. (a) 10^3, (b) 2×10^2, (c) 5.5×10^3, (d) 1×10^{-1}, (e) 1×10^{-3}, (f) 4×10^{-3}, (g) 3×10^{-12}, (h) 7.5×10^{11}, (i) 2.006×10^6, (j) 2.04×10^{-3}.
12. (a) 4.26×10^4, (b) 4.1075×10^{-1}, (c) 3.75×10^{-5}, (d) 7.287, (e) 4.65×10^{-1}, (f) 6.022×10^{23}, (g) 3.0103×10^4, (h) 6.25×10^3.
13. (a) 10^6, (b) 10^4, (c) 4.3×10^4, (d) 131, (e) 10^3, (f) 10^2, (g) 10, (h) 8×10^{11}, (i) 6.2×10^6, (j) $\pm 5 \times 10^3$, (k) $\pm 4 \times 10^2$, (l) $\pm 6 \times 10^2$, (m) 1.4×10^4, (n) $\pm 2x^3yz$.
14. $P = kt$, (b) $K.E. = \frac{1}{2}mv^2$, (c) $Z = \dfrac{k}{\lambda^{1/2}}$, (d) $f = k\dfrac{e_1e_2}{d^2}$, (e) $V = V_0\left(1 + \dfrac{1}{273}t\right)$.
15. (a) $\dfrac{m_1}{m_2} = \dfrac{\bar{v}_2^2}{\bar{v}_1^2}$, (b) $V_{SO_2} = 2 \times 10^5$ cm/sec.
16. (a) 2.1667, (b) 0.8697, (c) -2.4339, (d) 1.70, (e) 23.7798, (f) -4.74.
17. (a) 3.040×10^2, (b) 11.63, (c) 3.583×10^4, (d) 7.000×10^{-1}, (e) 5.508×10^{-4}, (f) 4.241×10^{-2}.
18. (a) 231 ml, (b) 1.7×10^{21} molecules, (c) 85 ml.
19. (a) 4.2×10^{-1}, (b) 5.27×10^7, (c) 30.44, (d) 5.000, (e) 58, (f) 1.574×10^{-1}, (g) 9.0×10^{-3}, (h) 2.10×10^3.
20. (a) 1.6, or -7.6, (b) 1.43 or -4.93, (c) $\pm 4.1 \times 10^{-3}$, (d) 5.0×10^{-4}, or -1.05×10^{-2}.

ANSWERS TO EXERCISES FROM APPENDIX B

1. (a) 25 miles, (b) 500 sec, or 8.3 min, (c) 10.8 yd/sec, (d) 1000 cm^3, 10 cm, (e) 6 boxes, (f) 10 moles.
2. (a) $27.8 \pm 0.4°C$, (b) 175.79 ± 0.03 g, (c) 43.9167 ± 0.0002 g, (d) 510.0 ± 0.3 ml.
3. Using significant figures: (a) 13.6 g/ml, (b) 13.6 g/ml; using percentage uncertainty: (a) 13.6 g/ml $\pm 1.2\%$, (b) 13.6 g/ml $\pm 0.5\%$.
4. Be sure to estimate the uncertainty of each measurement, taking into account the quality of the ruler used and the reproducibility of the readings you can obtain.
5. Weight of water: 80.6 ± 1.0 g; temperature increase: $9.1 \pm 0.4°C$; calories: 733 ± 41 cal.
6. 20 ± 1 g/cm^3.
7. 40.211 ± 0.003

Freeman Laboratory Separates in Chemistry

LABORATORY STUDIES IN GENERAL CHEMISTRY, 1001-1062
30¢ each
by FRANTZ and MALM
(These studies are included in the FREEMAN LIBRARY OF LABORATORY SEPARATES IN CHEMISTRY, Volume I.)

1001. Student Handbook
1002. Mass and Volume Relationships
1003. Weighing Operations. Gravimetric Techniques
1004. Physical Properties of Substances
1005. Some Elementary Chemical Properties of Substances
1006. Analysis for Substances in the Atmosphere
1007. The Chemical Separation of a Mixture
1008. The Packing of Atoms and Ions in Crystals
1009. The Chemistry of Oxygen. Basic and Acidic Oxides and the Periodic Table
1010. The Formation of Salts
1011. The Activity of Certain Metals and the Stability of Their Oxides
1012. The Experimental Determination of a Chemical Formula
1013. The Formula of a Hydrate
1014. Fundamental Weight Relationships 1: Equivalent and Atomic Weights by Oxidation of the Metal
1015. Fundamental Weight Relationships 2: Equivalent and Atomic Weights by Reduction of the Metal Oxide
1016. The Properties of Gases. Relationships Among the Variables Volume, Pressure, Temperature, and Amount. *A Study Assignment*
1017. Gas Analysis Based on the Molar Volume
1018. The Molecular Weight of a Gas
1019. Analysis Based on the Equivalent Weight of a Metal. Ionic Valence
1020. Stoichiometric Calculations Based on Chemical Reactions. *A Study Assignment*
1021. Group Relationships in the Periodic Table
1022. Ionic and Covalent Compounds. Ionic Reactions
1023. Oxidation-Reduction
1024. Common Oxidizing and Reducing Agents. Balancing of Oxidation-Reduction Equations. *A Study Assignment*
1025. The Relation of Oxidation-Reduction to Electric Cells and Electrolysis
1026. The Halogen Elements: Their Halide Salts and Acids
1027. The Chemistry of Chlorine: Oxidation States −1 to +7
1028. The Chemistry of Sulfur: Oxidation States −2 to +6
1029. The Chemistry of Nitrogen: Oxidation States −3 to +5
1030. The Chemistry of Chromium in Oxidation States 0 to +6 and of Manganese in Oxidation States 0 to +7
1031. Units of Concentration. The Volumetric Titration of Acids and Bases

1032. Volumetric Analysis —The Equivalent Weight of a Solid Acid
1033. Selected Volumetric Analyses
1034. The Equivalent Weight of Oxidizing and Reducing Agents
1035. The Selective Crystalization of Salts
1036. The Molecular Weight of a Soluble Substance by Freezing-Point Depression
1037. The Preparation and Properties of Colloidal Dispersions
1038. Thermochemistry. The Heat of Reaction
1039. The Rate of Chemical Reactions. Chemical Kinetics
1040. The Rate of Chemical Reactions. Activation Energy
1041. Reversible Reactions and Chemical Equilibrium
1042. The Equilibria of Water, Weak Acids, and Weak Bases. Indicators, pH
1043. The Ionization Constant of Acetic Acid
1044. The Ionization Constant of Some Weak Acids
1045. Equilibria Involving Volatility, Solubility, Degree of Ionization
1046. The Equilibria Between Slightly Soluble Salts and Their Ions. The Solubility Product Constant
1047. The Equilibria of Coordination Compounds —Hydration, Complex Ions. Slightly Soluble Metal Hydroxides
1048. Acid-Base Equilibria. Hydrolysis, Buffers
1049. Equilibria Involving Hard Water and Its Softening
1050. The Equilibria of Carbonic Acid and Its Salts
1051. Recapitulation of Problems on Ionic Equilibria. *A Study Assignment*
1052. The Qualitative Analysis of Some Common Negative Ions
1053. An Introduction to the Qualitative Analysis of the Metal Ions. *A Study Assignment*
1054. The Silver Group
1055. The Hydrogen Sulfide Group
1056. The Ammonium Sulfide Group
1057. The Alkaline Earth and Alkali Groups
1058. The Analysis of General Unknown Inorganic Substances. The Solution of Solids
1059. Some Inorganic Preparations
1060. Introductory Organic Chemistry. Types of Compounds and Reactions
1061. Simple Biochemical Tests
1062. Some Organic Syntheses
Teacher's Manual (Sent free to teachers)

LABORATORY STUDIES IN ORGANIC CHEMISTRY, 1063-1082
30¢ each
by HELMKAMP and JOHNSON
(These studies are included in the FREEMAN LIBRARY OF LABORATORY SEPARATES IN CHEMISTRY, Volume I.)

1063. Melting Points and Crystallization
1064. Boiling Points and Distillation
1065. Chromatography

1066. Spectra and Structure
1067. Chemical and Physical Equilibria
1068. Nucleophilic Substitution Reactions
1069. Reactions of Functional Groups: Aliphatic Hydrocarbons, Halides, Alcohols, and Esters
1070. Elimination Reactions
1071. Multistep Organic Syntheses I. Identification of Compounds
1072. Polymers and Polymerization
1073. Stereochemistry
1074. Triphenylcarbinol and Related Compounds
1075. Carbonyl Compounds
1076. Bimolecular Reduction of Acetone: Compounds Derived from Pinacol
1077. Aromatic Compounds: Derivatives of Nitrobenzene
1078. Heterocyclic Compounds
1079. Friedel-Crafts Reactions and Rearrangements
1080. Multistep Organic Syntheses II. Tetraphenyldihydrophthalic Anhydride
1081. Reactions of Carbonyl Compounds and Amines. Identification of a General Unknown
1082. Free Radical Reactions
Teacher's Manual (Sent free to teachers)

LABORATORY STUDIES IN GENERAL CHEMISTRY, 1083-1122
30¢ each
BIRDWHISTELL and O'CONNOR, Editors
(These studies are included in the FREEMAN LIBRARY OF LABORATORY SEPARATES IN CHEMISTRY, Volume II.)

1083. FRANTZ and MALM, Supplement to the Student Handbook, Analytical Methods and Equipment GESSER, BADER, JAGROOP, and LITHOWN
1084. Molecular Structure
1085. Densities of Pure Liquids and Solutions
1086. Gravimetric Analysis: Quantitative Analysis for Sulfate
1087. Quantitative Volumetric Analysis
1088. Volumetric Analysis: Determination of Sodium Carbonate in an Unknown
1089. Sources of Error—Measurement of the Gas Constant and the Molar Volume of Oxygen
1090. The Ideal Gas Law and the Molecular Weight of Gases
1091. Charles' Law and Absolute Zero
1092. *(No longer available)*
1093. Viscosity
1094. Surface Tension and Contact Angle
1095. Boiling Points of Binary Systems (Siwoloboff's Method)
1096. Distribution of Solute Between an Immiscible Solvent Pair
1097. Determination of Molecular Weight by Freezing-Point Depression
1098. The Enthalpy of Fusion of Naphthalene
1099. Heat of Solution
1100. Heat of Neutralization of an Acid and a Base
1101. Equilibrium